MOBILE CLOUD COMPUTING

Architectures, Algorithms and Applications

MOBILE CLOUD COMPUTING
Architectures, Algorithms and Applications

Debashis De
West Bengal University of Technology
Kolkata, India

CRC Press
Taylor & Francis Group
Boca Raton London New York

CRC Press is an imprint of the
Taylor & Francis Group, an **informa** business

A CHAPMAN & HALL BOOK

CRC Press
Taylor & Francis Group
6000 Broken Sound Parkway NW, Suite 300
Boca Raton, FL 33487-2742

© 2016 by Taylor & Francis Group, LLC
CRC Press is an imprint of Taylor & Francis Group, an Informa business

No claim to original U.S. Government works

Printed on acid-free paper
Version Date: 20150825

International Standard Book Number-13: 978-1-4822-4283-6 (Hardback)

Library of Congress Cataloging-in-Publication Data

De, Debashis, author.
  Mobile cloud computing : architectures, algorithms and applications / Debashis De.
    pages cm
  Includes bibliographical references and index.
  ISBN 978-1-4822-4283-6 (alk. paper)
  1. Cloud computing. 2. Mobile computing. I. Title.

QA76.585.D425 2016
004.67'82--dc23   2015029285

Visit the Taylor & Francis Web site at
http://www.taylorandfrancis.com

and the CRC Press Web site at
http://www.crcpress.com

This book is dedicated to
my beloved twin sons
Neelabhro and Neelangshu

Contents

Forewords

Mobile cloud computing has gained popularity because of its potential to minimize power consumption and enhance user experience by outsourcing resource/computation-intensive applications/operations from mobile devices to clouds. *Mobile Cloud Computing: Architectures, Algorithms, and Applications* is the first complete reference book on mobile cloud computing.

Mobile devices suffer from poor battery life and limited resource and storage capacity. To overcome these constraints, mobile cloud computing has introduced offloading, where data storage and computations are performed inside the remote cloud instead of the mobile device. With this emergence, mobile cloud computing has become a vital issue that requires an energy-efficient mobile network and green cloud environment. Resource allocation and security are also important issues discussed in detail in this book. Mobile cloud computing–based health monitoring, gaming, learning, and commerce are feasible due to high-speed 4G/5G mobile networks. The issues of mobile cloud computing are discussed in the 14 chapters with future open research problems.

Overall, this is an excellent book that serves not only as a resource for teaching purposes with a clear and detailed view of the various aspects of mobile cloud computing, but also as a complete research reference.

I wish success for the book.

Prof. Rajkumar Buyya
Fellow of IEEE
Director, Cloud Computing and Distributed Systems (CLOUDS) Laboratory
Department of Computing and Information Systems, The University of Melbourne, Parkville,
Victoria, Australia
Editor-in-Chief, Software: Practice and Experience
CEO, Manjrasoft Pty Ltd, Melbourne, Victoria, Australia

Mobile cloud computing (MCC) represents a most significant shift in IT, recently emerging as one of the buzzwords in the information and communications technology industry. MCC is a promising computation model that integrates the power of cloud data centers with the portability of mobile computing devices. The unabated flurry of research activities to augment various mobile devices by leveraging heterogeneous cloud resources has potentially given birth to a new research domain called MCC. In the core of such a nonuniform environment, MCC is a rich mobile computing technology that leverages unified elastic resources of varied clouds and network technologies toward unrestricted functionality, storage, and mobility. It can serve a multitude of mobile devices anywhere, anytime, through the channel of Ethernet or the Internet, regardless of heterogeneous environments and platforms, based on the pay-as-you-use principle. MCC enables mobile users to go beyond their hardware restrictions and take advantage of the rich applications offered by the cloud. While the economic case for mobile cloud computing is compelling, the challenge it poses to facilitate interoperability, portability, and integration among heterogeneous platforms is equally striking. Despite all the hype surrounding the MCC and its exponential growth, exploiting its full potential is demanding. Apart from its inherent problems, such as resource scarcity, frequent disconnections, and mobility, security, being the key concern, reduces the growth of MCC. Even though the processing and data-storage capabilities of a mobile device have grown exponentially, they are still limited in terms of resources compared to high-end computers. CPU and memory, such as RAM, are the main limiting factors.

This book comprehensively debates on the emergence of mobile cloud computing from cloud computing models. Various technological and architectural advancements in mobile and cloud computing have been reported. It has meticulously explored the design and architecture of computational offloading solutions in cloud and mobile cloud computing domains to enrich mobile user experience. Furthermore, to optimize mobile power consumption, existing solutions and policies toward green mobile computing, green cloud computing, green mobile networking, and green mobile cloud computing are briefly discussed. The book also presents numerous cloud and mobile resource allocation and management schemes to efficiently manage existing resources (hardware and software). Recently, integrated networks (e.g., WSN, VANET, MANET) have significantly helped mobile users to enjoy a suite of services. The book discusses existing architecture, opportunities, and challenges, while integrating mobile cloud computing with existing network technologies such as sensor and vehicular networks. It also briefly expounds on various security and privacy concerns, such as application security, authentication security, data security, and intrusion detection, in the mobile cloud computing domain. The business aspects of mobile cloud computing models in terms of resource pricing models, cooperation models, and revenue sharing among cloud providers are also presented in the book. To highlight the standings of mobile cloud computing, various well-known, real-world applications supported by mobile cloud computing models are discussed. For example, the demands and issues while deploying resource-intensive applications, including face recognition, route tracking, traffic management, and mobile learning, are discussed. This book concludes with various future research directions in the mobile cloud computing domain to improve the strength of mobile cloud computing and to enrich mobile user experience.

This book provides an introduction to the emerging computing paradigm of mobile cloud computing for the readers. It also enables mobile cloud application engineers and cloud service providers to leverage the appropriate features that can mitigate communication and computation latencies in order to increase the quality of service for mobile cloud users. Furthermore, discussion presented on various technological and architectural advancements in mobile cloud computing enables the reader to comprehend the mechanisms in mobile cloud computing. In particular, the highlighted open research challenges provide research directions to domain researchers for further investigations and improvements in mobile cloud computing.

Abdullah Gani
Professor and Dean
Faculty of Computer Science and Information Technology
University of Malaya
Kuala Lumpur, Malaysia

Preface

Mobile cloud computing is essential for high-speed fifth-generation mobile networks. The book consists of 14 chapters on the recent developments in mobile cloud computing. Chapter 1 deals with mobile computing, which is computation within a mobile device. The chapter discusses the evolution, architecture, and different generations of mobile networks along with the operating systems used in mobile devices. Various applications of mobile computing are also discussed with several challenges in this area.

In Chapter 2, the evolution, architecture, and applications of cloud computing are discussed in detail. Various aspects of cloud computing, such as security, data management, and energy-efficiency, are discussed, and different deployment schemes of the cloud are described.

Chapter 3 discusses mobile cloud computing in terms of the integration of mobile computing and cloud computing. The revolution of mobile cloud computing is described with its architecture, advantages, and applications. Various issues—and solutions—of mobile cloud computing are studied.

Mobile devices suffer from poor battery life and limited resource and storage capacity. To deal with these constraints, offloading is performed. In Chapter 4, offloading strategies are illustrated. Offloading refers to a mechanism in which data storage and computations are done inside the remote cloud instead of the mobile device. Consequently, the battery life of the device is increased and the difficulties of storage and resource limitations are overcome. In this chapter, we discuss offloading and its applications toward energy efficiency.

Green mobile cloud computing is an emerging research area. Green mobile network refers to energy-efficient mobile networks that consume low power. Small cell networks can be considered green mobile networks. By offloading computation inside the cloud, power consumption by a mobile device can be reduced. But this can cause more power, and thus cost, consumption inside the cloud. Therefore, there should be a trade-off between energy-efficient mobile networks and green cloud environment. In Chapter 5, the existing approaches for green mobile networks and green cloud computing are described. Based on the comparative study, a discussion is presented on how green mobile cloud computing can be achieved by merging green cellular networks with cloud environment.

Chapter 6 discusses the various resource allocation schemes of mobile cloud computing, including energy-aware resource management. Different task scheduling methods are also discussed. Challenges faced in the field of resource allocation are explored.

Sensor mobile cloud computing, an integration of wireless sensor networks with mobile cloud computing, is another emerging research area. In Chapter 7, the architecture and applications of sensor mobile cloud computing are studied. A life-cycle model of this architecture is developed, and different challenges of sensor mobile cloud computing are also discussed.

With the enlarging pervasive nature of social networks and cloud computing, users are exploring new methods with which to interact by utilizing these growing paradigms. A social network allows users to split information and build connections for generating dynamic virtual organizations. Massive use of mobile technologies such as laptops, smartphones, etc., is also drawing attention to the cloud for processing power, storage space, and energy savings, which in turn leads to a new concept known as mobile social cloud, illustrated in Chapter 8.

Chapter 9 presents the security and privacy issues in mobile cloud computing. Trust management, referring to the goodness, honesty, reliability, and faithfulness of an object, is also important in mobile cloud computing. We tend to trust a system that works according to our expectations. In mobile cloud computing, trust is a vital parameter because in this case long-distance personal data storage and data processing happen remotely.

In Chapter 10, different types of trust in mobile cloud computing are described. It is also discussed how the trust of the entire system can be increased by eliminating malicious users.

Research on mobile cloud computing studies several mobile agents such as vehicles, robots, and the like. All these mobile agents collaborate and interact to feel the environment, process the data, propagate the outputs, and mostly share resources. The vision of vehicular mobile cloud computing (VMCC) is a nontrivial argumentation regarding different dimensions with conventional mobile cloud computing. Chapter 11 provides a discussion of VMCC, in which underutilized resources of vehicles, such as storage, Internet connectivity, computing power, and the like, are shared among the drivers and are rented over the Internet for other customers/users.

Mobile cloud computing business management depends on the quality of service of the cloud service providers to the mobile users. A smile on the customer's face and retaining premium customers are the primary goals of service providers for business development. In Chapter 12, various economic and efficient business models are discussed, which have competitive advantages over each other on various quality-based parameters.

Mobile cloud computing has several applications including mobile learning, vehicles monitoring, digital forensic analysis, biometric application, and so forth. Through the help of mobile cloud computing, data can be stored and processed outside the mobile device and inside the cloud. This overcomes several challenges, for example, low bandwidth, limited speed, limited storage of traditional mobile learning, mobile health monitoring, and mobile gaming. Various applications of mobile cloud computing are presented in Chapter 13.

Mobile cloud computing has overcome several of the disadvantages of mobile computing mentioned earlier. Yet, various issues where research is needed remain. In Chapter 14, a variety of challenging applications of mobile cloud computing are discussed, including energy efficiency, latency minimization, efficient resource management, billing, and security. The chapter also recommends solutions to such challenges.

Debashis De

Acknowledgments

I acknowledge my past and present MTech students of the West Bengal University of Technology for their continuous support and enthusiasm. I am thankful to my PhD student Anwesha Mukherjee (gold medalist, MTech, IT), who is pursuing her research work as a DST-INSPIRE fellow under my supervision, and all the members of the mobile cloud computing laboratory of the West Bengal University of Technology. I am grateful to my brother Subhashis De for providing an excellent photograph for the cover page of this book. I thank my wife Swati De for supporting me all the way to completion of this project. The forewords written by Prof. Rajkumar Buyya, the cloud computing expert, and Prof. Abdullah Gani have added another dimension to this book. I am grateful for the invaluable feedback from all of the experts, reviewers, and the editorial team. I also thank Aastha Sharma, Hayley Ruggieri, Sarah Gelson, and Alex Edwards of CRC Press for their patience, enthusiasm, and support during my writing of this first edition of the book and at the same time for keeping me on schedule.

Finally, I am grateful to the Department of Science and Technology (DST) for sanctioning a research project entitled "Dynamic Optimization of Green Mobile Networks: Algorithm, Architecture and Applications" under the Fast Track Young Scientist scheme reference no. SERB/F/5044/2012-2013, under which this work was completed.

Debashis De

Author

Dr. Debashis De earned his MTech from the University of Calcutta, Kokata, West Bengal, India in 2002 and his PhD (engineering) from Jadavpur University, Kolkata, West Bengal, India in 2005. He is a senior member of the IEEE and member of the International Union of Radio Science. He previously worked as R&D engineer for Telektronics. Presently, Dr. De is the head of department and associate professor in the Department of Computer Science and Engineering of the West Bengal University of Technology, India and also adjunct research fellow at the University of Western Australia, Australia. He was awarded the prestigious Boyscast Fellowship by the Department of Science and Technology, Government of India, to work at the Herriot-Watt University, Scotland, UK. During 2008–2009, Dr. De received the Endeavour Fellowship Award from DEST, Australia, to work at the University of Western Australia. He also received the Young Scientist award both in 2005 at New Delhi and in 2011 at Istanbul, Turkey, from the International Union of Radio Science, Belgium. His research interests include location and handoff management, mobile cloud computing, traffic forecasting, green mobile networks, and low-power nanodevice designing for mobile applications. He has published in more than 60 peer-reviewed international journals, 50 conference papers, 2 research monographs, and 10 books. He can be reached at dr.debashis.de@gmail.com.

1

Mobile Computing

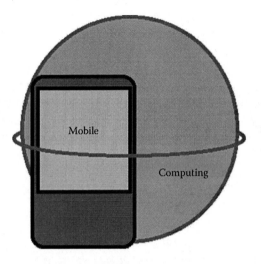

ABSTRACT   Mobile computing is computation within a mobile device. This chapter traces the evolution, architecture, and different generations of the mobile network. Different operating systems used in mobile devices are discussed. Various applications of mobile computing are described, along with several challenges in this area.

KEY WORDS: *mobile computing, Mobile network, 2G, 3G, 4G, 5G, mobility.*

1.1 Introduction to Mobile Computing

Mobile communication is the process of executing computations on a mobile device and transmission of data to and from one or more devices [1]. It is the technique of getting connected and making use of centrally located information and application software with the deployment of small, portable wireless communication and computing devices [1–3]. Mobile communication facilitates the execution of a number of applications on a single device. In this ultramodern world, everything is exceedingly reliant on technology. With the increase in the number of mobile users day by day, the need to provide better quality of service at very low power and cost also increases. This chapter deals with an overview of mobile computing and its challenges.

Mobile computing is the process of distributed computation on diversified mobile devices and hybrid networks interconnected by mobile communication protocols. Mobile computing focuses on the key technical issues related to the following:

1. *Mobile architectures*: Mobile networks and hosts, agents and proxies, wired and wireless systems integration, new development and standardization, mobility and location management, load balancing, mobile agent, and proxy architectures.

2. *Mobile support services*: Mobility and roaming, multimedia operation, operating system support, power/energy management, green mobile computing. Focusing on seamless incorporation of wired and wireless access networks to provide both increased mobility and bandwidth for broadband services.

3. *Algorithm for mobile network design*: Innovative dynamic optimization algorithm for mobile networks, bio-inspired algorithms for mobile networks.

4. *Protocol design and analysis*: Mobile environments, protocol design, efficient bandwidth utilization, intermittent connectivity, mobile IP protocol design for seamless access on mobile Internet, all-time hangout with social networks.

5. *Mobile environments*: Data and knowledge management, performance modeling and characterization, security, scalability and reliability, design, management and operation, systems, and technologies.

6. *Mobile communication systems*: Data encoding and compression, spread-spectrum technologies, multiuser access and multichannel processing, and channel coding.

7. *Applications*: Location-dependent and sensitive, nomadic computing, wearable computers and body area networks, multimedia applications and multimedia signal processing, pervasive computing, and wireless sensor networks.

8. *Emerging mobile technologies*: Opportunistic computing, urban sensing, Internet of things.

9. *Pervasive and mobile computing*: Pervasive/ubiquitous computing bringing in ground-breaking paradigms for computing models. Remarkable developments in mobile communications and networking, embedded designs, wearable body area networking, sensors, radio frequency identification (RFID) tags, smart spaces, middleware, and software agents have led to the evolution of pervasive computing platforms. The target of pervasive computing is to create ambient intelligence, where network devices embedded in the environment provide connectivity and services all the time, thus improving the human quality of experience. In this environment, the world around us, appliances, human body, cars, homes, offices, and cities are interconnected as a pervasive network of intelligent devices that cooperatively and autonomously collect, process, and transport information, in order to adapt to the associated context and activity. Various characteristic of *pervasive computing* are the following:
    a. Pervasive computing architectures and protocols
    b. Autonomic computing and communications
    c. Ambient intelligence, invisible and adaptive computing
    d. Mobile peer-to-peer computing
    e. Algorithmic paradigms, models and analysis of pervasive computing systems
    f. Smart spaces and intelligent environments

g. Bluetooth, body area networks, personal area networks
h. Embedded systems and wearable computers
i. Wireless sensors networks and RFID technologies
j. Multiple interconnected hybrid networking technologies
k. Positioning and tracking technologies
l. Auto-configuration and authentication
m. Context-aware computing and location-based services and applications
n. Service creation, discovery, management, and delivery mechanisms
o. Middleware and agent technologies
p. Application layer protocols and services
q. Programming paradigms for pervasive and ubiquitous computing applications
r. User interfaces and interaction models
s. Runtime support for intelligent, adaptive agents
t. Innovative applications requirements, performance, and benchmarking
u. Security, privacy, fault tolerance, and resiliency issues

1.2 Architecture of Mobile Network

Mobile networks can be classified into three different categories:

1. Cellular networks
2. Mobile ad hoc networks
3. Mobile wireless sensor network

In the following sections, we discuss the architecture of these three types of mobile networks.

1.2.1 Architecture of Cellular Network

The architecture of early mobile systems was intended to realize a huge coverage area using a single transmitter operating with high-power consumption and mounted on an elevated tower [1,2]. Though this architecture provided good coverage, it did not facilitate frequency reuse. To support frequency reuse and provide good coverage, a cellular architecture was sought for mobile networks [1]. In this architecture, the high-power-consuming transmitters used previously were replaced with many low-power-consuming transmitters [2]. The cellular architecture is such that each geographic coverage area is split into hexagonal cells, with each cell being served by a base station [2]. The cellular architecture is pictorially depicted in Figure 1.1.

The cells are surrounded by adjoining cells everywhere, and the boundaries of adjacent cells touch each other. The boundary specifies the area of coverage. Each cell is served by a base station. The base stations are assigned a portion of the net accessible frequency channel owed to the system. The base stations of the neighboring cells are allocated dissimilar

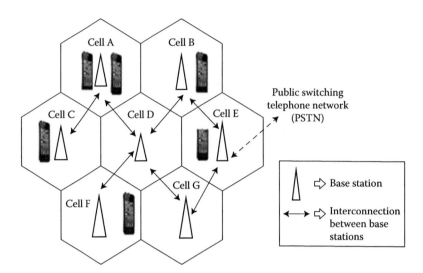

FIGURE 1.1
Cellular architecture of mobile communication.

frequency channels to prevent interference [1]. Let us consider cell D is using a frequency f_x; then none of the cells A, B, C, E, F, and G can use frequency f_x. If cell A uses frequency f_y, then cells C, B, and D cannot use this frequency, as they are adjacent to A, but F, G, or E can use this frequency. This concept is termed frequency reuse.

Let us assume cell D is using frequency f_x, cell A is using frequency f_y, and cell B is using frequency f_z; then cell E can reuse frequency f_y, cell C can use f_z, cell F can use frequency f_y, and cell G can use frequency f_z. In this case, if frequency reuse is employed, only three frequencies are needed. Thus, the frequency reuse factor is 1/3 [1,2]. The frequency reuse distance is computed as [1]

$$d = r \cdot \sqrt{(3 \times n)} \tag{1.1}$$

where
   r is the distance between cell center and cell boundary
   n is the number of adjacent cells surrounding the concerned cell [1,2]

In case the cells are divided into x sectors, then frequency reuse factor is x/n [1]. The concept of frequency reuse prevents interference among adjacent cells efficiently.

1.2.2 Architecture of Mobile Ad Hoc Network

Mobile ad hoc network (MANET) [1] is an infrastructureless network consisting of mobile devices communicating via wireless means. In such a network, the position of the switches, routers, or hubs are not fixed. In MANET, the organization is dependent on the position of nodes, connectivity among them, and their ability to locate service and send or receive messages to and from neighboring or peer nodes. In Figure 1.2, the architecture of MANET is depicted.

Figure 1.2 shows a MANET consisting of five nodes, out of which two are mobile nodes designated as Mobile node 1 and Mobile node 2, one is a handheld PC node, one is a

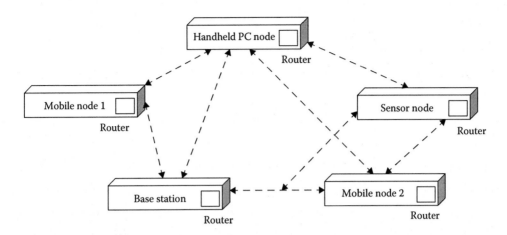

FIGURE 1.2
Architecture of MANET.

sensor node, and one is the base station. Each of these nodes behaves like a router and routes messages among the connected devices. The architecture is not static and may change with the movement of the nodes. Thus, it can be said that the MANET is a self-organizing network [1].

1.2.3 Architecture of Mobile Wireless Sensor Network

Mobile wireless sensor network (MWSN) is a MANET composed of sensor nodes having computational as well as communicational abilities [1]. MWSNs are flexible compared to static sensor networks as they can efficiently deal with topological changes. The architecture of the MWSN is depicted in Figure 1.3.

The architecture of a MWSN consists of sensor nodes acting as routers to route information between the nodes. The nodes communicate in an ad hoc manner with other MANETs as well as cellular networks. The benefit of using mobile sensors is the expansion in the

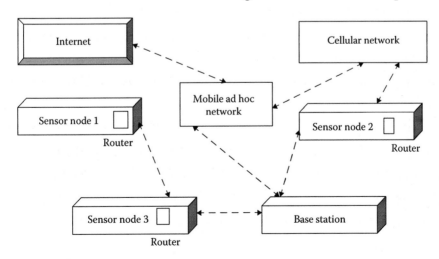

FIGURE 1.3
Architecture of mobile wireless sensor network.

number of applications, which would not be possible with static WSNs. Sensors can be employed for health monitoring such as heart rate, blood pressure, and so on [1]. Sensors can be used to track animal movements to get information related to their migration behavior, food habits, and many other purposes.

1.3 Generations of Mobile Communication

Budding technologies such as fourth generation (4G) and fifth generation (5G) attract a huge portion of engineering and commerce. However, second-generation (2G) and third-generation (3G) networks remain highly important and support nearly 80% of the mobile phone industry [1]. With the passage of time and improvements in technologies, mobile communication and computing have evolved from first generation (1G) to fifth generation (5G). The first generation of mobile communication was capable of providing only voice services. In addition to this capability, the second generation of mobile communication was capable of providing data services as well. The third generation provided multimedia services in addition to the services provided by the second generation. The fourth generation of mobile communications supported 3D audio and video in addition to the features supported by the third generation. The fifth generation of mobile communications will support complete multimedia using cloud-based offloading. In this section, we present an analysis of the evolving generations of mobile communications and computing.

1.3.1 1G Mobile Communication

Developed during 1970–1980, the first generation of mobile communications comprised the analog radio systems [1–5]. These systems function with the employment of frequency division multiple accesses (FDMA), which indicates that every channel makes use of a frequency band for voice calls [2–4]. The 1G mobile system has a small traffic capacity [3], and has a speed of 2.4 kbps [6]. The examples of the 1G mobile system include advance mobile phone service (AMPS), a total access communication system (TACS), Nordic mobile telephony (NMT), Japanese TACS (JTACS), and the C-system [4]. The features of these systems are presented in Table 1.1 [3].

TABLE 1.1

Features of 1G Mobile Communication Systems

System	Range of Frequency (MHz)	Channel Spacing (kHz)	Number of Channels
AMPS	824–949	30	832
NMT-450	453–457.5	25	180
NMT-900	890–915	12.5	1999
TACS	935–960	25	1000
JTACS	915–925	25	400
C-450	460–465.74	12.5	573

The major disadvantages of these systems include the following [4]:

- Low data rate
- Vulnerability to security attacks
- Unavailable roaming facilities

This generation continued to dominate the field of mobile communications until it was replaced by 2G digital cell phones.

1.3.2 2G Mobile Communication

Second generation mobile devices were introduced in 1990 [1]. 2G mobile communications makes use of digital modulation and provides voice and limited data services. Global system for mobile communications (GSM) forms the basis of 2G mobile communication. GSM provides roaming and short message service (SMS) facility [4]. GSM was developed to provide a mobile telephone standard, but it was readily honored and accepted universally [3]. GSM standards operate in the frequency ranges of three bands: 900, 1800, and 1900 MHz [2]. The 900 and 1800 MHz bands make use of the same base band signals but operate using different carrier frequencies. The separation of radio frequency between the uplink and downlink carrier for 900 MHz is 45 MHz while that of 1800 MHz is 90 MHz [5]. The 1900 MHz band is used primarily in North America. The frequency separation between the uplink and downlink frequencies is 80 MHz. The GSM features are as follows [3]:

- Adjacent carrier frequencies are separated by 200 kHz.
- GSM uses Gaussian minimum shift keying modulation.
- The transmission rate is 270 kbps.

The GSM architecture consists of the following components [1–3]:

- Mobile station (MS)
- Base station subsystem (BSS)
  - Base transceiver station (BTS)
  - Base station controller (BSC)
- Network switching subsystem (NSS)
  - Mobile services switching center (MSC)
  - Gateway mobile services switching center (GMSC)
  - Home location register (HLR)
  - Visitor location register (VLR)
- Operating subsystem
  - Operation and maintenance center (OMC)
  - Authentication center (AuC)
  - Equipment identity register (EIR)

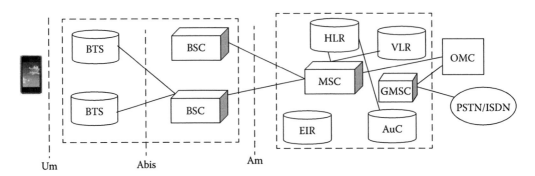

FIGURE 1.4
GSM architecture.

The architecture is pictorially depicted in Figure 1.4. The interface U_m connects the mobile station with the BTS. The BTS is connected with the BSC via the A_{bis} interface. The BSC in turn connects with the MSC via the A interface.

The functioning of all the components of the GSM architecture is presented in Table 1.2.

GSM provides three types of services, which are represented in Figure 1.5.

GSM services are classified as follows:

1. *Teleservices*: This includes voice calls, Fax, SMS, emergency calls, MMS (multimedia message service), and so on.

2. *Supplementary services*: This includes call forwarding, caller line identification, closed user group formation, call barring or waiting, call charge advice, and so on.

3. *Bearer services*: This includes data transmission and reception over network interfaces, and so on.

TABLE 1.2

Functioning of Components of GSM Architecture

* *Mobile station*: This represents the mobile device and consists of the subscriber identity module (SIM). The international mobile equipment identity (IMEI) uniquely identifies the mobile device. The SIM holds a secret key that helps in authentication and security measures.
* *Base station subsystem*: This unit is composed of BTS and BSC. BTS represents the transceivers and antennas used in the cells. BSC administers radio resources for the BTSs [1]. It handles channel setup, frequency hopping, and handover procedures when the users are moving [2].
* *Network switching subsystem*: This unit consists of MSC, GMSC, HLR, and VLR [1–3]. The functioning of each of these parts are as follows [4]:
  * MSC manages processing of signals, setting up and termination of calls, monitoring calls made to and from a mobile device, call charging, call forwarding, and so on.
  * HLR maintains databases of MSs in a GSM network. It is responsible for storing subscriber data such as a mobile subscriber ISDN (integrated services digital network) number, and details of subscription permissions such as call forwarding, roaming, user's location area, and user's current VLR status.
  * VLR is a database that stores both permanent and temporary subscriber data, which are required for communication between MSs in the coverage area of MSC associated with the HLR.
  * GMSC handles connections to fixed networks such as PSTN (public switched telephone network) or ISDN.
* *Operating subsystem*: This unit consists of OMC, AuC, and EIR [1–4]. The functioning of these parts are as follows:
  * OMC is accountable for controlling the functioning of the component units of GSM architecture.
  * AuC is used by the HLR to authenticate the users.
  * EIR is responsible maintaining a list of all mobile numbers that distinguish the devices stolen, currently in use, and nonfunctioning devices based on the IMEI number.

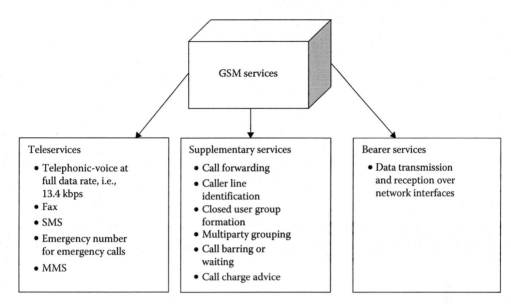

FIGURE 1.5
GSM services.

Figure 1.6 shows a GSM spectrum. The GSM spectrum shows the resolution bandwidth (RBW) and the video bandwidth (VBW) with the sweep and span. RBW is a band-pass filter in an IF (intermediate frequency) path. It is the minimum bandwidth over which two signals can be separated. The video bandwidth is used to filter noise. The center frequency is set to 935.1991 MHz as observed from the figure.

1.3.3 2.5G Mobile Communication

The 2.5G mobile communications technologies were developed in 1995 and supported a data rate of 64 kbps [3–5]. This generation of mobile communication was primarily dominated by the general packet radio services (GPRS). GPRS is an upgraded version of the GSM system in terms of the speed of data transmission. GPRS operates using the packet switching mode for both data transmission and Internet access. The GPRS architecture is shown in Figure 1.7.

The architectural entities are described in Table 1.3 [3–5].

GPRS supports a data rate of ~115 kbps [4]. It allows users to make voice calls and data transmission at the same time. The development of GPRS is an important step toward the development of 3G technologies.

1.3.4 3G Mobile Communication

3G mobile communication systems support data rates of over 153.6 kbps [1–3]. They provide better quality of experience for the users and support multimedia data transfers such as transfer of audio, video, text, and pictures. In 1997, WCDMA was selected as the 3G radio interface [4–7]. CDMA2000 was brought to the market in 2000 [5]. In Table 1.4, an analysis of the characteristics of WCDMA and CDMA2000 is presented.

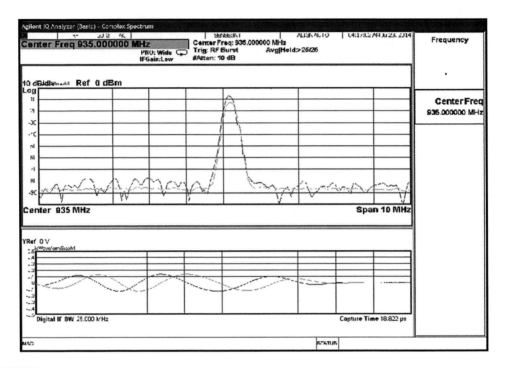

FIGURE 1.6
GSM spectrum.

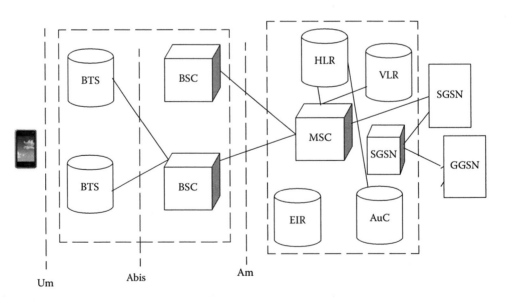

FIGURE 1.7
GPRS architecture.

TABLE 1.3

Functioning of Components of GPRS Architecture

- The radio subsystem (RSS) consists of MS, BTSs, and BSCs.
- An MS with GPRS features holds a cipher key sequence number (CKSN), which is used for user authentication and security purposes.
- The network subsystems are composed of serving GPRS support node (SGSN) and MSC. The SGSN interfaces with the BSCs as well with other SGSNs.
- The GPRS context-containing information related to MS status, data compression, and routing details is stored in the MSs as well the SGSN.
- EIR helps preserve security issues and maintenance activities.
- The gateway subsystem consists of gateway GPRS support node (GGSN) and serving GPRS Support Node (SGSN) and provides Internet connectivity.

TABLE 1.4

Characteristics of WCDMA and CDMA2000

Features	WCDMA	CDMA2000
Frequency spectrum	1.920–1.980 GHz for reverse and 2.110–2.170 GHz for forward	Can operate at 400, 800, 900, 1700, 1800, 1900 MHz
Bandwidth	5 MHz	3.75 MHz
Chipping rate	3.84 Mchips/s	3.6864–3.84 Mchips/s
Data transfer rates	Greater than or nearly equal to 2 Mbps for short distances and 384 kbps for long distances	Greater than or nearly equal to 2.05 Mbps for short distances and 614 kbps for long distances
Forward link	From BTS to MS	From BTS to MS
Reverse link	From MS to BTS	From MS to BTS
Modulation type	Quadrature pulse shift keying (QPSK) modulation for both forward and reverse link	QPSK modulation for forward link and binary pulse shift keying modulation (BPSK) for reverse link
Multi-rate transmission	Use single code while transmitting small amounts of data and multiple codes while transmitting large amount of data	Use variable spread factors between 4 and 256 for data transfer rates of 307–5 kbps

Source: Kamal, R., *Mobile Computing*, Oxford University Press, Inc., Oxford, U.K., 2008.

The ability to transfer both voice data such as a phone call and nonvoice data such as uploading and downloading information, e-mail exchange, and instant messaging is provided by 3G.

Figure 1.8 presents the WCDMA spectrum with RBW and VBW with sweep and span. The center frequency is set to 999.989 MHz, as seen from the figure. In case of the WCDMA spectrum, the RBW and VBW are both 20 kHz, as seen from Figure 1.8.

1.3.5 4G Mobile Communication

4G mobile communication was deployed during 2010–2012 [3–8]. It supports high-speed multimedia data transfer, high-definition television (HDTV) content, and high-speed Internet access [8]. It provides seamless Internet access anytime, anywhere. 4G provides a

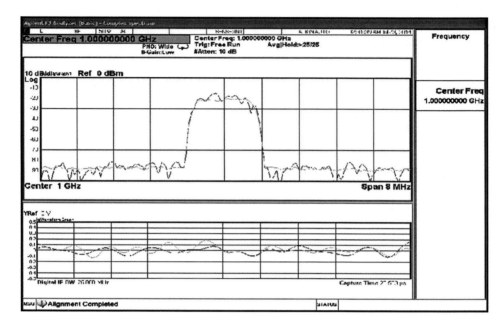

FIGURE 1.8
WCDMA spectrum.

trustworthy network, better quality service, and enhanced mobility and high security [8]. The technologies dominating 4G are the following [8]:

- Worldwide interoperability for microwave access (WiMAX) advanced
- Long-term evolution (LTE) advanced

The features of WiMAX advanced are presented in Table 1.5. The features of LTE advanced are presented in Table 1.6.

4G is a fully IP-based integrated system and the Internet work is accomplished with the union of wired and wireless networks including computers, consumer electronics, communication technology, and the capability to provide 100 Mbps and 1 Gbps, respectively, in outdoor and indoor environments with better quality of service (QoS) and improved security, facilitating any kind of services anytime, anywhere, at affordable cost and single billing [8].

TABLE 1.5

Features of WiMAX Advanced

Features	Definition
Frequency spectrum	Can be 450–470 MHz, 698–960 MHz, 1.710–2.025 GHz, 2.110–2.200 GHz, 2.300–2.400 GHz, 2.500–2.690 GHz or 3.400–3.600 GHz
Multiplexing	Frame duplex division (FDD) and time duplex division (TDD)
Modulation	Quadrature pulse shift keying (QPSK)
Channel bandwidth	5–20 MHz
Type of encoder	Channel encoder
Uplink power control	Both closed-loop AND open-loop power control are used
Applications	Mobile wireless Internet access, streaming multimedia, video, HDTV, Data, broadband Internet access

TABLE 1.6

Features of LTE Advanced

Features	Definition
Frequency spectrum	100 MHz
Uplink data rate	500 Mbps
Downlink data rate	1 Gbps
Spectral efficiency	Uplink 15 bps/Hz
	Downlink 20 bps/Hz
Uplink power control	Fractional path loss compensation
Latency	10 ms user plane and 50 ms control plane
Applications	Streaming multimedia, video, HDTV, internet access

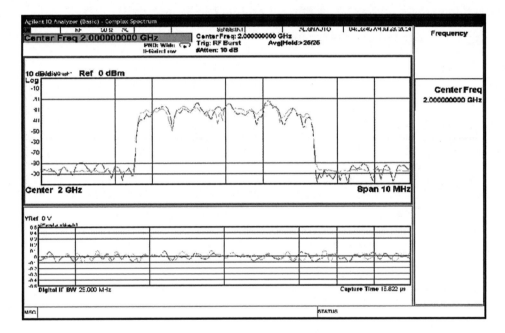

FIGURE 1.9
LTE spectrum.

Figure 1.9 presents the LTE spectrum with RBW and VBW with the sweep and span. The center frequency is set to 2.001091 GHz, as observed from the figure. In the case of LTE spectrum, the RBW and VBW are both 180 kHz.

1.3.6 5G Mobile Communication

To fulfill the user demand for high-speed mobile communication, fifth generation (5G) mobile communication was introduced as a promising research area. It is evident that it should be valid for all sorts of radio access technologies while taking into consideration a smooth transition to 5G, so that 5G can generate better profits for current global operators; in addition, interoperability will become more feasible. There should be a general platform unique for all the technologies to build 5G practical for all sorts of radio access technologies (RAT). The fifth generation of mobile communication is a wireless race that will be sustained by large area synchronized code division multiple access (LAS-CDMA), orthogonal frequency-division multiplexing (OFDM), multicarrier code division multiple access (MC-CDMA), ultrawide band (UWB), local multipoint distribution system (network-LMDS), and *Internet Protocol version 6* (IPv6) [9,10]. IPv6 is the elementary protocol for running on both 4G and 5G [10]. The 5G mobile communication system is an all-IP based system. In 5G, the IP-based mobile applications and services such as mobile portals, mobile commerce, mobile health care, mobile government, and others, are obtainable by means of cloud computing resources (CCRs) [9]. Cloud computing is a technology for on-demand network access to configurable computing resources such as networks, servers, storage, applications, and services. Cloud computing facilitates users to get their data stored and applications run in a distant cloud server. The importance of 5G can be understood once the 4G technology and its limitations become clear and practiced. Thus, the typical 5G concept should be introduced somewhere around 2020.

1.3.7 Comparison of the Generations of Mobile Communication

A comparative analysis of the generation of mobile communication is presented in Table 1.7 [1–10].

1.4 Mobile Operating Systems

The mobile operating system, or the mobile OS, offers the platform to operate a smartphone, tablet, or other mobile devices [11]. The present-day mobile OS sustains features of a personal computer together with those of touch screen, Bluetooth, Wi-Fi, GPS, camera for still and moving images, video and audio recorders, video and audio player, and many other attributes. The development of mobile OS over generations has led to the enhancement of features of the mobile devices and evolution of smartphones from ordinary phones. A brief history of the generations of mobile OS is presented here [11].

- The period 1979–1992 saw mobile phones using embedded systems for controlling their functioning.
- In 1993, the first smartphone was launched with touch screen, e-mail, and personal digital assistant (PDA) facilities.
- The year 1996 saw the introduction of the Palm Pilot 1000 PDA operating with the Palm OS. This year was also marked by the introduction of the Windows CE portable PC devices.
- In 1999, the Nokia S40 OS was introduced to the market.
- In 2000, Symbian OS was introduced.
- In 2001, Kyocera 6035 was launched as the first smartphone with Palm OS.
- In 2002, the first Blackberry smartphone was introduced.
- In 2005, Maemo OS was released by Nokia for the first Internet tablet N770.
- In 2007, Apple iPhone with iOS was introduced.
- In 2008, Android 1.0 was introduced by HTC.
- In 2009, the webOS366 was introduced by Palm, and the Bada OS was introduced by Samsung.
- In 2011, the first mobile device with Linux operating system was introduced.

TABLE 1.7

Mobile Communication History and Status

Generation	1G	2G	2.5G	3G	4G	5G
Starting time	1985	1992	1995	2002	2010–2012	Will be in the market by 2020
Driven technique	Analogue signal Processing	Digital signal processing	Packet switching	Intelligent signal processing	Intelligent software auto configuration	Packet switching
Representative standard	AMPS, TACS, NMT	GSM, TDMA	GPRS, I-Mode, HSCSD, EDGE	IMT-2000 (UMTS, WCDMA, CDMA2000)	OFDM, UWB	OFDM, MC-CDMA, LAS-CDMA, IPv6
Radio frequency (Hz)	400 M–800 M	800 M–900 M	1800 M–1900 M	2G	3G–5G	Greater than 4.9G
Bandwidth (bps)	2.4 kbps–30 kbps	9.6 kbps–14.4 kbps	171 kbps–384 kbps	2 Mbps–5 Mbps	10 Mbps–20 Mbps	Higher than 1 Gbps
Multiaddress technique	FDMA	TDMA, CDMA	TDMA, CDMA	CDMA	FDMA, TDMA, CDMA	LAS-CDMA, MC-CDMA,
Cellular coverage	Large area	Medium area	Medium area	Small area	Mini area	
Core networks	Telecom networks	Telecom networks	Telecom networks	Telecom networks, some IP networks	All-IP networks	All-IP networks
Service type	Voice Mono-service Person-to-person	Voice, SMS Mono-media Person-to-person	Data service	Voice, data Some multimedia Person-to-machine	Multimedia Machine-to-machine	Dynamic access to information Wearable devices with artificial intelligence capabilities

- The year 2012 was marked by the introduction of the Firefox OS by Mozilla.
- In 2013, the Ubuntu Touch version of Linux was introduced, and Blackberry introduced Blackberry 10 as its new OS for smartphones.

The mobile OS executes resource allocation and management and provides a platform to run the software that operates the mobile devices [11].

1.4.1 Windows CE Operating System

Windows CE OS was introduced by Microsoft [1]. It is officially known as Windows Embedded Compact. It is a real-time operating system suitable for handheld devices and embedded systems. Windows CE is suitable for devices with reduced storage and is run under a megabyte of memory. Many platforms have been developed on the Windows CE platform, which include Microsoft's AutoPC, Pocket PC 2000, Pocket PC 2001, Windows Mobile 2003, Windows Mobile 5.0, Windows Mobile 6, Smartphone 2002, Smartphone 2003, Windows Phone, and many other embedded and handheld devices. Some of the features of the Windows CE operating system are as follows [1]:

- It is a 32-bit operating system.
- It is well suited with a diversity of processor architectures.
- It is compiled for a precise set of hardware.
- The source code is contained in the kernel including the hardware abstraction layer.
- Memory space is partitioned.
- It supports large storage media.
- It includes a file system that supports larger file sizes.
- Files and data are saved in the flash file system when the battery is exhausted.
- It supports connectivity to cellular networks.

1.4.2 Mac OS X

Mac OS X is used in Apple iPhones [1]. There are four layers in this OS. The first layer facilitates the basic services, the second layer is responsible for the core services, the third layer supports media features, and the fourth layer is designated as the touch layer. The OS size is 500 MB but may vary depending on the OS version [1]. They support highly efficient and speedy user interfaces.

1.4.3 Symbian OS

Symbian OS is used in a large range in smartphones [11]. It is very similar to the desktop operating systems [1]. It is an open source operating system designed for smartphones. It supports preemptive multitasking, multithreading, and memory management. It puts huge stress on memory conservation. Symbian OS supports event-based processing, and when applications are not related to events, the processor is switched off. This technique helps in reducing power consumption and enhancing battery lifetime. Some of the features of the Symbian OS are as follows [11]:

- It supports multimedia and multimodal applications.
- The graphics of Symbian support 3D rendering.

- Application development languages for Symbian include Personal Java, Java JVM, Java micro edition, JavaPhone, Symbian C/C++, Adobe Flash, Web Runtime, and others.
- Symbian OS supports Linux, Mac OS, MeeGi, and Maemo.
- It supports maintenance of contact lists, push-to-talk, slideshows, e-mail download, address book, spreadsheets, calendar, memo, Internet browsing, word processor, text-to speech converter, Adobe Reader, music, and video payer.

1.4.4 Android OS

Android OS was introduced as an open source operating system, but a large proportion of software running on Android devices such as Play Store, Google Search, Google Play Services, Google Music, and others are licensed [1]. Android versions 1, 1.5, and 1.6 are solely used for mobile phones, Android 2.x was used for tablets in addition to mobile phones, but Android 3.0 was solely developed for tablets and did not run on mobile phones [11]. The latest Android version used is Android 4.4. Android holds a share of 52.5% in the global market [11]. Some of the features of the Android OS are as follows [1]:

- Supported by platforms such as ARM, MIPS, and x86
- Supports HTML 5, open source web browser, Google Maps
- Supports pictures to be directly uploaded to Picasa via mobile devices
- Supports upgraded search and navigation
- Supports animation between screens and 2D and 3D graphics
- Supports multitasking
- Supports GPS, accelerometer, magnetometer, picture camera, FM tuner, touch screen facility, live wallpapers, and many other features

1.4.5 Blackberry 10

Blackberry 10 is a proprietary operating system developed by Blackberry Limited for the Blackberry smartphones and tablets [1]. Blackberry 10 OS is supported by devices such as Q5, Q10, Z10Z30, and the P'9982 smartphones. This OS uses a revolutionary grouping of motion and touch for navigation and control [11]. The back button is functionless with this OS, and any command or data can be entered without the need to press any key. The only need to press a key is to turn the device on/off. Blackberry 10 OS supports multitasking and provides facilities such as Twitter, Facebook, Blackberry Messenger, LinkedIn, and so on. Blackberry 10 supports capturing multiple frames of every single photo. The features of the clicked images can be easily corrected with the software. They also support video chats and voice over IP calls [11].

1.5 Applications of Mobile Communication

The improvement in the field of mobile communications has led to latest technologies and devices [1]. The passing generations of mobile communications and computing have seen the evolution of smartphones from mobile phones. These phones encapsulate the capacities of uploading and downloading data to and from a PC, FM radio, television viewing,

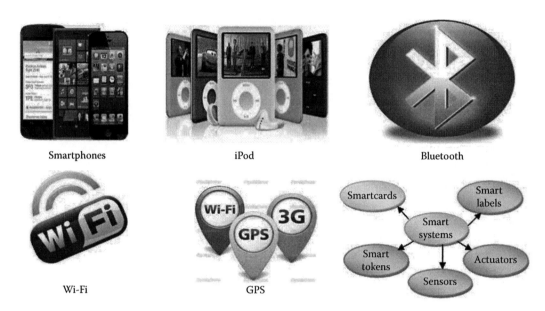

FIGURE 1.10
Applications of mobile communication.

TABLE 1.8

Features of Smartphone

---

1. It provides GSM or CDMA interface with the core network made available by the mobile service provider.
2. It comes with a keypad which allows the facility to enter text data by pressing a numeric keypad a multiple number of times.
3. It comes with a high-quality LCD screen.
4. It provides features like sending and receiving SMS and MMS, contact list, call logs, calculator, task-to-do list, alarm clock, calendar, memo, location detection services, voice and video chat services.
5. These devices are WAP, Wi-Fi, or Internet enabled for accessing web pages and downloading data.
6. Provides features like games, e-business and e-ticketing.
7. Provisioning of features like MP3 audio and MP4 video formats. Cameras are employed for viewing still pictures as well video recording. Picture editing features are also available with these devices.

---

sending and receiving e-mails, camera, and Internet connectivity [1]. These devices bring the world to the hands of the users with the help of providing information related to anything, anytime, and anywhere. Some of the applications of mobile communication are pictorially depicted in Figure 1.10.

1.5.1 Smartphones

A smartphone is a handheld device with computation capabilities [1]. It provides features of an ordinary mobile phone and has various other advanced features. The features of smartphones are tabulated in Table 1.8 [1].

1.5.2 Digital Music Players

The vast range of applications provided by mobile communications has transformed the field of user entertainment and the way of listening to music in the form of the digital music players [1]. These players consist of software and portable music players that can play music files encoded in formats like MP3 and Real-media on mobile devices [1]. They make use of flash memory. They have a storage capability of about 128 MB to 80 GB and can store more than 15,000 songs. The players with flash memory can be used as a storage device. The media players commonly used these days support the playback of both video and audio. The leading device in the player market is the iPod. iPods are players with the flash memory enabled. They include both video and audio players. Present day iPods incorporate the features of transferring, storing, managing, and playing videos and audios as well as displaying photos. Any media file downloaded in iPod or PC can be transferred one way or both ways.

1.5.3 Bluetooth and Wi-Fi

Bluetooth versions 1.2, 2.1, and 3.0 are used in mobile devices [1]. Bluetooth 1.2 and 2.1 have a data throughput of 1 and 3 MHz, respectively, and can be used to transfer contact information and files [1]. Bluetooth 3.0 has a data throughput of 24 MHz and is capable of transferring video files [1]. Wi-Fi versions popular these days are 802.11b and 802.11n [1]. They support a data rate of about 54 Mbps to 600 Mbps. It is used in the frequency bands 2.4 GHz or 5 GHz [1]. Bluetooth technology is constructive when transporting information between two or more devices that are close to each other when the speed of data transfer is insignificant, such as that of phones, printing devices, modems, and headphones. It is excellently suited to low-bandwidth applications such as transferring audio with Bluetooth headset or files with handheld computers or keyboard and mice. Wi-Fi is suitable for functioning full-scale networks because of speedy connection, improved coverage from the base station, and enhanced security services compared to Bluetooth [1].

1.5.4 GPS

Current mobile devices have a global positioning system (GPS) receiver, which provides the location information, or it may receive the GPS location from the service provider [1,12]. GPS is a satellite-based system that gives information related to the location and time irrespective of the weather conditions. The system is highly beneficial in military, civil, and commercial world [12].

1.5.5 Smart Systems

Smart systems with embedded computational abilities are one of the major applications of mobile communication. Smart systems provide efficient and remote access to devices [1]. These devices are used over a wide range and have numerous applications in daily lives. Some of the smart systems are as follows:

- Smartcards
- Smart labels
- Smart tokens
- Sensors
- Actuators

1.5.5.1 Smartcards

Smartcards are small cards with computational capabilities embedded. These cards are capable of storing and updating data. There are two types of smartcards: contact smartcards and contactless smartcards [1]. The gold-plated pins on the chips of contact smartcards facilitate contact with the circuit of the card reader on inserting the card [1]. Contactless smartcards communicate with the card reader via radio frequency technology on holding it close to the reader. These cards consist of a fabrication key, a personalization key, and a utilization lock. The fabrication key is used to uniquely identify the card. Personalization key is used by the server to activate the card for different transactions. The utilization card is used to lock or unlock the card. The applications of smartcards include the following [1]:

- Monetary transactions like a credit card or an ATM card
- Storing bank account balance after any transaction
- Storing information like personal ID or photo or other personal information
- Storing medical information of the card holder
- Used by doctors to get faster access to patients' medical information
- Used by the employees in an organization to open security locks and logging
- Issued to students for borrowing library books

1.5.5.2 Smart Labels

Smart labels are used to identify the contents of a package. For example, the barcode labels on a book pack identify the publisher, the title, the author, the date of publication, and the reprint edition of the book. These labels can also be used for identifying the product details and the price. Smart labels consist of the processor, memory, transmitter, receiver, and antenna [1]. These cards are charged by the received signals and use wireless communication.

1.5.5.3 Smart Tokens

Smart tokens are used for authentication prior to entry into any restricted area. It consists of processor and memory embedded in the chip. These are small-size systems, almost the size of a button or pen nib, and can be used for the following purposes [1]:

- To authenticate employees before entry to an office
- To open car doors remotely
- To authenticate parcels sent to the defense departments

1.5.5.4 Sensors

Sensors are electronic devices capable of sensing physical environment such as temperature, pressure, light, dust, smoke, and distance from any object [1]. Smart sensors consist of a processor and memory. In mobile devices, sensors provide communication with other devices as well as the surrounding environment. Some applications of the sensors are the following [1]:

- Background noise detection and voice intensification at the time of a call
- Controlling the brightness of an LCD screen

- Measuring signal strength and strengthening signals
- Measuring angular velocities
- Turning a device on and off

1.5.5.5 Actuators

Actuators operate by receiving a signal from a central controller and activating a physical device. A smart actuator receives a signal from mobile devices, computers, or controllers. Sensor and actuator together can be used to manage other systems such as, for example, a sensor actuator pair for controlling an oven temperature or pressure [1].

1.6 Challenges of Mobile Communication

Mobile communication is gaining utmost importance with the increasing usage of portable computers and the need to access the Internet irrespective of the location of the user [1]. The effectiveness of upcoming technologies lies in their capability to offer more and more innovative applications and contribute to their growth. Mobile devices are expected to meet the goals of effectiveness and convenience. Although mobile devices have attained heights of success, they still have some drawbacks. The drawbacks or challenges faced by mobile communication are divided into the following sections [1,13]:

1. Wireless communication
   a. Network disconnection
   b. Network bandwidth
   c. Network optimization for confined areas with high user concentration
   d. Variable network conditions
   e. Security issues
2. Mobility
   a. Changing network address
   b. Locality migration
   c. Location management
   d. Mobile network traffic forecasting
   e. Mobile call admission control and handover management
3. Resource limitation
   a. Data storage
   b. Power consumption
4. Mobile channel models
5. Disaster management
6. Mobile data mining
7. Quality of service

1.6.1 Wireless Communication

Mobile devices have either wired or wireless communication with the Internet. When the mobile device is stationary, wired communication provides cheaper and better Internet access [13]. But due to mobility, these devices need wireless communication too. Wireless communication comes with several disadvantages such as frequent disconnection due to network error, low bandwidth accessibility, and the varied network conditions. Moreover, security of wireless communication is more vulnerable than wired communication.

1.6.1.1 Disconnection

Network failure can result in disconnection of mobile devices, which affects the quality of service and hampers the quality of experience by the mobile users. Methods involved in the prevention of network disconnection involve expenditure of extra resources to deal with the disconnection scenario effectively [13]. Round-trip delays are expensive because of wasted clock cycles in case of disconnection [1]. Network disconnections are better handled by stand-alone machines that can execute the operation entirely by itself rather than splitting the operations. For a network with disconnections, a stand-alone machine is the optimal solution.

1.6.1.2 Low Bandwidth

Wireless networks provide lower bandwidths compared to wired networks. Thus mobile communication is greatly susceptible to low bandwidth. The bandwidth that can be delivered to each user depends on the number of users sharing a cell [13–15]. For improving network capacity, techniques such as data compression, splitting large applications into smaller ones (known as logging), pre-fetching and write-back caching are commonly used [15].

1.6.1.3 Network Optimization for Confined Areas with High User Concentration

Wireless communication, from broadband Internet access to basic voice calls, is becoming an ever-more vital part of people's existence globally. Regrettably, wireless communications can experience technical hitches in restricted areas with high user density, such as convention centers, airports, train cars, airplane cabins, special events locations, and stadiums. Commonly experienced problems include insufficient capacity and high interference [16], and these constitute a major challenge for mobile communication.

1.6.1.4 Variable Network Conditions

Mobile devices encounter varied network conditions as a result of their mobility. In their path of leaving the range of one network transceiver and entering the range of another, they may experience environments where they can access multiple network transceivers operating with variable frequencies [13]. There is even a need to switch interfaces while moving from one region to another, which requires changing the access protocol for different networks [14]. Thus these lead to complexity of mobile communication compared to traditional computing.

1.6.1.5 Security Issues

Wireless communication is more vulnerable to the risk of security hazards due to the ease of breaking into the wireless links. Kerberos are often implemented to preserve security of a system. Kerberos allow users to be authenticated without their passwords being exposed over the network [13]. They allow mobile users to authenticate to foreign networks where they are unknown [15]. However, security provided by Kerberos is also limited and is susceptible to offline attacks [13–15].

1.6.2 Mobility

Mobile devices are able to change locations unlike static devices. The risk of information loss increases with mobility. The challenges associated with mobility are discussed in the following sections.

1.6.2.1 Changing Network Address

As their location changes, mobile devices use different access points and different addresses. When in active mode, devices cannot be moved to a new address [13]. The address for a host name is cached in the system memory for a long time without the provision of validating the data stored. In the IP, for example, a host name is embedded with its network address. Thus, changing the location means acquiring a new IP name. Some methods for solving address migration are as follows [13]:

- *Broadcasting*: The broadcast method facilitates a message to be sent to all mobile hosts in the network asking for their current address. But this method is cost-inefficient owing to the increase in the number of messages being sent over the network [13]. To solve this problem, selective broadcast can be advantageous, where broadcast messages are sent only to the hosts that know the sending device previously. Thus with selective broadcasting, the number of messages being sent over the network is reduced.
- *Central service*: The central service method facilitates the current address of each mobile device in the network to be maintained in a centralized database [13]. The information in the database is periodically updated. Although centralized, this method is advantageous with the usage of database replication, which reduces the possibility of loss of information.

1.6.2.2 Locality Migration

Mobile communication prompts the locations to migrate as the user moves; for example, if the mobile device finds the nearest server over time, there is no restriction that this server will continue to be nearest to the mobile device. Thus, additional cost, power, and time are required to select a new server each time the device moves. In order to eliminate this disadvantage, the service connections may be dynamically moved to servers that are closer to the device [13].

1.6.2.3 Location Management

Location management is an important area of mobile communication [17–20]. In a personal communication services (PCS) network, the service area is divided into a number of location areas. The process of tracking the location of a mobile terminal (MT) is referred to as location management. Basic operations involved in location management are location update and paging. Whenever a mobile terminal enters a location area, a location update is performed. On arrival of an incoming call, the network searches the called MT; this process of searching the called MT is referred to as paging [21–23].

1.6.2.4 Mobile Network Traffic Forecasting

Network traffic forecasting is a significant component of early network deployment and dynamic optimization of network capacity. Advanced network traffic forecasting can lessen the initial network investment costs by better approximation summation and intracell network demand. Short-term traffic prediction results in more optimal beam forms within a cell and better allocation of both intracell and intercell network capacity. Long-term network traffic forecasting is critical for the optimal deployment of network equipment throughout a mobile operator's coverage area. Short-term network traffic forecasting helps the operators to dynamically optimize capacity within a cell and among cells [21,23–25].

1.6.2.5 Mobile Call Admission Control and Handover Management

A base station blocks new calls and call handoffs during peak times when its reserve capacity, called the guard channel, for new calls and call handoffs is completely filled [24]. By utilizing spare capacity, a cell could accommodate new requests without negatively affecting the service levels of existing requests. With the growth in other types of call classes, or applications, such as e-mail and file downloads, which are less sensitive to latency, a mobile operator might be able to shift capacity from one call class to another without significantly affecting the users' experience levels during peak times [25]. Future hardware and software advancements could enable channel borrowing between neighboring base stations, thereby increasing a mobile network's overall utilization. Efficient handoff management prevents call dropping and avoids unwanted interruption of service. By dynamically adjusting the call admission thresholds for new calls and call handoffs, a base station can better utilize its available capacity and potentially increase revenues during peak periods [25,26].

1.6.3 Resource Limitation

Mobile devices are resource-constrained in terms of storage capacity and battery lifetime. These constraints are discussed in the following subsections.

1.6.3.1 Data Storage

The memory of mobile devices is finite. Thus, these devices are not capable of storing an unlimited volume of information. This is a major constraint of the mobile devices because, with the increasing usage of mobile devices, the need to store various data in the device memory also increases. Solutions proposed for handling this shortcoming includes offloading the storage to some other device as facilitated by cloud computing, or, more specifically, mobile cloud computing [26].

1.6.3.2 Power Consumption

In the absence of any power outlet or portable generator, mobile devices rely on battery power for functioning [13]. Battery is one of the main alarms for mobile devices. Quite a few solutions have been proposed to enhance the CPU performance and to manage the disk and screen in an efficient manner to reduce power consumption [15]. However, these solutions require changes in the structure of mobile devices, or they require new hardware, which results in an increase of cost and may not be feasible for all mobile devices [15]. The computation offloading technique is proposed to migrate large computations and complex processing from resource-limited mobile devices to resource-rich cloud servers. This avoids lengthy application executions on mobile devices that consume a large amount of power [26]. On the other hand, base stations in a network consume high power. To reduce the power consumption by base stations, various schemes are proposed by Mukherjee and De [27]. To achieve a low-power congestion control scheme, Hung et al. proposed a traffic-based cellular network deployment method [28].

1.6.4 Mobile Channel Model

Modeling of a mobile communications channel is one of the most fundamental areas of mobile technology and is essential for the design, evaluation, and testing of mobile systems. Understanding the channel state is also critical for smart antennas [16,21–24]. Among other benefits, more sophisticated channel models could help mobile operators optimize base station deployment and periodically tune base stations' transmission patterns, which would allow them to increase the usable capacity those stations [26]. To achieve higher mobile network performance levels, channel models need to more accurately estimate the actual channel conditions for initial network planning and continuously update the channel state to adjust a base station's beam forms [16,21–27].

1.6.5 Disaster Management

Because of the sudden occurrence of a natural disaster or an accident, the number of users in a cell within that devastated area increases rapidly. Therefore, the number of call arrivals and call generations increases suddenly in that confined area.

1.6.6 Mobile Data Mining

Highly developed data mining could provide a means for better understanding mobile users, enabling mobile operators and other mobile service providers to provide services tailored to specific customer segments and ultimately to individual users. Such efforts could stimulate greater user spending and increase customer satisfaction, thus decreasing customer defection [29].

1.6.7 Quality of Service

New QoS measurement tools will help service providers, such as mobile telephone vendors, to better measure and monitor users' experience with various applications, including voice calls, SMS, email, web browsing, video clips, streaming video, and GPS. The growth of mobile data applications is expected to hurt the capacity of mobile networks and could degrade the quality of experience during periods of peak demand. Acceptance of new mobile services will depend on the quality perceived by end users; this will require providers to have effective tools and methods to monitor and measure services.

1.7 Conclusion

Mobile communication is one of the primary areas of science and technology that has been focusing on the way information exchanges between individuals and organizations via mobile devices. The evolution of mobile communication replaced the usage of telegrams and letters with mobile phones and emails. Mobile communication has recently become the spine of civilization. The technology of mobile communications has improved the art of living and made lives easier. But it comes with certain challenges. In this chapter, we presented a discussion on the details of mobile communication such as the architecture of mobile communication, the evolution of mobile communication from 1G to 5G, different operating systems supporting mobile communication, and the applications of mobile communications, and highlighted a few challenges of mobile communication.

Questions

1. What are the differences between mobile communication and mobile computing?
2. Compare between 2G, 3G, and 4G mobile networks.
3. Compare between TDMA, FDMA, and CDMA networks.
4. Draw and explain the GSM block diagram.
5. What is a GPRS network? What are advantages of SGSN and GGSN?
6. Draw and explain the architecture of a mobile ad hoc network.
7. Explain the architecture of a mobile wireless sensor network.
8. What are the challenges of mobile WSN?
9. Explain the functions of HLR and VLR.
10. What is LTE? Explain the LTE standard.
11. What is a mobile operation system? Explain different types of OS used in mobile devices.
12. What are the advantages of 5G network over 3G and 4G networks?
13. What are the features of a smartphone?
14. What are the challenges of mobile communication?
15. What are the challenges of mobile computing?
16. Explain the working principle of GPS.

References

1. R. Kamal, *Mobile Computing*, Oxford University Press, Inc., Oxford, U.K., 2008.
2. T. S. Rappaport, *Wireless Communications: Principles and Practice*, vol. 2, Prentice Hall PTR, Upper Saddle River, NJ, 1996.
3. L. S. Ashiho, Mobile technology: Evolution from 1G to 4G, *Electronics for You*, 94–98, 2003.

4. K. B. Moses, Mobile communication evolution, *International Journal of Modern Education and Computer Science*, 1, 25–33, 2014.
5. P. Kanani, K. Shah, and V. Kaul, A survey on evolution of mobile networks: 1G to 4G, *International Journal on Science and Research*, 3(2), 802–810, 2014.
6. S. Ohmori, Y. Yamao, and N. Nakajima, The future generations of mobile communications based on broadband access technologies, *IEEE Communications Magazine*, 38(12), 134–142, 2000.
7. X. Li, G. Abudulla, S. Rosli, and Z. Omar, The future of mobile wireless communication networks, in *IEEE International Conference on Communication Software and Networks*, Macau, pp. 554–557, 2009.
8. T. Miki, T. Ohya, H. Yoshino, and N. Umeda, The overview of the 4th generation mobile communication system, *International Conference on Communications*, 2(3), 1551–1553, 2005.
9. P. Sharma, Evolution of mobile wireless communication networks-1G to 5G as well as future prospective of next generation communication network, *International Journal of Computer Science and Mobile Computing*, 2(8), 47–53, 2013.
10. K. K. Sharma, C. Kumar, and D. Kumar, An overview on 5G technologies, *International Journal of Information Technology and Knowledge Management*, 6(2), 171–174, 2013.
11. D. Johansen, R. Van Renesse, and F. B. Schneider, Operating system support for mobile agents, in M. N. Huhns and M. P. Singh (eds.) *Readings in Agents*, pp. 263–266, Morgan Kaufmann Publishers, San Francisco, CA. 1997.
12. B. Hofmann-Wellenhof, H. Lichtenegger, and J. Collins, *Global Positioning System: Theory and Practice*, Springer Science & Business Media, 2013.
13. G. H. Forman and J. Zahorjan, The challenges of mobile computing, *Computer*, 27(4), 38–47, 1994.
14. M. Satyanarayanan, Fundamental challenges in mobile computing, in *15th Annual ACM Symposium on Principles of Distributed Computing*, Santa Barbara, CA, pp. 1–7, 1997.
15. G. Deepak and B. S. Pradeep, Challenging issues and limitations of mobile computing, *International Journal of Computer Technology and Applications*, 3(1), 177–181, 2012.
16. A. Alexiou, C. Bouras, and A. Papazois, A study of multicast congestion control for UMTS, *International Journal of Communication Systems*, 22(6), 739–754, 2009.
17. A. Mukherjee and D. De, DAS: An intelligent three dimensional cost effective movement prediction of active users in heterogeneous mobile network, *Journal of Computational Intelligence and Electronic Systems*, 1(1), 31–47, 2012 (American Scientific Publishers).
18. D. De and A. Mukherjee, A cost-effective location management strategy based on movement pattern of active users in a heterogeneous system, in *IEEE URSI General Assembly and Scientific Symposium*, Istanbul, Turkey, pp. 1–4, 2011.
19. A. Mukherjee and D. De, A cost-effective location tracking strategy for femtocell based mobile network, in *IEEE International Conference on Control, Instrumentation, Energy and Communication*, Calcutta, India, 2014.
20. D. De and A. Mukherjee, A novel cost-effective and high-speed location tracking scheme for overlay macrocell-femtocell network, in *IEEE URSI General Assembly and Scientific Symposium*, Beijing, China, 2014.
21. S. Tabbane, An alternative strategy for location tracking, *IEEE Journal in Selected Areas in Communications*, 13(5), 880–892, 1995.
22. R. Khalil, W. Elkilani, N. Ismail, and M. Hadhoud, A cost efficient location management technique for mobile users with frequently visited locations, in *Fourth Annual Communication Networks and Services Research Conference*, Moncton, NB, Canada, pp. 259–266, 2006.
23. S. Wu, T. Chow, K. T. Ng, and K. F. Tsang, Improvement of borrowing channel assignment for patterned traffic load by online cellular probabilistic self-organizing map, *Neural Computing and Applications*, 15(3–4), 298–309, 2006.
24. T. S. Kim, M. Y. Chung, and D. K. Sung, Mobility and traffic analyses in three-dimensional PCS environments, *IEEE Transactions on Vehicular Technology*, 47(2), 537–545, 1998.
25. H. T. Dinh, C. Lee, D. Niyato, and P. Wang, A survey of mobile cloud computing: Architecture, applications, and approaches, *Wireless Communications and Mobile Computing*, 13(18), 1587–1611, 2013.

26. A. Mukherjee, S. Bhattacherjee, S. Pal, and D. De, Femtocell based green power consumption methods for mobile network, *Computer Networks*, Elsevier, 57(1), 162–178, 2013.
27. A. Mukherjee and D. De, Congestion detection, prevention and avoidance strategies for an intelligent, energy and spectrum efficient green mobile network, *Journal of Computational Intelligence and Electronic Systems*, 2(1), 1–19, 2013 (American Scientific Publishers).
28. A. I. Zreikat, K. Al-Begain, and K. Smith, A comparative capacity/coverage analysis for CDMA cell in different propagation environments, *Wireless Personal Communications*, 28(3), 205–231, 2004.
29. S. Y. Hung, D. C. Yen, and H. Y. Wang, Applying data mining to telecom churn management, *Expert Systems with Applications*, 31(3), 515–524, 2006.

2

Cloud Computing

ABSTRACT   Cloud computing is a recent trend in the area of computation. This chapter discusses the evolution, architecture, and applications of cloud computing. Various aspects of cloud computing such as security, data management, energy efficiency, etc., are discussed. Different deployment schemes of cloud are also described.

KEY WORDS: *cloud, virtualization, utility, resource, service.*

2.1 Introduction

Cloud computing is an evolution in the field of computer science and technology. In the twenty-first century, computer users access Internet services via lightweight portable devices because powerful desktop machines are going through a phase of drought. Cloud computing emerged as a solution to this problem. Cloud is a distributed computing paradigm. It is a collection of interconnected and virtualized computers, which are provisioned and presented dynamically as unified computing resources offered on a pay-per-use basis [1]. Cloud computing is defined as applications that are delivered as Internet services: the hardware and system software in the data centers are used to provide these services [2]. Cloud computing is an advanced technology that focuses on the way of designing computing systems, developing applications, and leveraging existing services for building software [3]. It is based on dynamic provisioning [3]. In cloud computing, resources are offered in an on-demand and pay-per-use basis from the cloud computing vendors [3]. In this chapter, we will discuss the evolution, architecture, applications, and other issues related to cloud computing.

2.2 Evolution of Cloud Computing

The shift in computing paradigm over six distinct phases ranges from simple dummy terminals to powerful and efficient grids and clouds [4]. The birth of the concept of cloud computing dates back to 1950 when people used terminals to connect to powerful mainframes simultaneously shared by many users. In the next phase, almost everyone started possessing a personal computer (PC) for daily usage, and there was no need to share the mainframe with anyone else. It was during the third phase that the computer networks ushered into the world of computing. We could work on a PC and connect to other computers via local networks to share resources. The fourth phase was the era of the Internet where local networks could connect to other local networks to establish a global network. Users started accessing remote applications and resources via the Internet. Grid computing emerged during the fifth phase as an evolution. Grid computing is the collection of computer resources from multiple locations to reach a common target. The grid can be thought to be a distributed system with noninteractive workloads that involve a large number of files. Grids are a form of distributed computing. PCs were used to access a grid of computers in a transparent manner.

With the gradual advancement in the world of computing, the sixth phase witnessed the exploitation of available resources on the Internet in a simple and scalable way. This came to be known to us as "cloud computing." Cloud computing, in general, means a large number of computers connected through a real-time communication network, typically the Internet. A cloud provides a standard interface but hides all the available resources, hardware and software, and services, which ideally are similar to the transparency property of distributed systems. Our PCs are simply used as lightweight terminals having access to the powerful Internet cloud allowing us to utilize the cloud.

But unlike the mainframe computer, which is a physical machine providing finite computing power, cloud computing provides us with the entire power of the Internet. Figure 2.1 depicts the evolutionary cycle or the birth of this new genre of computing, cloud computing [4].

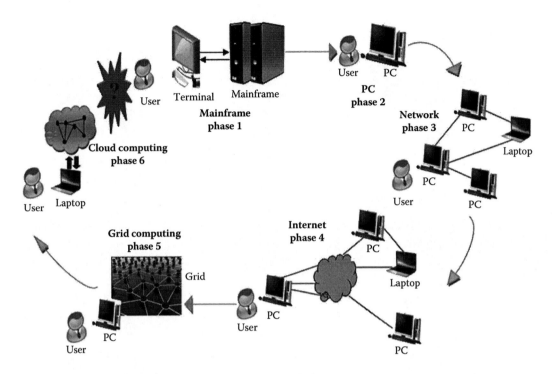

FIGURE 2.1
Birth of cloud computing.

2.3 Cloud Computing: What Is It Actually?

In the simplest terms, cloud computing is said to be the next stage in the evolution of the Internet. The cloud provides the means through which anything and everything—from computing power to computing infrastructure, applications, business processes, and personal collaboration—can be delivered to consumers as a service whenever needed [5].

Internet services are the most popular applications with lots of users. Websites such as Facebook, Yahoo, and Google are accessed by millions every day, as a result of which a huge volume of valuable data (in terabytes) is generated, which can be used to improve online strategies of advertising and user fulfillment. Storage, real-time capture, and analysis of that data are general needs of all applications. To trace these problems, some cloud computing strategies have recently been implemented. Cloud computing is a style of computing where virtualized resources are provided to the customers as a service, which is dynamically scalable, over the Internet. The cloud refers to the data center hardware and software that a client requests from remotely hosted applications, often in the form of data stores. Those companies are using these infrastructures to cut costs by eliminating the call for physical hardware, which allows them to outsource data and on-demand computations. The function of large-scale computer data centers is the main focus of cloud computing. These data centers benefit from the economies of scale, allowing for decrease in the cost of bandwidth, operations, electricity, and hardware [6].

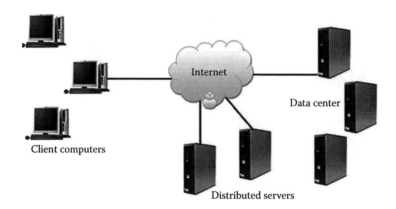

FIGURE 2.2
Components of cloud computing solution.

The term "cloud computing" has been coined as a metaphor for the Internet. Generally, the Internet is represented in network diagrams as a cloud, as shown in Figure 2.2.

A cloud computing system mainly consists of clients, data centers, and distributed servers. Each of these has a specific purpose and plays an important role in delivering a functional cloud-based algorithm. The components of the cloud computing solution are depicted in Figure 2.2.

In cloud computing architecture, clients are the devices with which the end users interact to manage information on the cloud. Mobile, thin, and thick clients are different types of existing clients. Mobile clients include personal digital assistants (PDAs) or smartphones, such as Blackberry or iPhone. Thin clients are computers that do not have internal hard drives. They let the server do all the work and then display the result. A thick client is a regular computer using a web browser such as Firefox or Internet Explorer to connect to the cloud. The collection of servers where the applications to which the customers subscribe is termed a data center. But these servers are not located at the same site. They are present in geographically different locations, but this does not appear to the end users. This cloud computing architecture solves the problem of site failure. For instance, Amazon has its servers present all over the world. If any one site of Amazon fails, the entire system will continue functioning efficiently as many other Amazon servers function perfectly across the globe. The concept of distributed servers is the backbone behind the success of the cloud computing system.

2.3.1 Virtualization as a Component of Cloud Computing

Cloud computing is actually meant for the delivery of services to an end user. In a lucid way, virtualization can be defined as the creation of a virtual, rather than actual, version of something, such as a hardware platform, operating system, a storage device, or network resources. VMware provides us with the platform to implement the concept of virtualization. According to experts, the architectures of today's servers allow only a single operating system to run on them at a time [7]. Server virtualization unbolts the traditional one-to-one architecture of servers currently available in the market by abstracting the operating system (OS) and application software from the physical hardware, enabling a more gainful agile and basic server situation. Using server virtualization, multiple OSs can run on a single physical server as virtual machines, each with access to the original server's computing resources.

2.3.1.1 Characteristics of Virtualized Environments

Virtualization can create the artificial view that many computers are a single computing resource or that a single machine is really many individual computers. It can make a single large storage resource appear to be many smaller ones or make many smaller storage devices appear to be a single device. The three major components of virtualized environment are guest, host, and virtualization layer. The original environment is represented by the host where the guest is supposed to be managed. The guest represents the system component that interacts with the virtualization layer rather than with the host. The virtualization layer is responsible for recreating the same or a different environment where the guest will operate. The common characteristic of all these different implementations is that the virtual environment is created by means of a software program. Our present technologies allow profitable use of virtualization, and such advantages have always been characteristics of virtualized solutions.

2.3.1.2 Taxonomy of Virtualization Techniques

A taxonomy model provides an overview of the different types of virtualization technologies. The taxonomy model illustrates the main virtualization domains and their relation. Each domain is further divided into subtypes.

- *Machine reference model*: Virtualization methodologies substitute one of the layers and intercept the calls directed toward it. Hence, a clear separation between the layers simplifies their implementation, only requiring emulation of the interfaces and proper interaction with the underlying layer.

- *Execution virtualization*: Implementation of execution virtualization takes place on the top of the hardware by the OS, application, or libraries linked to an application image dynamically or statically. This type of virtualization includes the methods with an objective to imitate an execution environment separate from the one that hosts the virtualization layer.

- *Hardware-level virtualization*: The virtualization of computers or OSs, computer hardware virtualization hides the physical characteristics of a computing platform from users, instead showing another abstract computing platform. Hardware-level virtualization gives an abstract execution environment by means of computer hardware on top of which a guest OS can run. In this model, the host is represented by the physical computer hardware, the guest by the OS, the virtual machine by its emulation, and the virtual machine manager (VMM) by the hypervisor. Hardware-assisted virtualization means a situation in which the hardware gives an architectural support to build a VMM, which can run a guest OS in complete isolation. The hardware virtualization techniques are described as follows:

  - *Full virtualization*: This method virtualizes the main physical server entirely to support applications and software to operate on virtualized divisions in a similar pattern. Full virtualization supports the unmodified guest OSs where unmodified denotes OS kernels that have not been modified to run on a hypervisor and, therefore, still execute privileged operations. Full virtualization has the following advantages:

    - The existing system is combined with the newer one along with efficiency and well-organized hardware.

    - The overall performance of a company is augmented, and the need for physical space is reduced by this technique.

- *Paravirtualization*: In paravirtualization, the kernel of the guest OS is modified to run on the hypervisor. So, the hypervisor has the responsibility to do the task for the guest kernel. Paravirtualization has the following advantages:
  - Paravirtualization reduces the number of VMM. So, it increases the performance significantly.
  - Many OSs can run on a single server through paravirtualization.
- *OS-level virtualization*: OS-level virtualization is considered in order to get the best performance and measurability. This method works on OS layer. The instant of the OS and the physical server is virtualized. This virtualization takes place in multiple isolated partitions.

- *Hypervisors*: The "hypervisor" forms the backbone of the virtualization concept. A hypervisor is nothing but a computer software, firmware, or hardware to create and run virtual machines. A host machine is a computer on which a hypervisor runs one or more virtual machines. Each of the instances installed on the host machine is called a guest machine. The hypervisor presents the guest OS with a virtual operating platform and handles the execution of the guest OSs. The virtualized hardware resources may be shared by multiple instances of different OSs. Hypervisors have been classified into two categories [8]: type 1 (native/bare metal) hypervisor and type 2 (hosted) hypervisor. The type 1 hypervisors are installed directly on the hardware of the host machine on which various OSs can be installed, such as Windows, Linux, and Unix. Hence, a guest OS runs on another level above the hypervisor, whereas type 2 hypervisors run within a conventional OS environment. An OS is installed on the hardware of the host machine on which the type 2 hypervisor should be installed. The instances of guest OSs are then installed on the hypervisor. Hence, with the hypervisor layer as a distinct second software level, guest OSs run at the third level above the hardware. These two categories of hypervisors are shown in Figure 2.3a and b.

- *Programming language-level virtualization*: Language is implementation to cover different underlying machine architectures. The support software is compiler, OS, and system runtime software. Programming language-level virtualization has a long path in computer science history and was originally used in 1966 for BCPL. One of the important elements of the Common Language Infrastructure is to support multiple programming languages. It has the following advantages:
  - It provides uniform execution environment across different platforms.
  - It can be made more secure by filtering the I/O operations.

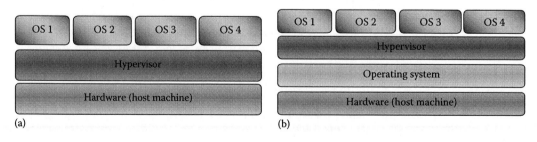

(a)   (b)

FIGURE 2.3
(a) Type 1 and (b) type 2 hypervisors.

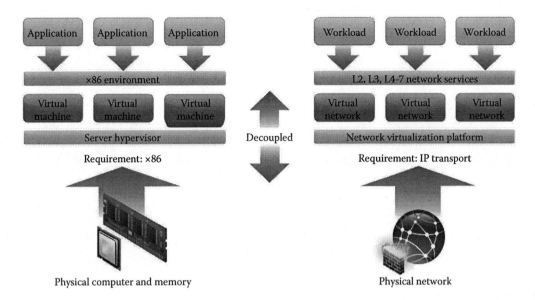

FIGURE 2.4
Network virtualization.

- *Application-level virtualization*: Application virtualization is a virtualization technique that isolates applications from the OS. With application virtualization, a set of files can be distributed independently from the OS. Two common types of application virtualization are sandbox application and application streaming.

- *Storage virtualization*: Storage virtualization is often used at locations with many storage systems. It consists of a set of technologies that creates an abstraction layer between logical storage and physical storage systems. Storage virtualization allows harnessing a wide range of storage facilities and representing them under a single logical file system. It is commonly used in storage area networks and is a form of block virtualization.

- *Network virtualization*: Network virtualization is a set of technologies that hides the true complexity of the network and separates it into manageable parts, as shown in Figure 2.4. With network virtualization, multiple networks can be combined into a single network, or a single network can be logically separated into multiple parts. The well-known network virtualizations are virtual LAN, virtual IP, and virtual private network.

- *Desktop virtualization*: Often called client virtualization, desktop virtualization is a virtualization technology used to separate a computer desktop environment from the physical computer, as shown in Figure 2.5. Each user retains his or her own instance of desktop OS and applications, but that stack runs in a virtual machine on a server, which users can access through a low-cost thin client similar to an old-fashioned terminal.

2.3.1.3 Virtualization and Cloud Computing

Cloud computing is possible without virtualization [9]. Some hardware, OS, and application clusters are able to deliver cloud services, although they are complicated and costly. They provide only limited features but require a lot of work. Virtualization and cloud

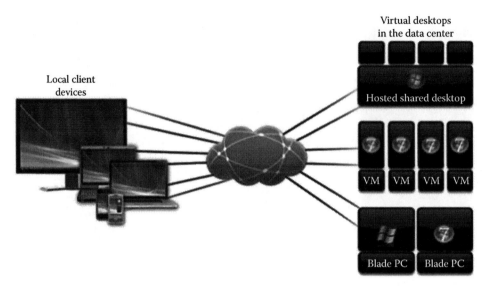

FIGURE 2.5
Desktop virtualization.

computing are interrelated because the major hypervisor vendors such as VMware, Microsoft, and Citrix Systems highlight the concept of cloud. They have closely allied their products with tools and complementary technologies that endorse the acceptance of private cloud computing. Cloud computing is an emerging trend, and it will reshape the world of smart computing. It closely allies with virtualization even though coming up with many other technologies.

2.4 Characteristics of Cloud Computing

The characteristics of cloud computing are indispensable for the clear understanding of the concept of cloud computing.

- *Broad network access*: Capabilities are accessed over the network using standard methodologies that promote the use by heterogeneous thin or fat client platforms such as laptops, PDAs, and mobile phones.

- *On-demand self-service*: A user can provision computing capabilities such as server time and network storage automatically whenever needed without human intervention with service provider.

- *Resource pooling*: The service provider's computing resources such as storage, memory, network bandwidth, processing, and virtual machines are pooled to serve multiple users based on a multi-tenant model. Different physical and virtual resources are rapidly assigned and reassigned according to user demand. In this case, the subscriber has no knowledge or control over the accurate location of the provided resources though the location at a higher level of abstraction can be specified; this is an example of location independence.

- *Rapid scalability and elasticity*: Capabilities are rapidly, elastically, and sometimes automatically provisioned to quickly scale out and released to quickly scale in. Consumers can purchase the capabilities at any time in any amount, which are available for provisioning and often appear to be unlimited.

- *Measured service*: Leveraging a metering capability at some level of abstraction suitable to the service type, cloud systems automatically manage and optimize resources. Resource usage can be observed, controlled, and reported by offering transparency with respect to both users and service providers.

2.5 Related Technologies

Different technologies related to cloud computing are discussed as follows [10,11]:

Grid computing: It is a distributed computing system. To attain a common computational objective, it coordinates networked resources. Grid computing is developed on computation-intensive scientific applications. Cloud computing is quite similar to grid computing as it also employs distributed resources to fulfill objectives at application level. In cloud computing, virtualization technologies are leveraged at multiple levels for the purpose of dynamic resource provisioning and resource sharing.

Utility computing: On-demand resource provisioning and resource usage–based billing are represented by utility computing. For economic reasons, cloud computing takes on a utility-based pricing scheme. With utility-based pricing and provisioning resources on demand, service providers can maximize resource utilization by minimizing their operating costs.

Autonomic computing: Autonomic computing intends to build self-managing computing systems to overcome the management complexity of existing computer systems. Although cloud computing possesses several autonomic features such as automatic resource provisioning, its aim is to reduce the resource cost rather than to reduce system complexity.

Virtualization: Virtualization is a technology that offers virtualized resources for high-level applications by abstracting the details of physical hardware. A virtualized server is generally referred to as a virtual machine. Virtualization forms the base of cloud computing, as it provides the capability of pooling computing resources from clusters of servers and on-demand dynamically provisioning or releasing virtual resources to applications.

To attain the objective of providing computing resources as a utility, cloud computing leverages virtualization technology and shares certain aspects with autonomic computing and grid computing.

2.6 Cloud Computing Architecture

2.6.1 Cloud Computing Service Models

Different service layers of cloud are presented in Figure 2.6. The details of the services are shown in Figure 2.7.

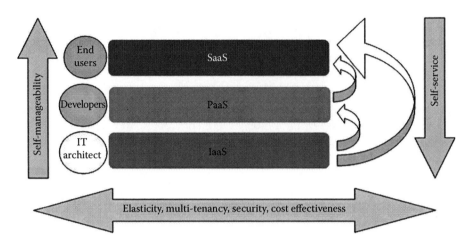

FIGURE 2.6
Different service layer of cloud.

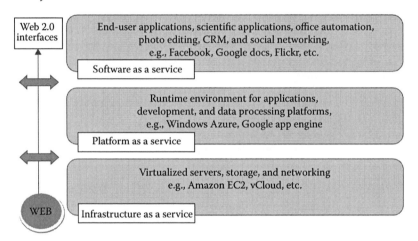

FIGURE 2.7
Different cloud services.

2.6.1.1 Software as a Service

Software as a service (SaaS) is a service-oriented model used for long-term purposes. SaaS is implemented to provide application- and process-oriented services to users, which are implemented into the cloud and hosted in an infrastructural view of the cloud. SaaS provides domain-specific service to registered users for using applications into cloud as a service over the Internet. But users can take these services on a pay-per-usage basis [12–15].

2.6.1.2 Platform as a Service

Platform as a service (PaaS) is a service model that provides all kinds of required software development life cycle model such as design tools, development tools, debugging tools, testing tools, and deployment tools. So, the main consumers of a PaaS are testers, designers, debuggers as well as software developers. Most PaaS cloud service providers fix developers into particular development platforms, and debugging and testing tools are not allowed direct link with lower computing infrastructures that are provided, although programming application program interfaces (APIs) might be provided with restricted functionalities of road and rail network control and organization [1–3].

2.6.1.3 Infrastructure as a Service

Infrastructure as a service (IaaS) gives users infrastructure support (i.e., compute, storage, OS, networking) as a service. The IaaS model allows clients to start a new project quickly by renting computing assets. The key characteristics of an IaaS cloud are *scalability and elasticity, enabling computing resources to level up and down*. Most IaaS cloud service providers offer scalability under customers' control with straight self-service interfaces, through which consumers can request to control and manage computing as well as scale the resources. An IaaS is also referred to as *Rcloud*. According to the different types of resources on hand, Rcloud can be further separated into three subcategories [12–15]:

1. *Computing as a service (CaaS)*: It offers clients raw power for computing on virtual cloud servers or virtual machine instances. CaaS gives users self-service interface for on-demand dynamic provisioning and management (i.e., start, stop, destroy, and reboot) of virtual machine instances. A CaaS contributor may also supply interfaces that are self-management oriented for autoscaling and automatable management services.
2. *Storage as a service*: This is an on-demand online storage service from cloud service providers to end users. The service is provided on a pay-per-GB basis.
3. *Database as a service (DaaS)*: It is a sub-service model of IaaS, which standardizes processes for controlling, manipulating, and accessing (i.e., read, update, write, and delete). Data in the database are accessed by the cloud in the cloud storage and provided as DaaS to users.

2.7 Cloud Computing Deployment Models

A cloud environment (IaaS, PaaS, and SaaS) can be deployed using the following three main models [1,12–15], as shown in Figure 2.8: public cloud, private cloud, and hybrid cloud.

2.7.1 Public Cloud

A public cloud is shared and used by customers via the Internet; for example, Amazon Web Services is the leading public cloud provider. The benefits of public cloud are as follows [16]:

- Very low-cost because all the virtual resources, whether hardware, data, or applications, are enclosed by the cloud service providers.
- Storage efficiency and computational services.
- Easy to connect to the cloud servers and to partake of information.

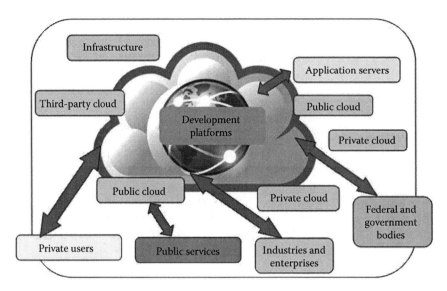

FIGURE 2.8
Different types of cloud deployment scenarios.

- Assures suitable use of resources because customers only pay for the services they need.
- Widely spread availability irrespective of geological area.
- Public cloud empowers workers and enables them to be productive even outside the office. The SaaS model makes sure that while delivering the flexibility of output software on the cloud, corporations save on IT expenditures.

The disadvantages include the following:

- Variety of applications, such as Microsoft–Amazon incompatibility
- Security issue

2.7.2 Private Cloud

A private cloud is a network of all services or a data center that stores hosted services for a restricted number of people [16]. When public cloud resources are used by a service provider to create a private cloud, it is called a virtual private cloud. Whether it is a private or public cloud, the objective of cloud computing is to provide scalable and easy-to-access computing resources and IT services.

2.7.3 Hybrid Cloud

Sometimes, it may happen that private clouds run out of capacity. To overcome this problem, hybrid cloud is introduced [12–16]. It is an integration of two or more clouds that remain distinctive entities but are bound together. The benefits and shortcomings of different types of cloud are presented in Figure 2.9.

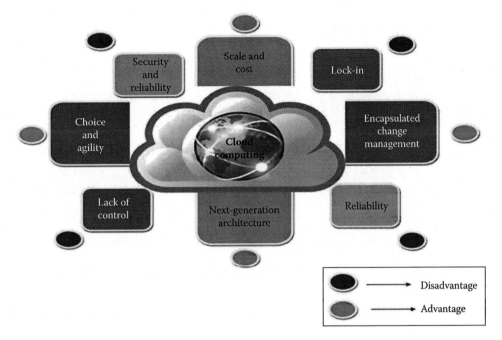

FIGURE 2.9
Different types of cloud advantages and disadvantages.

2.7.4 Public Cloud versus Private Cloud

Rationale for private cloud:

- Service-level agreements are required for reliability and real-time performance.
- Security and privacy of valuable data are huge issues.
- Cost savings of shared model achieved.
- Prospective for vendor lock-in.

2.8 Issues of Cloud Computing

2.8.1 Scheduling

In the cloud environment, different types of job-scheduling algorithms are applied with suitable modification. The main aim of a job-scheduling algorithm is to improve the performance and quality of service and, at the same time, maintain the efficiency and fairness among jobs as well as reduce the execution cost [17]. Traditional job-scheduling algorithms cannot provide scheduling in the cloud environment. Job-scheduling algorithms in cloud computing can be categorized into two main classes [17]:

1. Batch-mode heuristic algorithms
2. Online-mode heuristic algorithms

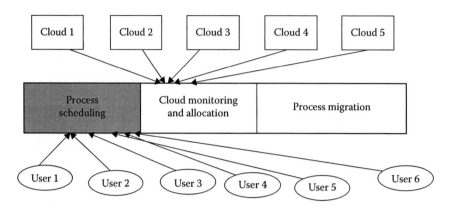

FIGURE 2.10
Architecture of job-scheduling system.

In batch-mode heuristic scheduling, jobs are collected when they arrive in the system. The scheduling algorithm starts working after a fixed amount of time, such as a Round Robin algorithm. In online-mode heuristic algorithms, jobs are scheduled individually as soon as they arrive in the system. Online-mode algorithms are more suitable for a cloud environment because of the cloud environment's heterogeneity and varying processor speed, such as the Most Fit Task scheduling algorithm.

Figure 2.10 shows the middle layer architecture of a job-scheduling system, which performs cloud allocation in the case of underload and overload conditions. The process migration handles the overloading. The middle layer exists between the cloud and the client. The middle layer accepts requests coming from the users and analyzes the cloud servers. There are three main tasks performed by the middle layer [17]:

1. Perform the allocation of process and monitor the cloud server's capabilities
2. Process migration in overloading situations
3. User requests scheduling

2.9 Security and Trust

Security is a major issue in the field of cloud computing. Personal data of users are sensitive. Storing personal information inside the cloud requires high security to protect it from hackers. Moreover, protection is needed during data transmission to the cloud. Offloading is an important aspect of cloud computing. Most IT people are keen to offload their data to cloud in order to save the costs of storing data and to reduce the overhead of private servers. But still many people are unwilling to use offloading for fear of losing their confidential information. However, for secured communication, AAA (Authentication-Authorization-Accounting) is used and for data security, various encryption techniques are available, such as AES, DES3, and Twofish.

Trust is certainly based on past experience; it is the firm belief in the reliability of others. We trust a system or person if that system or person behaves or works as we want. In the field of cloud computing, trust is an important parameter since interaction takes place over the network. There must be a trustworthy relationship between the cloud service provider and the cloud consumer. If there is a cloud manager, who maps the cloud consumer to the cloud service provider, then the cloud manager must also be trustworthy.

2.9.1 Types of Trust

2.9.1.1 Direct Trust

If one entity trusts another entity through direct association or communication, then this type of trust is called direct trust.

2.9.1.2 Indirect Trust

An entity can indirectly trust another based on the recommendations of others. If two entities are not directly associated to one another and know each other through a third entity, then this type of trust is called indirect trust [18]. Indirect trust is required when there has not been prior interaction between two entities. Hence, the trust value will be calculated based on observation and recommendation.

2.9.1.3 Hybrid Trust

Hybrid trust [18] is computed based on both direct and indirect trust. In a private cloud, all the services are provided exclusively to trusted consumers/users. Essentially, an organization's data center delivers cloud computing services to clients who may or may not be in the premises. In contrast, public cloud services are consumed by those users/consumers who do not want to pay for the overhead cost of maintaining in-house infrastructures; rather, they prefer to pay the rent for using any other organization's cloud infrastructure. The users of public cloud are by default treated as untrustworthy. In the case of hybrid cloud, the users get combined services of both the private and public cloud.

2.10 Energy Efficiency

Figure 2.11 shows the energy impact of a cloud-based system over the traditional system.

Migration from the traditional solution to a cloud-based solution directly affects energy consumption in the following three ways: servers, cooling, and network.

2.10.1 Reduction in Direct Energy

An IT company must have a redundancy policy to protect its data from loses or a redundant storage location, which requires more server installation. This, in turn, creates the necessity for more servers. It is common that most of the companies do not use these servers 24 × 7. Thus, these servers remain underutilized. But the amount of energy drawn by the servers is approximately the same whether they are in working mode or in idle mode. A cloud-based solution makes it possible to reduce energy consumption by aggregating the demand of global users. In this case, only those servers can be utilized properly.

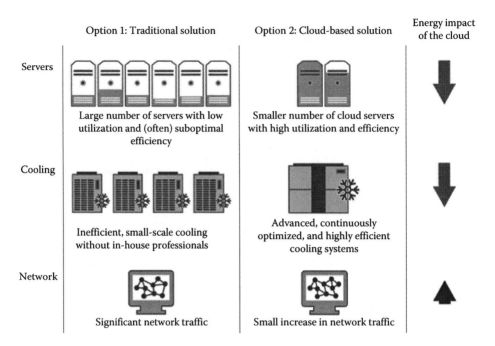

FIGURE 2.11
Energy saving by using cloud.

2.10.2 Reduction in Energy for Cooling Server

The energy consumed by a server produces heat in the location/building where the server is located. It increases the need for more powerful air-conditioning systems. Through significant R&D and innovative technologies, Google's data centers need just 0.13 W of cooling for each watt of direct power [19].

2.10.3 Increase in Energy Consumption for Increase in Network Traffic

Since communicating with the cloud is possible only through the Internet, network traffic is increased, causing additional energy consumption. But we can ignore this additional energy consumption because the aforementioned two points keep the cloud-based result a bit ahead of the traditional solution.

2.11 Interactivity, Real-Time Streaming

Interactivity is the communication process that takes place between human and computer software. In computer games, it is found that the gamer is always interacting with the game. This is a real-life example of interactivity. Database applications and other financial, engineering, and trading applications are also good examples of interactivity. All background applications are not interactive applications, since there is no need for user interaction with them. The most common interactive program is a web browser. When a person does an online transaction, the interactive application is the web browser. Cloud computing has become the daily need of common people too. If you want to video chat with your nearest contact, there are Skype, Yahoo Messenger, etc. In these cases, real-time streaming of images is done to provide seamless connectivity. Cloud computing gives a new approach in which the required bandwidth is supplemented to the cloud nodes for real-time communication. This implies that a specific user might be served media from different sources. Multipath RTP (real-time transport protocol) [20] can be used to optimize transmission.

2.12 Data Management

The major issue of cloud computing is providing security to users who store their data on the cloud. The privacy policies should be maintained by the cloud-hosting company. Data management in the cloud is a vital issue in cloud computing. Storing personal data in remote cloud servers brings various questions regarding security, privacy, trust, data portability, and interpretability [21].

2.12.1 Data Storage and Access on Cloud

Cloud makes it possible to store, retrieve, and access data anywhere anytime. As in Gmail, it has a 15 GB capacity of storage and users can access all data from anywhere through the Internet. Dropbox, SkyDrive, and iCloud are examples of storage space in cloud. In private cloud, data storage capacity is bigger. People can store large amounts of data by investing much less than traditional systems (pay-per-use concept) and access the data whenever required by multiple application servers [22]. Sometimes, it faces bandwidth and connectivity problems, but it can be removed by locally storing parts of data or full data on cache-based nonvolatile memory when accessing those data on cloud.

2.12.2 Data Portability and Interoperability

Migrating data among different application programs, computing environments, or cloud services is called portability of data. Customers of cloud do not want to be bounded within a single cloud provider. They would love the freedom of moving among clouds, from private to public and back to private. Interoperability gives cloud users the freedom to switch from one cloud service provider to another as their needs for computation may grow or shrink from time to time, and as a result, migration is needed to take place to cut the cost.

2.13 Quality of Service

Quality of service (QoS) has a special role in the service-oriented distributed system to effectively reserve resources. QoS is the most important topic in a distributed system and is often regarded as the resource reservation technique to achieve a certain level of performance and availability of services. Simply QoS provides priorities to different applications to guarantee the expected level of performance. The required QoS can be achieved by managing packet loss, delay, delay jitter, and bandwidth parameters on a network. Dell provides software, Scrutinizer NetFlow Traffic Analyzer [23], which quickly pinpoints the source of jitter, packet loss, latency, or a misconfigured network, visualizes real-time or archived application traffic data, and analyzes traffic in high-throughput environments.

2.14 Resource Utilization

A computer's own resources are specific to it. On the other hand, the resources owned and controlled by other networked computers are referred to as global or remote resources. In a cloud computing environment, resources could be storage space, processors, and the like. Resource management encompasses space management, process synchronization, and time management.

Resource sharing: A computer can request an appropriate service from the cloud by sending a request message over the communication network. Both hardware and software resources can be shared efficiently.

Modular expandability: In a cloud computing system, modular expansion of resources without replacing the entire system is the built-in feature.

Resource utilization: Resource utilization includes how to make both local and global resources available to users in an efficient and effective manner and this process is done in the following ways:

- *Data migration*: In this process, data are brought and computed to the location of computation through the network, and after the computation, the resultant sets of data are brought back to the original location from where the data had come. This is a technique by which the remote resources are utilized efficiently. But security and trust are important factors in this case because these data may be confidential and need to be protected from malicious users.
- *Computation migration*: In this technique, computation, instead of data, migrates from one location to another. In computation migration, part of the current thread is moved to the processor where the data reside. At the remote processor, the thread portion executes on the data, which are also in the cloud or in remote storage. This, in turn, makes subsequent access in the thread portion local.
- *Distributed scheduling*: In this technique, processes are migrated from one location to another. When the first machine or processor where the process originated is overloaded or it does not have necessary resources to execute that process, the process is transferred to another location for execution.

2.15 Applications of Cloud Computing

Cloud computing, in the simplest way, is said to be the next stage in the evolution of the Internet. Cloud computing can make it possible to access applications from anywhere. In cloud computing, the virtualization technique provides good support to resource utilization.

2.15.1 Mobile Cloud Computing

Simply stated, mobile cloud computing refers to an infrastructure where data storage and data processing happen outside the device. Mobile cloud application gives this kind of opportunity to computing power and storage of data away from mobile phones.

2.15.2 Healthcare

E-health, which employs wireless and mobile technologies for the attainment of health-related goals, can prove to be the next big thing in the field of delivering health services all over the world. A number of aspects are responsible for bringing about this transformation. These comprise rapid breakthrough in the field of mobile applications, the increase in opportunities for the amalgamation of present e-health services with mobile health, and persistent advancements in the expansion of mobile cellular network. Figure 2.12 shows the scenario of an e-healthcare system.

2.15.3 Cloud Gaming

Cloud gaming, also called gaming on demand, is a kind of online gaming. Presently, there are two varieties of cloud gaming: cloud gaming involving video streaming and cloud gaming involving file streaming. Cloud gaming focuses on providing continuous and undeviating play ability to end users across different devices. Very soon gaming technologies such as NVIDIA GRID, which is cloud based, will enable end users to play video games directly from the web. The cloud servers of GRID provide 3D games, encode the frames immediately, and send the outcome to the wired/wireless Internet connection.

Gaming as a service provided by NVIDIA GRID has huge untapped advantages over conventional gaming systems, as shown in Figure 2.13. Cloud gaming has superiority with respect to quality as well as economy. Cloud gaming employs a system in which the games execute on cloud servers, and the end users connect via networked slim clients. The slim clients are portable and are easily ported to resource-restricted platforms such as cellular devices. Cloud gaming makes gaming ubiquitous, and developers can improve the games according to varying configurations of PCs.

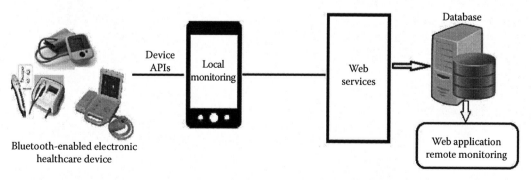

FIGURE 2.12
Electronic healthcare system.

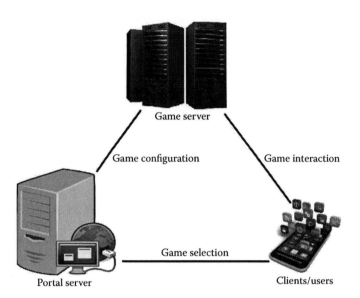

FIGURE 2.13
Gaming using cloud computing.

2.15.4 Storage

APIs provide easy access to cloud storage, but through superior features, power and flexibility can be attained. APIs can be accessed via XML or JSON or other programming languages. Cloud storage can be used to access static objects as well as for data storage. Big Query and App Engine are some of the popular cloud storage services. Authentication, speed, and flexibility are among the advantages of accessing data using cloud storage services.

2.16 Conclusion

Cloud computing has brought a revolution in the field of computation. Cloud is a cluster of computers with a virtually huge storage capacity, unlimited resources, and processing ability. Cloud computing is related to utility computing where services are provided as utility to users who use them on a pay-per-use basis. In this chapter, we have discussed the evolution, architecture, and applications of cloud computing. A case study of creating a cloud environment is also described in this chapter.

Questions

1. What does "cloud computing" mean? What are the components of cloud computing?
2. What is virtualization? How is cloud computing related to virtualization?
3. Explain the service models of cloud computing.
4. Discuss the characteristics of cloud computing.

5. Explain the deployment models of cloud.
6. Discuss the different types of trust in cloud computing.
7. How can energy efficiency be achieved in cloud computing?
8. Describe the data management and resource utilization schemes of cloud computing.
9. Describe the service-oriented architecture of mobile cloud computing.
10. What do you mean by agent–client architecture of mobile cloud computing? Explain with an example.
11. Discuss the various applications of cloud computing.

References

1. R. Buyya, J. Broberg, and A. M. Goscinski, *Cloud Computing: Principles and Paradigms*, John Wiley & Sons, Hoboken, NJ, 2010.
2. B. Sosinsky, *Cloud Computing Bible*, John Wiley & Sons, Indianapolis, IN, 2010.
3. R. Buyya, C. Vecchiola, and S. T. Selvi, *Mastering Cloud Computing*, Tata McGraw-Hill Education, New Delhi, India, 2013.
4. J. Voas and J. Zhang, Cloud computing: New wine or just a new bottle? *IT Professional*, 11(2), 15–17, 2009.
5. J. Hurwitz, R. Bloor, M. Kaufman, and F. Halper, *Cloud Computing for Dummies*, John Wiley & Sons, Hoboken, NJ, 2009.
6. M. Armbrust, O. Fox, R. Griffith, A. D. Joseph, Y. Katz, A. Konwinski, G. Lee et al., Above the clouds: A Berkeley view of cloud computing, Technical Report No. UCB/EECS-2009-28, Berkeley, CA, 2009.
7. K. Barr et al., The VMware mobile virtualization platform: Is that a hypervisor in your pocket? *ACM SIGOPS Operating Systems Review*, 44(4), 124–135, 2010. http://www.vmware.com/virtualization/.
8. G. J. Popek and R. P. Goldberg, Formal requirements for virtualizable third generation architectures, *Communications of the ACM*, 17(7), 412–421, 1974.
9. R. Uhlig, G. Neiger, D. Rodgers, A. L. Santoni, F. C. Martins, A. V. Anderson, S. M. Bennett, A. Kagi, F. H. Leung, and L. Smith, Intel virtualization technology, *Computer*, 38(5), 48–56, 2005.
10. L. Badger, T. Grance, R. Patt-Corner, and J. Voas, Draft cloud computing synopsis and recommendations, NIST Special Publication, 800-146, National Institute of Standards and Technology, Gaithersburg, MD, 84 pages, May 2011. http://citeseerx.ist.psu.edu/viewdoc/download?doi=10.1.1.232.3178&rep=rep1&type=pdf.
11. Q. Zhang, L. Cheng, and R. Boutaba, Cloud computing: State-of-the-art and research challenges, *Journal of Internet Services and Applications*, 1(1), 7–18, 2010.
12. M. Armbrust, A. Fox, R. Griffith, A. D. Joseph, R. Katz, A. Konwinski, G. Lee et al., A view of cloud computing, *Communications of the ACM*, 53(4), 50–58, 2010.
13. M. D. Dikaiakos, D. Katsaros, P. Mehra, G. Pallis, and A. Vakali, Cloud computing: Distributed internet computing for IT and scientific research, *IEEE Internet Computing*, 13(5), 10–13, 2009.
14. K. Chard, S. Caton, O. Rana, and K. Bubendorfer, Social cloud: Cloud computing in social networks, *IEEE Third International Conference on Cloud Computing*, Miami, FL, pp. 99–106, 2010.
15. D. Durkee, Why cloud computing will never be free, *Queue*, 8(4), 20, 2010.
16. B. Sotomayor, R. S. Montero, I. M. Llorente, and I. Foster, Virtual infrastructure management in private and hybrid clouds, *Internet computing, IEEE*, 13(5), 14–22, 2009. https://www.eucalyptus.com/news/security-software-giant-f-secure-selects-eucalyptus-systems-private-and-hybrid-cloud.

17. S. Ghanbari and M. Othman, A priority based job scheduling algorithm in cloud computing, *Procedia Engineering*, 50, 778–785, 2012.
18. K. Govindan and P. Mohapatra, Trust computations and trust dynamics in mobile adhoc networks: A survey, *IEEE Communications Surveys & Tutorials*, 14(2), 279–298, 2012.
19. Google, Google apps: Energy efficiency in the cloud, pp. 1–6, 2012. http://static.googleusercontent.com/media/www.google.com/en//green/pdf/google-apps.pdf.
20. S. Ahsan, J. Ott, T. Karkkainen, V. Singh, and L. Eggert, *Multipath RTP*. Internet draft, IETF, Toronto, Canada, July, 23, 2014.
21. D. J. Abadi, Data management in the cloud: Limitations and opportunities, *IEEE Data Eng. Bull.*, 32(1), 3–12, 2009. http://www.ipms.fraunhofer.de/en/press-media/press/2014/2014-06-16.html.
22. D. Agrawal, S. Das, and A. El Abbadi, Big data and cloud computing: Current state and future opportunities, in *Proceedings of the 14th International Conference on Extending Database Technology*, ACM, Gwangju, South Korea, pp. 530–533, 2011.
23. G. Huang, A conceptual analysis on the taxation system of highly virtual enterprises, 2010. http://www.sonicwall.com/us/en/products/Scrutinizer-Netflow-Analyzer.html.

3

Mobile Cloud Computing

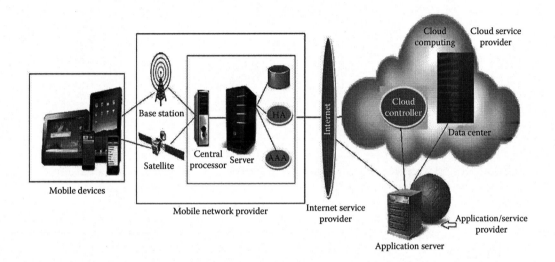

ABSTRACT Mobile cloud computing has gained popularity recently because of less power consumption and overhead of mobile devices. Mobile cloud computing-based health monitoring, gaming, learning, and commerce are gaining importance day by day. This chapter discusses the revolution of mobile cloud computing including its architecture, advantages, and applications. Various issues of mobile cloud computing are studied with solutions. Research challenges of mobile cloud computing are also discussed.

KEY WORDS: *mobile cloud computing, cloud, bandwidth, resource, energy, security, latency.*

3.1 Introduction

In recent years, as a result of the tremendous developments in mobile networks and technologies, mobile computing (MC) has become an emerging area of research. In the past decades, people used computers for computing purposes. According to the recent surveys, people are willing to use mobile devices such as laptops, Smartphones, personal digital assistants (PDAs), tablets, i-Pads, and so on, rather than the immobile desktop computers.

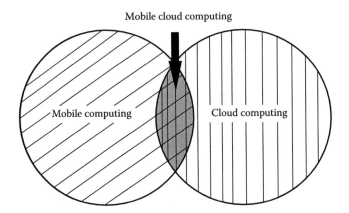

FIGURE 3.1
Origin of MCC from MC and CC.

The current smartphone user base already has reached the 1 billion mark. So, computing through mobile devices has become a more feasible concept than the conventional approach. But still some drawbacks such as lack of storage, computational power, and limited battery life of mobile devices have become the challenges for MC technology.

To overcome these challenges, cloud can be a useful solution. Cloud is the combination of virtualization of a high amount of resources with a distributed computing paradigm incorporated with software as a service (SaaS), platform as a service (PaaS), and infrastructure as a service (IaaS). Various cloud providers such as Microsoft Azure and Amazon EC2 provide seamless elastic storage and processing in an "On demand," "Pay as you use" manner. So, integration of mobile computing with cloud computing (CC) has given birth to a newer and better technological approach called mobile cloud computing (MCC), as shown in Figure 3.1. In a simple sense, MCC is nothing but cloud computing in which mobile devices are involved as the thin clients. Here, data will be offloaded into cloud from mobile devices for computation or storage.

In this chapter, we discuss the framework for mobile social cloud with related applications and issues along with the future scope in this field.

3.2 Motivation to Mobile Cloud Computing

Mobile cloud computing is defined as a rich mobile computing technology that controls integrated elastic resources of different clouds and network technologies toward unlimited functionality, mobility, and storage in order to serve a large number of mobile equipment anywhere and at anytime through the Ethernet channel or Internet in spite of heterogeneous environments and platforms on the basis of the pay-as-you-use principle [1]. MCC is an infrastructure where the data storage and the data processing are performed outside the mobile device but inside the cloud. In MCC, the computing power and data storage are moved away from mobile devices and performed in the cloud, bringing mobile cloud applications and mobile computing not only to smartphone users but also to a wider range of mobile subscribers [2]. So, MCC is an infrastructure that combines the mobile computing and cloud computing domains where both data storage and data processing happen outside the mobile device. It is not always that offloading will be to a remote cloud, but it can be to a local cloud-let or to the collective resources of mobile devices in the local vicinity.

The motivation behind MCC is simply to remove the existing drawbacks of mobile computing. There are several limitations of mobile computing discussed, which are as follows [1–3]:

1. Limited battery life of mobile devices: Because of the mobility of the device, it is impossible to find an external power source every time. Mobile devices have to rely on the internal battery, which has a charge life of only a few hours, in most cases. If computation is continuous or various applications are running continuously, battery will drain soon.

2. Limited storage capacity of mobile devices: Each mobile device has a limited amount of internal memory. A well-configured smartphone has only 8 GB of internal memory, and a laptop has 500 GB of storage. Though they can be expanded using external memory, in case of organizational data backup, they are insufficient.

3. Limited processing power of mobile devices: Smartphones have ARM processors, which are capable of running only small and a limited number of applications. Though laptops have i3, i5, and i7 type 3G high processing units, often they are not affordable due to their high cost. Processors are an irreplaceable part of a mobile device, so if anyone wants to upgrade it, it may not be possible.

4. Low bandwidth: In mobile computing, EDGE, GPRS, and GSM technologies have very low bandwidth. Though 3G and 4G technologies such as HSPA, WCDMA, LTE, and so on, are popular, they are available only in a limited number of cities/towns, at too high a cost.

All these problems can be solved only by a cloud platform with MC. Cloud is a very large, virtualized, shared resource or infrastructure that has the capability of computing, analyzing, and warehousing a large amount of data. Cloud serves its clients on an "on demand," "pay as you use" basis, in much the same way as using electricity from a service provider. Because of the elastic nature of cloud, a client can get the desired amount of service as his or her requirements change, and with a high-speed Internet connection, the user can get seamless service from cloud providers. Hence, MCC has become much more feasible than MC.

3.3 Architecture of Mobile Cloud Computing

3.3.1 Service-Oriented Architecture

The service-oriented architecture (SOA) of MCC consists of three layers [3], as shown in Figure 3.2.

SOA of MCC consists of the following components [3]:

1. Mobile network
2. Internet service
3. Cloud service

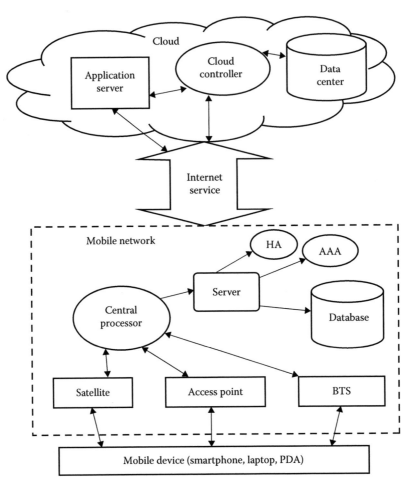

FIGURE 3.2
Service-oriented architecture of mobile cloud computing.

1. *Mobile network*: A mobile network contains mobile devices and network operators. Mobile devices may be smartphones, PDA, satellite phones, laptops, and so
   on. They are connected to the network operator via the BTSs (base transceiver
   stations), access points, or satellites. They establish and control the connection
   between the functional interface between mobile device and network operator.
   A mobile device's request and information, such as ID and location, are transmitted to the central processor and servers of the network providers. Here, operators provides various services such as AAA (authentication, authorization, and
   accounting) based on the HA (home agent) and subscriber data stored in database.

2. *Internet service*: Internet service plays the role of a bridge between the mobile
   network and cloud. Subscriber requests are delivered to the cloud via a high-speed
   Internet service. Using wired connections or advance 3G or 4G technologies such
   as HSPA, UMTS, WCDMA, LTE, and so on, the user can get seamless service from
   cloud.

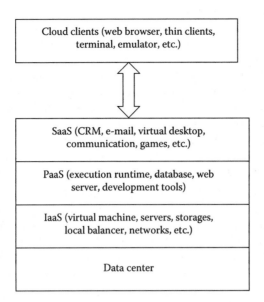

FIGURE 3.3
Service model of cloud computing.

3. *Cloud service*: After getting all the requests from the users, the cloud controller processes the requests and provides service to them according to their requirements. Cloud has a few service providing layers, as shown in Figure 3.3. These service layers are discussed as follows:

a. Data center layer: Data center provides the hardware facilities and infrastructure for the cloud. In a data center, there are several servers connected with high-speed networks and high power supply. Normally, they are built in less populated places with a low risk of disasters.

b. Infrastructure as a service: IaaS resides on the top of the data center layer. It provides storages, servers, networking components, and hardware to its clients on a "pay as you use" basis. It has an elastic nature, so infrastructures can be expanded or shrunk dynamically according to user demands. Amazon EC2 and S3 are examples of IaaS.

c. Platform as a service: PaaS provides an integrated environment or platform for users to build, test, and deploy several applications. Any kind of platform such as Java, .NET, PHP, and so on, are available. Google App Engine, Microsoft Azure are examples of PAAS.

d. Software as a service: SaaS is a software delivery model provided by application service providers (ASPs). Software and the associated data are centrally hosted on the cloud. SaaS can provide numerous kinds of software solutions such as CRM, ERP, MIS, HRM, and so on, on demand without any dedicated installation in client site.

This way, in mobile cloud computing, data storage and computations are moved into the cloud, and the user gets seamless, on-demand service without having to worry about battery life or the processing power of mobile devices.

3.3.2 Agent–Client Architecture

In this architecture, mobile devices are not connected to the cloud directly. They are connected to the cloud via some agents such as femtocell [4], cloudlet [5], or both, as shown in Figure 3.4. MCC is all about wide area network (WAN) and cloud. Usually, clouds are situated at a long distance from the users, so there are chances of delay or cost inefficiency. These agents can fulfill the user demand with high bandwidth, low latency, and low cost. Only when the demands are not fulfilled by the agents will the request go to the cloud. The architecture is shown in Figure 3.4. Here, M1, M2, M3, and M4 are the mobile devices that are not directly connected to cloud but via agents such as femtocell, cloudlet, or both.

Cloudlet: Cloudlet [5] is a resource-rich computer or cluster of computers that is trusted and well connected to high-speed Internet and available to mobile devices. When a user does not want to offload any task directly to cloud due to delay or cost, he or she can offload it to the nearest available cloudlet. If the device cannot find any cloudlet available, then it will send its request to the cloud or, in the worst case, complete the task with its own resources. Thus, a user gets real-time response by low-latency, one-hop, high-bandwidth, and low-cost access to cloudlet.

Femtocell: Femtocell [4], well known as the "home base station," is the smallest version of traditional macrocells. Femtocells are deployed indoors to provide good coverage [6–8]. Mobile devices are connected to femtocells, and femtocells

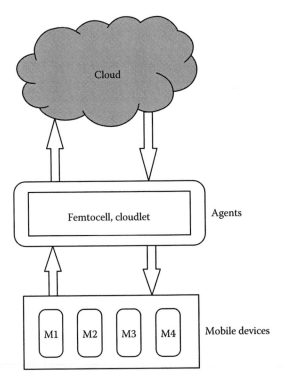

FIGURE 3.4
Client–agent architecture of MCC.

are connected to the mobile network with residential DSL, cable broadband, optical fibers, or wireless last-mile technologies [9–11]. Femtocell access points can implement cellular technologies such as UMTS/HSPA/LTE and mobile WiMax. So, they can provide seamless 3G and 4G service to the user and can be used to connect mobile devices with a cloudlet or cloud. As it gives higher bandwidth, the user will face very little latency to offload tasks to a cloud or cloudlet.

3.3.3 Collaborative Architecture

Nowadays, smartphones are operated independently using their local computing, sensing, networking, and storage capabilities. When data are shared with other devices through a centralized server or cloud, it requires expensive upload and download. It can be avoided by collaborative computing [3]. In this architecture, resources of a mobile device are used by considering the device as a part of a cloud. The cloud server may be the controller and scheduler for collaboration among the devices. By combining smartphone data and computing, a smartphone cloud can be generated. Mobile applications can utilize these resources of the smartphone cloud, so processing of mobile data in a smartphone cloud can remove the bottlenecks of global network and the limitation of offloading data to a remote server. In Figure 3.5, the collaborative architecture of MCC is shown, where M1, M2, M3, and M4 are the mobile devices, and M2, M3, and M4 have formed a smartphone cloud and M1 is using it.

Smartphones such as androids or i phones support cloud computing. Hyrax [12] is a platform based on Hadoop, and supports cloud computing on an Android smartphone. Hyrax allows client applications to conveniently utilize data and execute computing jobs on networks of smartphones and heterogeneous networks of phones and servers. By scaling with the number of devices and tolerating node departure, Hyrax allows applications to use distributed resources abstractly and exactly as the cloud. Using an x86 virtual machine, different applications can run from the BOINC server to an apple i-Phone, which integrates the grid computing framework. By this technique, an i-Phone can support the CC approach.

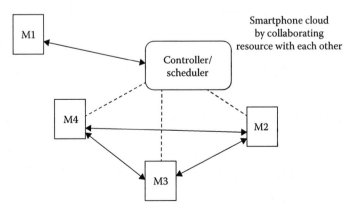

FIGURE 3.5
Collaborative architecture of MCC.

3.4 Platform and Technologies

3.4.1 Platform of MCC

Nowadays, tablets and smartphones are the most popular and useful platforms of MCC [3]. BlackBerry, Nokia, Samsung, and Google are popular developers of smartphones. Smartphones use different operating systems such as Research in Motion (RIM), BlackBerry, Windows™ Mobile®, Nokia Symbian platform, as well as UNIX® variations such as Google Android and Apple iOS made by Samsung, Motorola, and Acer, to support MCC.

3.4.2 Enabling Technologies of MCC

The most useful and enabling technologies for improving bandwidth and network latency of MCC are 3G and 4G [3]. HTML5 and CSS3 can improve mobile web applications by allowing specification of offline support, which makes local storage possible and solves the problem of connectivity interruptions. For Internet application, Web-4.0 is also used. Another enabler for cross-platform applications is the embedded hypervisor, which allows a web application to run on any smartphone.

3.5 Mobile Augmentation Approaches

Resource poverty is the main disadvantage of mobile devices. Mobile computation augmentation, or augmentation in brief, is the process of increasing, enhancing, and optimizing the computing capabilities of mobile devices by leveraging varied feasible approaches, hardware, and software. Using MCC capacity and storage, processing can be increased significantly, which is called mobile augmentation [13].

Augmentation approaches shown in Figure 3.6 are classified as follows:

3.5.1 Hardware Approaches

Hardware approaches simply generate high-end resources by installing a better processor, RAM, and so on, in the mobile device. However, in many cases, the installation is not easy and it becomes a dedicated installation, so the hardware approach is not viable.

3.5.2 Software Approaches

There are several software approaches for leveraging the cloud:

1. *Remote execution and storage*: This approach conserves the native resource of mobile devices by migrating [14] the resource-hungry part of the application fully or partially into the nearby cloudlet, remote cloud, or a resource-rich computing machine that is connected to a seamless power source and the Internet.

2. *Fidelity adaption*: Fidelity adaption [14] provides a tradeoff between application quality and energy conservation of the device. Quality of the application is altered here; for example, videos on YouTube have a number of copies of the

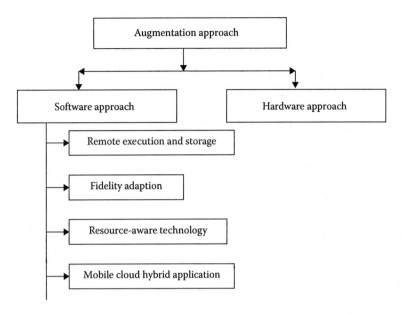

FIGURE 3.6
Mobile augmentation approaches.

same video of different quality. A low quality of a video consumes less resource and less energy in a mobile device.

3. *Resource-aware technology*: By creating a resource-aware application [14], resource consumption can be reduced significantly. Mobile RAM and phase-change memory (PCM) belong to this type of technology. In mobile RAM, the power management unit maintains multiple power states such as "Self Refresh" and "Power Down" to minimize power consumption, whereas PCM leverages three states of I/O, on and off, to store data with enhanced energy consumption. Proximity sensors are used in smartphones to adjust screen brightness automatically, which is also a resource-aware technique.

4. *Mobile cloud hybrid application*: Mobile cloud hybrid application [14] reduces resource and energy consumption of an application without trading off the quality of the application. Cloud resources run the application quickly, and mitigated communication overhead reduces the energy consumption and the overall execution time. μCloud is a kind of application framework that leverages the cloud with the least component-level communication and dependency with the promise of high functionality and resource efficiency.

3.6 Issues of Mobile Cloud Computing

As MCC is composed of two different technologies, namely mobile computing and cloud computing, it has to face several issues and challenges such as resource management, network and security management, and operational, end-user, and service-level management, as shown in Figure 3.7.

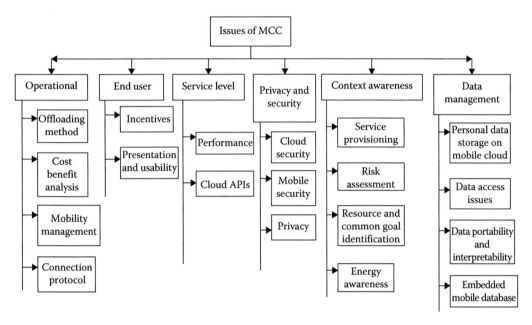

FIGURE 3.7
Issues of mobile cloud computing.

3.6.1 Operational Issues

Operational issues [13] include offloading methods and computation of offloading cost. It computes whether offloading will be cost effective or not. Also, mobility management and connection protocols exist.

3.6.1.1 Offloading Methods

3.6.1.1.1 Client-Server Communication Method

In this offloading process, communication is done using remote procedure call (RPC), remote method invocation (RMI), and sockets between the offloader and the surrogate device. But when the ad hoc or mobile nature of the device is considered, it becomes a disadvantage because those services need to be preinstalled in the device. Thus, it restricts the mobility of the device. Spectra [13] and Chroma [13] are kinds of systems that use offloading computation via RPC by invoking functionalities in local and remote Spectra servers. When offloading is needed from a device, Spectra client consults a database, which stores information such as current availability and the CPU load of the Spectra server, and, depending on the resource pool, decides at runtime and does offloading.

Hyrax [12] is a smartphone application based on Hadoop ported on Android platform, distributed both in terms on data and computation. It uses a cluster of mobile devices as a resource provider and mobile cloud. HyraxTube is such an application used for distributed mobile multimedia search and sharing. It allows users to search through multimedia files according to the time, quality, and locations. Hyrax has a central server that has NameCode and JobTracker instances and access to each client mobile device. The central server does not process a job, but coordinates data and jobs. Hadoop distributed file system (HDFS) stores the multimedia data, and threads run on the mobile devices.

Another Hadoop-based approach was presented [15] to use mobile devices as a resource provider. Here, the offloading manager module performs the sending and receiving jobs from and to mobile devices and creates virtual machines on the surrogates. On a surrogate device, the tasks are executed on a virtual machine acting as a protected space, thus ensuring the security of device data.

The "Cuckoo" [16] framework is based on java stub/proxy model. It offloads the task to any resource that runs a java virtual machine such as Amazon EC2 or local mini cloud, cluster of laptops, or mobiles. Cuckoo's offloading objectives are to enhance performance and reduce battery usage. Here, applications have to be written in such a way that they support remote as well as local execution. There is an advantage of parallelism, as Cuckoo generates the same code for the remote version of the application and can run on multicore computers, for instance. If the remote resource or network connectivity is not available, then the application can run on the local device entirely. There are two applications, "eye-Dentify" and "PhotoShoot," based on Cuckoo, which have gained speedup of 60% and reduce battery consumption to 40%.

3.6.1.1.2 Virtual Machine Migration

A live migration is where memory image of a virtual machine (VM) from a source server is transferred to another server without any interruption in execution.

MAUI [17] uses virtual machine migration and code partitioning, both approaches saving energy. Here, offloading is done from the device to a local or remote server. Code partitioning is done in MAUI dynamically and during runtime. MAUI decides whether an application will be offloaded or not if the resource is available in the remote server.

Clone cloud [18] is another VM migration technique where part of the application is offloaded to a resourceful server through 3G or Wi-Fi. Here, mobile applications are unmodified and there is no need of any annotating method because clone devices are used. "Clone" is a mirror image of the smartphone created on cloud and running on a virtual machine. Compared to the device, the clone has a better resource and computational environment. A cost analysis model is used to compare the cost between the offloading task in cloud and in monolithic execution. Clone cloud has been tested over applications such as virus scanner, image search, and privacy-preserving advertising on Android phones using clone laptops running on the Ubuntu platform. Clone cloud architecture is shown in Figure 3.8.

MobiCloud [19] is another service-oriented approach that uses cloud computing and a mobile ad hoc network (MANET). Here, each node is considered as service node used as service provider or service broker depending on its computational capabilities and available resources. Every service node is incorporated onto the cloud as a virtualized component and is mirrored in the cloud. The main objective of MobiCloud is to provide a secure service architecture and a virtual trusted and provisioning domain.

3.6.1.2 Cost–Benefit analysis

A cost–benefit analysis is very important in MCC to measure the cost of offloading and determine whether offloading a task will be beneficial or not. In Kumar and Lu [20], a model has been proposed to calculate the cloud cost, which is divided into two parts as follows:

1. Total cost of ownership (TCO): TCO is a financial estimate to determine the costs associated with owning and managing an IT infrastructure. But from a cloud perspective, it is used to estimate commercial value of cloud investment and also

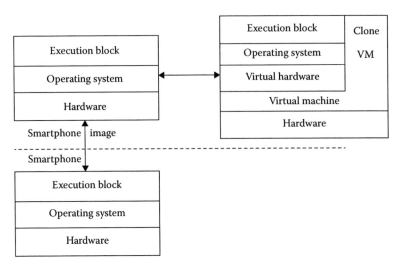

FIGURE 3.8
Architecture of clone cloud.

> to take care of service cost, network cost, software cost, cooling cost, power cost, facilities cost, maintenance cost, and real estate cost.

2. Utilization cost: Utilization cost is the actual cost of using a resource by a user. It is dynamically elastic.

Spectra [13] and Chroma [13] are two methods that employ cost versus benefit offloading to the cloud. In Spectra, there is a tradeoff between the offloading application and energy consumption, performance, and quality. Spectra continuously monitors resources such as the CPU, battery, network, cache, and so on. For best resource placement, "self-tuning" is used to maintain user demand with available resource, observe execution, and maintain profile history of surrogates for future studies. Chroma also uses a similar "tactics," and does resource monitoring and history-based prediction for cost estimation. Chroma trades off between attributes such as power consumption and speed using a utility function.

Scavenger's [21] framework, using a component called the scheduler, also decides whether offloading should be done. It considers various factors for cost analysis, such as the relative speed and current utilization of surrogate, network bandwidth and latency to the surrogate, task complexity, and input/output size.

MAUI [17] does the cost–benefit analysis of an application through serialization. It does the analysis based on device's energy consumption, application characteristics, and network characteristics.

Clone cloud [18] uses a "dynamic profiler" to collect data, which are then used to analyze cost and benefit. These data are fed into the "optimization solver" to decide whether offloading should be done or not. Costs are analyzed in terms of execution time, energy consumption, or resource footprint.

A cost model using parametric analysis [22] compares energy usage between cloud and mobile devices. The parameters are the network bandwidth (B), the number of bytes to be transferred (D), the speed of the remote cloud (S) and the mobile device (M), the number of instructions of the computation (C) assuming that both mobile and cloud versions have the same number of instructions, and the energy consumed by the mobile device in computing (P_c), communicating (P_{tr}), and idling (P_i) states. If the cloud is T times faster than the mobile device, then the energy saving will be $(C/M)\{P_c - (P_i/T)\} - P_{tr}(D/B)$. The ratio D/B should be than C/M, and T should be sufficiently large. Then, the equation gives a value greater than zero, which implies that energy saving is possible.

Cost models using stochastic methods, a semi-Markov decision process (SMDP), have been described in Puterman [23]. There are three states in a mobile cloud: new service request or an inter-domain request, intra-domain transfer service request, and the service leaves domain. SMDP is based on these three states. Here, the cloud accepts a migration request if there is an overall system gain, which is based on maximizing cloud profit and reducing the expenses of the mobile user. There is ongoing debate whether an intra-domain service transfer from one service node to another would usually generate more profit than a new service migration from a mobile device, or an inter-domain transfer in which the transfer happens from another cloud service provisioning domain. The overall system gain also takes care of CPU cost in the cloud server due to virtual image occupation. To calculate the cost of system states and their actions, a model named "rewarded" is used.

3.6.1.3 Mobility Management

Mobility is one of the main issues of MCC, as mobile devices are present here. One particular position may be suitable for a device but, due to change of location, services should not be interrupted. Mobility is one of the reasons for disconnection. In mobility management, localization is very important and it can be achieved using two techniques: infrastructure-based and peer-based. Infrastructure-based techniques use GSM, Wi-Fi, ultra sound RF, GPS, and RFID, which are less suitable for the needs of mobile cloud devices. On the other hand, peer-based techniques are more suited to manage mobility, considering that relative location is adequate and can be implemented with low-range protocols such as Bluetooth. Escort [24] represents a peer-based technique to localize without using GPS or Wi-Fi, which are power-consuming applications. Here, social encounters between users are monitored by audio signaling and the walking traits of individuals by phone compasses and accelerometers. Here, various routes are created by various encounters. For example, if X wants to locate Y and if X had met Z recently and Z had met Y, the route is first calculated to the point where X met Z, and then to the place where Z met Y. There will be many possible paths but the optimal one is chosen. Thus a mobile device can be localized when it is in a mobile cloud.

Virtual compass [25] is another approach in which Bluetooth and Wi-Fi short-range protocols are used to construct a two-dimensional reorientation of the nearby device. Here, peer-to-peer messaging is used to measure the distance via signal strength and pass information about the device's neighbors and their distances. In mobile cloud, it will be helpful to know about devices that are newly joined or previous nodes that have returned to the cloud. Here, the mobile cloud is constructed based on Bluetooth or Wi-Fi, so passing information is not a big burden. Because of continuous scanning, there is a possibility of energy drain. To remove this drawback, a self-adaptive scanning technique can be used based on regulating the scanning interval with the change of neighbor graph. This regulation is done by a central server where, as in context of the mobile cloud, a decentralized approach is better. One decentralized approach is "Friends Radar" [26], a peer-to-peer technique using XMPP. Only "known" contacts or friends' locations are visible, and it uses GPS. All the participating devices are pre-known and trusted.

MoCA (mobile collaboration architecture) [27] uses component and proxy migration to support mobility management. Here, user mobility is supported by monitoring the locations of the users and switching to an application proxy more suited to the new location. Mobile clients query the service, which contains registered information on available application server proxies, and discover the means to access a collaborative service at their closest proxy determined by the location of the mobile device.

3.6.1.4 Connection Protocol

MCC is incorporated with a large number of protocols and technologies. Wi-Fi (wireless Ethernet 802.11b) and Bluetooth (2.4 GHz ISM band) are used for short-range connection. For longer range, there are 2G, 3G, and 4G technologies. Different protocols such as GSM, CDMA, EDGE, GPRS, WCDMA, HSPA, HSPDA, UMTS, and so on, are used.

3.6.2 End-User Issues

End-user issues are related to the direct mobile user. There are several questions that need to be tackled by a mobile user while using cloud: In which way will the cloud service be billed? How is credit represented in collaborative mobile cloud? How will the user be motivated to share these resources in cloud? There are many others.

3.6.2.1 Incentive Scheme

Incentives [28] are given to mobile users to motivate them to contribute their resources as cloud or into the cloud. Here, resource may be storage, computation power, information, important result, license key, and so on. Incentives may be tangible or intangible. Tangible incentives are mainly monetary, and intangibles are the name, fame, reputation, and so on. There are many resource sharing protocols available, such as volunteer, action, trophy, posted price, spot price, reciprocation, reputation, and so on. Each protocol has its own incentive schemes for the participants.

3.6.2.2 Presentation and Usability

Presentation and usability are related to the user interface of the mobile device. The size of the mobile screen is a big factor. A small screen is less power consuming but is difficult to use, whereas a bigger screen is easy to use but consumes high power. Touch screens have better usability than non-touch screen devices. In touch screen devices, feather touch is better than captive touch. On the other hand, nontouch devices are harder in nature.

3.6.3 Service-Level and Application-Level Issues

Service and application issues are related to the performance and quality of service of the system, which include the availability and fault tolerance of the system.

3.6.3.1 Fault Tolerance

Faults occur as a result of the mobility of the device from one network to another, network failure with cloud, running out of battery, hardware failure, and so on. Hyrax [12] is based on Hadoop, and it recovers task failure by re-execution and redundancy. If one node fails, then it replicates the code to another node and completes the execution.

By proxy migration, one faulty proxy node is replaced by another service node from the cloud and fault is resolved. The resource tracking model [29] uses JEL API for fault tolerance. It stands for "join elect leave." "Join" operation notifies when a new mobile node is connected to the distributed system. "Elect" operation elects a mobile node to coordinate with other node. When a node leaves by choice or fault, a "leave" operation is notified and another node is elected to fill it.

3.6.3.2 Supporting Performance at Service Level

There are several mobile applications, based on the REST (representational state transfer) [30] architecture, that are able to be connected and request services that are hosted on a remote cloud using interfaces. Limited computational resources, frequent connectivity interruptions, and low bandwidth are the constraints of web services. They can be resolved by caching and pre-fetching. This enables a user to do his or her job for a certain period in the offline mode also.

3.6.3.3 Cloud APIs

Cloud APIs are application programming interfaces used to build applications in cloud computing or to use cloud utility [31]. Funambol Cloud API is a cross-platform API that provides client-side and server-side SDKs for developing mobile cloud applications and services that use images, contacts, calendar, and so on, stored inside a Funambol server. Dropbox, OpenNebula, and Nimbus are good examples of famous cloud APIs.

3.6.4 Security and Privacy

Keeping each and everything in one's own mobile device is secure and trusted, but, still the use of the cloud for computational and storage purposes is essential. Offloading personal information, data, and application to the remote cloud as well as in communication channel raises various questions regarding security, privacy, and trust.

3.6.4.1 General Cloud Security

The security risks needed to be addressed when data are offloaded into the cloud are discussed as follows [32]:

1. *Regulatory compliance*: Cloud service providers should undergo external audits and security certifications.
2. *Privileged user access*: When sensitive data get offloaded to the cloud, it may appear that the data are no longer under the direct physical, logical, and personal control of the user.
3. *Data location*: Exact physical location of user's data is not transparent, and this may result in confusion in particular authorities and commitments on local privacy needs.
4. *Recovery*: Cloud providers should provide proper recovery management schemes for data and services when a technical fault or disaster arises.
5. *Data segregation*: As cloud data are usually stored in a shared space, each user's data should be separated from the others' with efficient encryption methodologies.

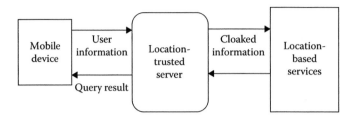

FIGURE 3.9
Cloaking architecture.

6. *Long-term viability*: It must be ensured that users' data would be safe and accessible even in the event the cloud company itself goes out of business.

7. *Investigative support*: For multiple customers, logging and data may be colocated. Thus, it may be vital, but hard, to predict any inappropriate or illegal activity.

3.6.4.2 Security for Mobile User

Mobile devices are usually affected by various kind of malicious codes such as viruses, worms, and so on. Installing and running security software such as AVG, Kasperkey, McAfee, and Avast can help get rid of these threats.

3.6.4.3 Privacy

Most mobile devices use GPS-enabled location-based service (LBS) nowadays. It reveals the user's current location every time. Sometimes it discloses the privacy of a user. A location-trusted server (LTS) [3] is used to solve this problem. Working between the mobile device and the LBS, it takes information from mobile user and creates a "cloaked region." This "cloaked region" is sent to the LBS. Thus, LBS can know only general information about users but cannot identify them. It is shown in Figure 3.9.

3.6.5 Context Awareness

Context awareness is one of the vital issues of MCC. MCC is all about mobility, heterogeneous environment, and technology, so, it is very important for a system to be very aware of its context or surroundings. The system should be able to reconfigure or change itself according to the change in its environment.

3.6.5.1 Context-Aware Service Provisioning

Mobile cloud provides services based on the context of a user. It can sense various attributes such as location, acceleration, and so on, of a mobile user. Based on this context information, cloud provides the most suitable service to the user. There are four layers of the context element [33]:

1. *Monitored context*: Referring to the current monitored context of a device, monitored context includes environmental and device settings, user-specific preference settings, service context information like QoS, and so on.

2. *Types of gaps*: When a user changes his or her service from the current service to a new one, there may be a gap between contexts of the two services. The gap may be functional or nonfunctional.

3. *Types of causes*: There may be gaps between different interfaces and different implementations of a single service. These gaps occur as a result of the mismatch of the service level, service interface level, service component level, and component instance level.

4. *Adapters*: Adapters are needed to remove all the previously mentioned causes.

Based on these factors, context-aware service provisioning has been built with three tiers. The user tier consists of mobile devices where the application runs, the agent tier adapts the services according to the context, and the service tier deploys the services.

In intelligent access [34], context information is provided by terminals, sensor nodes deployed with the users' environment, network nodes, and so on. Dynamic context information, such as user profile, terminal status, and sensor information, is provided to the mobile cloud. Context management architecture (CMA) is responsible for acquiring, processing, managing, and delivering context information. Context quality enabler (CQE) controls the supply of context information to the mobile cloud. CMA has the following three components, as shown in Figure 3.10.

1. *Context provider*: The context provider generates the context information. Then it is supplied to other components of the CMA. According to the context request, communication between the context provider and other components is established. Static data such as terminal capabilities, user preferences, or user information, and dynamic data such as user location and movement (i.e., speed and direction) and network conditions, and so on, are provided by the context provider.

2. *Context broker*: Context broker acts as a middleman between the context provider and the context consumer. There is a registry containing the information about the context provider's availability and capabilities. It provides a look-up service for the context provider and also forwards the data received from it. There are two

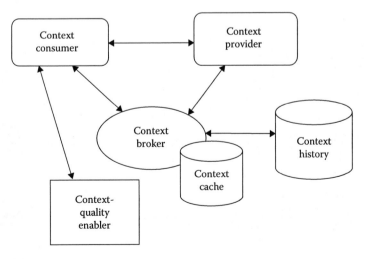

FIGURE 3.10
CMA architecture.

kinds of communication mode: asynchronous and synchronous. In the asynchronous mode, context information is forwarded if a specific condition becomes true. In synchronous mode, context information is answered instantly by the context broker. It maintains a context cache to store context that has not expired yet, and a context history to store expired context.

3. *Context consumer*: These are the entities that take context data as inputs for their actual functionality, for example, network services, applications for end-users, and service enablers.

3.6.5.2 Risk Assessment Using Context Awareness

MobiCloud [19] uses context information to facilitate risk assessment and routing decisions by virtual trusted and provisioning domain (VTaPD). It can isolate different information flows in different domains via programmable router technologies by creating separate flows and different domains. Thus, a user can run different applications in different secure domains based on the context. Various pieces of context information such as device sensing value, location, neighboring device status, and so on, are recorded at VTaPD and used for risk management and intrusion detection.

3.6.5.3 Resource and Common Goal Identification Using Context Awareness

The context manager can sense and store context information, which can be used by other components. Using the information context, the application manager can launch, intercept, and modify the current applications. The location and the number of devices in a vicinity are considered the main context information. Location context is used for tracing the mobility of the user, and a number of devices are used for forming the mobile cloud, so, the system is aware of when a new node enters or leaves the system and also can identify which node is stable. These are essential to deciding whether the new node is following a common movement pattern with other devices, leading to common activities.

3.6.5.4 Energy Awareness

Because of limited battery life in a mobile device, energy awareness has become one of the vital issues of MCC. Mobile devices have a finite power source, and running many applications constantly causes a large amount of battery drain. So, energy profiling and energy usage estimation of the device are very important. In MCC, the power consumption must be less than the benefit gained. Unnecessary software components should be unloaded, and the energy-intensive components should be redeployed to the more resourceful host.

PowerScope [35], which maps the power usage to a specific code component in the application and operating systems, is an energy profiling tool for mobile application. It allows the analysis of the power draining process, which helps the software developer to build energy-efficient software. An experiment was done by some laptops, minicomputers, and digital multimeters, and the result showed that 46% energy saving was possible in adaptive video streaming by using PowerScope.

PowerSpy [36] is a software approach for the energy profiling of mobile devices in the Windows platform. It has two stages: tracking and analysis. When an application is running, it tracks several factors such as the CPU time, energy consumption, and I/O activity. In the analysis stage, it filters the result by I/O activity. It estimates a certain energy consumption for an I/O operation specified by the manufacturer, which is then compared with the recorded energy consumption. The difference is noted as the energy taken up by CPU threads. Next, energy consumption is drilled down to the individual threads, on the assumption that CPU power usage is proportional to the number of CPU cycles spent on a thread.

Energy usage estimation is just opposite of energy profiling. It is the prior estimation of an application not during runtime but at the construction time. This estimation is done based on some context parameters such as the size of data exchanged over the network, inputs to the components' interfaces, invocation frequency of components' interfaces, and so on. The equation for energy cost estimation is as follows [37]:

$$\text{System energy cost} = \text{Computational energy cost} + \text{Communicational energy} \\ + \text{Overall infrastructure energy cost}$$

where (computational energy cost + communicational energy) refers to the overall energy cost.

3.6.6 Mobile Data Management

Data management in the cloud has become a vital problem for both the cloud user and provider. Storing personal data in a remote cloud has raised various questions regarding security, privacy, trust, data portability, interpretability, and so on.

3.6.6.1 Personal Data Storage on Mobile Cloud

A mobile device stores a lot of personal information such as photos, videos, chat logs, contact lists, passwords, financial documents, financial records, and so on. All these pieces of information are much safer to keep in the mobile cloud than in a conventional cloud. SkyDrive and iCloud are examples of mobile cloud in Apple and Windows platforms. SkyDrive provides 7 GB and iCloud provides 5 GB of initial free storage for the users. These mobile clouds are much safer and easier for managing data than the traditional cloud, as a limited number of users are involved in this case.

3.6.6.2 Data Access

In MCC, data are stored into cloud because of the limited storage, power, and computation power of the mobile device, but still it sometimes faces a bandwidth and connectivity problem. To remove this problem, an approach called "Pocket Cloudlet" has been proposed in Koukoumidis et al. [38]. Here, a local storage cache based on nonvolatile memory is used to store parts or full cloud services in the mobile device. Here, the device has to decide previously which portion of the cloud is needed to cache locally.

3.6.6.3 Data Portability and Interoperability

Mobile cloud is compatible with assorted mobile devices such as Android, BlackBerry, iPhone, and so on. These devices have various sensing capabilities, so mobile cloud has to store various sensor-specific data. The mobile cloud should have platform independence, as it communicates with typical cloud structures containing large-scale servers and with mobile devices. Considering data interoperability, a Palm Pre user [39] faces problems when Apple's iTune disables the Palm Pre's ability to sync its multimedia with iTunes software. In case of Funambol cloud service from iPhone, only syncing contacts are allowed by the apple SDK through the official client. Though it is possible to sync the calendar with SQLite database, the iPhone has to be unlocked, which makes the warranty void.

3.6.6.4 Embedded Mobile Databases

An embedded mobile database [40], or simply mobile database, cannot provide all the features of a traditional database to a user. The mobile devices must be lightweight, should be able to download data from a remote depositary and execute even in the offline mode, should be able to synchronize the modified data during the downtime with the enterprise whenever the network becomes available again, should have a quick start-up time, and so on. Mobile databases are usually integrated with the operating system and specific applications similar to traditional databases, but they are directly driven by procedure calls and designed for data storage for persistent media. SQLite is an example of such a database used in Android, iPhone, and web browsers such as Mozilla Firefox.

3.7 Advantages of Mobile Cloud Computing

There are several promising advantages of MCC, which are due to the use of a cloud environment as discussed here:

1. *Extending battery lifetime*: In MCC, data storage and processing happen outside the device and in the cloud, so it automatically increases the battery lifetime of the device. Any large computation drains the battery very quickly as it consumes a lot of power. It has been observed that offloading task into cloud like large-scale matrix computation can reduce the battery power by 45%, and in the case of chess game using cloud, a 45% energy saving is possible [17]. So offloading and task migration are effective solutions to extend the battery life of mobile devices.

2. *Extending storage capacity*: As was pointed out earlier, storage was a big constraint for a mobile device. But MCC provides a huge amount of storage. Amazon's simple storage service and Dropbox are examples of cloud that provides storage to the user. Flickr is an application for photo-sharing based on MCC. Even Facebook application is an example of image sharing based on MCC.

3. *Extending processing power*: Many applications such as transcoding, playing games, broadcasting multimedia service, and so on, require high-processing power, which can be made available by offloading tasks into the cloud.

4. *High reliability*: In MCC, data and applications are stored in multiple computers, so there is no chance of data loss. Disaster management has become faster because of multisite availability. Many times, the cloud provides copyright to digital content such as music and video, thus preventing unauthorized distribution. Apart from this, cloud provides security services such as virus scanning, authentication, malicious code detection, and so on. In this way, MCC has improved reliability.

5. *On-demand service*: In MCC, the user gets, on demand, seamless service from the cloud. Because of the elastic nature of the cloud, users need not install dedicated hardware or software in their device. Everything can be obtained from the cloud. As an example, an Android user can get plenty of mobile apps from Google Play Store any time and in any amount they want in a "pay as you use" fashion.

3.8 Applications of Mobile Cloud computing

Because of the massive use of smart mobile devices and cloud, many applications of MCC have emerged. These applications are very promising in terms of mobility and availability. They are discussed in the following sections.

3.8.1 Mobile Commerce

In simple language, mobile commerce is the mobile version of e-commerce. Each and every utility of e-commerce is possible through mobile devices using the computation and storage in the cloud. According to Wu and Wang [41], mobile commerce is "the delivery of electronic commerce capabilities directly into the consumer's hand, anywhere, via wireless technology." There are plenty of examples of mobile commerce, such as mobile transaction and payment, mobile messaging and ticketing, mobile advertising and shopping, and so on. Wu and Wang [41] further report that 29% of mobile users have purchased through their mobiles 40% of Walmart products in 2013, and $67.1 billion purchases will be made from mobile device in the United States and Europe in 2015. This statistics proves the massive growth of m-commerce. In m-commerce, the user's privacy and data integrity are vital issues. Hackers are always trying to get secure information such as credit card details, bank account details, and so on. To protect the users from these threats, public key infrastructure (PKI) can be used. In PKI, an encryption-based access control and an over-encryption are used to secure the privacy of user's access to the outsourced data. To enhance the customer satisfaction level, customer intimacy, and cost competitiveness in a secure environment, an MCC-based 4PL-AVE trading platform is proposed in Dinh et al. [3].

3.8.2 Mobile Learning

Mobile learning is e-learning with mobility, which is fully related to e-learning and distance education. Mobile learning refers to learning across multiple contexts through social and content interactions, through the use of personal mobile devices. It is a process of learning where the learner is not at a fixed predetermined location, and learning opportunities are offered by mobile technologies. It includes portable technologies such as handheld computers, MP3 players, video players, notebooks, and mobile phones, which increases interactivity and portability. Traditional mobile learning applications suffer from various limitations such as limited educational resources, lack of transmission rate, high cost of device and network, and so on. The use of cloud has removed these drawbacks and provided large storage, processing power, and longer battery life. GeoSmart mobile application [42] is a perfect example of mobile learning. It is an online education system using cloud in an Indonesian community. Dinh et al. [3] introduced the IMERA platform-based contextual mobile learning system to facilitate the learners to access learning resources remotely. MCC-based mobile learning using smartphone software based on the open source JavaME UI framework and Jaber for clients, which enhances the quality of communication between students and teachers, is presented in Dinh et al. [3]. These applications provide educational material online in videos with student–teacher interaction. In Rolim et al. [43], a real-time mobile learning scheme, in which cloud is used to enhance the performance of the distance education system, is proposed.

3.8.3 Mobile Gaming

Games have always been a very interesting feature of mobile devices. But many times, games need high processing power, storage, and battery consumption, so in mobile gaming (m-gaming), the game execution takes place completely or partially in the cloud, and gamers interact with the screen interface of the mobile devices. Offloading multimedia code into the cloud reduces the power consumption of the device. MAUI [17] is an energy-aware mechanism that offloads the multimedia code into the cloud. But it does not offload all the multimedia code to the cloud. To mitigate the offloading cost, it partitions the multimedia code at runtime. Experiments show that MAUI [17] saves 27% energy in video games and 45% energy in a chess game. Dinh et al. [3] proposed utilizing a rendering adaptation technique, a new cloud-based m-gaming in which the game rendering parameters are dynamically adjusted based on the gamers' demands and communication constraints. The rendering adaptation method focuses on the minimization of the number of objects in the display list and scaling of the complexity of rendering operations to maximize the user experience and, at the same time, communication quality and cost.

3.8.4 Mobile Health Monitoring

Appropriate and real-time health monitoring of a patient is a very challenging and important issue nowadays. In a conventional healthcare system, nurses monitor the patient's health, record the data, and forward to doctors and other medical staff. All these activities are manually done, which introduces delay. In case of an emergency, the patient's health condition can change rapidly, and thus real-time monitoring without delay is needed. To make this possible, sensor cloud computing [44–46], which is another application of MCC, is introduced in a conventional healthcare system. Because sensor cloud computing is a combination of wireless sensor networks and cloud computing, sensor nodes are attached to the patient's body or to various kinds medical devices such as x-ray, ECG, MRI, and so on. The sensor nodes will collect the data and send them immediately to the cloud. Then doctor and or medical staff can access these data without any delay using the Internet. Microsoft's HealthVault and IBM's SmartHealth provide a cloud solution for health monitoring on a very large scale throughout the world.

3.9 Research Challenges in Mobile Cloud Computing

MCC has come up with many brilliant ideas but still there are some challenges that need to be solved.

3.9.1 Connectivity between Mobile Device and Cloud

MCC is all about mobility. Here, a mobile device is connected to a distant cloud. Because of continuous mobility, disconnection becomes a vital problem. Suppose a device is connected through a 3G network with the cloud. Now, if the device moves to such a place where 3G network is not available, connection with cloud will be broken. Cloudlet [5] or Clone cloud [18] has reduced the connectivity problem to an extent, but it is maximum for stationary devices and not always for mobile devices. It is useful in places such as shopping malls, airports, and so on.

3.9.2 Cloudlet Deployment in MCC

Deployment of cloudlet [5] is a vital support in MCC. In agent–client architecture, cloudlet plays a vital role. But deploying cloudlet has raised various questions: How much processing, storage, and networking capacity should a cloudlet possess? How do these resource requirements depend on the specific applications supported? How do they vary over time in the short and long term, taking into account natural clustering of users? How do cloudlet resource demands vary across individual users and groups of users? How sparse can cloudlet infrastructure be, yet provide a satisfactory user experience? What management policies should cloudlet providers use to maximize user experience while minimizing cost? Also, what kind of trust policy should be included for trust establishment or reputation-based trust? There are many more. These are the various research challenges to be faced in deploying cloudlet in MCC.

3.9.3 Centralization in Collaborative Model of MCC

In case of collaborative architecture of MCC, Hyrax [12] provides the architecture for Android devices. Here, mobile devices are used as resource providers using a centralized server that controls all the sharing and movements. On the other hand, Hadoop is a decentralized approach. Hyrax is based on the Hadoop principles. So, there is a conflict between these two due to centralization.

3.9.4 Security in MCC

Security, privacy, and trust [32] are always vital challenges for MCC. Though several security mechanisms have been developed for MCC, still it is not safe enough for many users to offload their personal data to the cloud. A malicious VM manager may interfere with important secret business transaction, so trust in the cloud provider is very important. A token-based authorization scheme is proposed in Ahmad et al. [47] to provide secure access to mobile cloud resources. A secure cloud-based augmented reality scheme for book search is proposed in Rafiq and Ahsan [48].

3.9.5 Incentives in MCC

In case of collaborative mechanism of MCC, surrogate devices share their resources for some tangible or intangible benefits [28]. But in many cases, there may be lack of common goals. So users may not share their resources. Thus, there should be a good incentive mechanism to encourage users to share their resources every time. In the case of monetary incentives, there should be good transparency and easy user interface. For efficient resource allocation, an application migration scheme is proposed in Ahmed et al. [49].

3.9.6 Energy Efficiency in MCC

"Green MCC" has become an important terminology nowadays. The main goal of MCC is to reduce the energy consumption [37] of mobile device, making the device greener. But in many cases, offloading a simple task to the cloud causes more energy consumption and cost than running the task in the device itself. So, there must be a tradeoff between offloading and energy savings.

3.9.7 Business Model of MCC

MCC is composed of two vital service providers [3]: the mobile service provider (MSP) and the cloud service provider (CSP). When a user gets services from cloud, both entities are involved in the process. Suppose a user wants to play online games. Game is provided by cloud, and it runs in cloud, so this is the job of CSP. On the other hand, connectivity of the user with the cloud is done by MSP through the Internet. So it is a question of how the profit will be divided between these two service providers. So a good business model has to be developed for profit sharing between MSP and CSP.

3.9.8 Data Traffic Management

Data traffic management is another important issue in MCC. In Bharath and Priyadarsini [50], a low-power data traffic management method, in which cloud computing is used to increase the battery lifetime of mobile devices, is proposed. Cloud-assisted pre-fetching is used where the multimedia content are fetched.

3.10　Conclusion

In this chapter, we discussed the features, architecture, advantages, and applications of MCC. MCC is a mixture of mobile computing and cloud computing. MCC integrates cloud computing into the mobile environment to enable users to utilize resources in an on-demand fashion. MCC provides a simple infrastructure for mobile applications and services by performing both data storage and data processing outside the mobile devices and in the cloud. This, in turn, reduces the energy consumption of the mobile device. Moreover, using the Internet, different applications such as m-commerce, m-learning, m-healthcare, and m-gaming can be accessed by the mobile device with limited storage capacity and battery life. Resource poverty, latency, bandwidth, mobility management, security, QoS, and so on are the critical issues in MCC. To deal with these issues, different schemes have already been developed. But still these difficulties cannot be removed completely. We have discussed the existing challenges in this chapter. In the future, the existing approaches will be modified to solve these problems more efficiently and effectively.

Questions

1. Define mobile cloud computing. What are the limitations of mobile computing? How does mobile cloud computing help overcome these limitations?
2. Describe the service-oriented architecture of mobile cloud computing.
3. What do you mean by agent–client architecture of mobile cloud computing? Explain with an example.
4. Briefly explain the collaborative architecture of mobile cloud computing.
5. What are the platforms and technologies used in mobile cloud computing.

6. Discuss mobile augmentation approaches.
7. What is offloading? What are the offloading methods used in mobile cloud computing.
8. Explain mobility management in mobile cloud computing.
9. Discuss the security risks to be considered during offloading to the cloud.
10. Describe the context management architecture of mobile cloud computing.
11. Explain the advantages of mobile cloud computing.
12. Discuss the applications of mobile cloud computing.
13. Discuss the research challenges of mobile cloud computing.

References

1. Z. Sanaei, S. Abolfazli, A. Gani, and R. Buyya, Heterogeneity in mobile cloud computing: Taxonomy and open challenges, *IEEE Communication Surveys and Tutorials*, 16(1), 369–392, 2014.
2. N. Fernando, S. W. Loke, and W. Rahayu, Mobile cloud computing: A survey. *Future Generation Computer Systems*, 29(1), 84–106, 2013.
3. H. T. Dinh, C. Lee, D. Niyato, and P. Wang, A survey of mobile cloud computing: Architecture, applications, and approaches, *Wireless Communications and Mobile Computing*, 13(18), 1587–1611, 2013.
4. J. Zhang and D. L. Roche, *Femtocells: Technologies and Deployment*, Wiley, Chichester, U.K., pp. 1–13, 2010.
5. M. Satyanarayanan, P. Bahl, R. Caceres, and N. Davies, The case for VM-based cloudlets in mobile computing, *IEEE Pervasive Computing*, 8(4), 14–23, 2009.
6. A. Mukherjee, S. Bhattacherjee, S. Pal, and D. De, Femtocell based green power consumption methods for mobile network, *Computer Networks*, Elsevier, 57(1), 162–178, 2012.
7. A. Mukherjee and D. De, Congestion detection, prevention and avoidance strategies for an intelligent, energy and spectrum efficient green mobile network, *Journal of Computational Intelligence and Electronic Systems*, American Scientific Publishers, 2(1), 1–19, 2013.
8. A. Mukherjee and D. De, A cost-effective location tracking strategy for femtocell based mobile network, in *IEEE International Conference on Control, Instrumentation, Energy and Communication*, Kolkata, India, pp. 533–537, 2014.
9. A. Mukherjee, P. Gupta, and D. De, Mobile cloud computing based energy efficient offloading strategies for femtocell network, in *Applications and Innovations in Mobile Computing*, Kolkata, India, IEEE, pp. 28–35, 2014.
10. A. Mukherjee and D. De, Femtocell based green health monitoring strategy, in *URSIGA*, Beijing, China, IEEE, pp. 1–4, 2014.
11. A. Mukherjee and D. De, A novel cost-effective and high-speed location tracking scheme for overlay macrocell-femtocell network, in *URSIGA*, Beijing, China, IEEE, pp. 1–4, 2014.
12. E. E. Marinelli, Hyrax: Cloud computing on mobile devices using MapReduce, Defense Technical Information Center, Ft. Belvoir, VA, 2009.
13. N. Fernando, S. W. Loke, and W. Rahayu, Mobile cloud computing: A survey, *Future Generation Computer Systems*, 29(1), 84–106, 2013.
14. Z. Sanaei, S. Abolfazli, A. Gani, and R. H. Khokhar, Tripod of requirements in horizontal heterogeneous mobile cloud computing, in *Proceedings of the First International Conference on Computing, Information Systems, and Communications (CISCO'12)*, Singapore, May 2012.
15. G. H. Canepa and D. Lee, A virtual cloud computing provider for mobile devices, in *Proceedings of the First Workshop on Mobile Cloud Computing and Services: Social Networks and Beyond*, San Francisco, CA, ACM, p. 6, 2010.

16. R. Kemp, N. Palmer, T. Kielmann, and H. Bal, Cuckoo: A computation offloading framework for Smartphones, in *Mobile Computing, Applications, and Services*, Springer, Santa Clara, CA, pp. 59–79, 2012.

17. E. Cuervo, A. Balasubramanian, D. K. Cho, A. Wolman, S. Saroiu, R. Chandra, and P. Bahl, MAUI: Making smartphone last longer with code offload, in *Proceedings of the Eighth International Conference on Mobile Systems, Applications, and Services*, San Francisco, CA, ACM, pp. 49–62, 2010.

18. H. Qi, and A. Gani, Research on mobile cloud computing: Review, trend and perspectives, in *Second International Conference on Digital Information and Communication Technology and Its Applications*, Bangkok, Thailand, pp. 195–202, 2012.

19. D. Huang, X. Zhang, M. Kang, and J. Luo, MobiCloud: Building secure cloud framework for mobile computing and communication, in *Fifth International Symposium on Service Oriented System Engineering*, Nanjing, China, pp. 27–34, 2010.

20. K. Kumar and Y. H. Lu, Cloud computing for mobile users: Can offloading computation save energy?, *Computer*, 43(4), 51–56, 2010.

21. M. D. Kristensen, Scavenger: Transparent development of efficient cyber foraging applications, in *International Conference on Pervasive Computing and Communications*, Mannheim, Germany, pp. 217–226, 2010.

22. A. Berl, E. Gelenbe, M. Di Girolamo, G. Giuliani, H. De Meer, M. Q. Dang, and K. Pentikousis, Energy-efficient cloud computing, *The Computer Journal*, 53(7), 1045–1051, 2010.

23. M. L. Puterman, *Markov Decision Processes: Discrete Stochastic Dynamic Programming*, John Wiley & Sons, New York, vol. 414, 2009.

24. I. Constandache, X. Bao, M. Azizyan, and R. R. Choudhury, Did you see Bob?: Human localization using mobile phones, in *Proceedings of the 16th Annual International Conference on Mobile Computing and Networking*, Chicago, IL, ACM, pp. 149–160, 2010.

25. N. Banerjee, S. Agarwal, P. Bahl, R. Chandra, A. Wolman, and M. Corner, Virtual compass: Relative positioning to sense mobile social interactions, in *Proceedings of the Eighth International Conference on Pervasive Computing*, Helsinki, Finland, pp. 1–21, 2010.

26. R. Mayrhofer, C. Holzmann, and R. Koprivec, Friends radar: Towards a private p2p location sharing platform, in *Proceedings of the 13th International Conference on Computer Aided Systems Theory*, Las Palmas de Gran Canaria, Spain, pp. 527–535, 2012.

27. V. Sacramento, M. Endler, H. K. Rubinsztejn, L. S. Lima, K. Goncalves, F. N. Nascimento, and G. A. Bueno, Moca: A middleware for developing collaborative applications for mobile users, *Distributed Systems Online, IEEE*, 5(10), 2–2, 2004.

28. K. Chard, K. Bubendorfer, S. Caton, and O. Rana, Social cloud computing: A vision for socially motivated resource sharing, *IEEE Transactions on Services Computing*, 5(4), 551–563, 2012.

29. N. Palmer, R. Kemp, T. Kielmann, and H. Bal, Ibis for mobility: Solving challenges of mobile computing using grid techniques, in *Proceedings of the 10th Workshop on Mobile Computing Systems and Applications*, Santa Cruz, NY, pp. 171–176, 2009.

30. R. Khare and R. N. Taylor, Extending the representational state transfer (rest) architectural style for decentralized systems, in *Proceedings of the 26th International Conference on Software Engineering*, Edinburgh, United Kingdom, pp. 428–437, 2004.

31. D. Petcu, C. Craciun, and M. Rak, Towards a cross platform cloud API, in *Proceedings of Components for Cloud Federation*, Busan, Korea, pp. 166–169, 2011.

32. A. N. Khan, M. M. Kiah, S. U. Khan, and S. A. Madani, Towards secure mobile cloud computing: A survey. *Future Generation Computer Systems*, 29(5), 1278–1299, 2013.

33. H. J. La and S. D. Kim, A conceptual framework for provisioning context-aware mobile cloud services, in *Proceedings of the IEEE Third International Conference on Cloud Computing*, Miami, FL, pp. 466–473, 2010.

34. A. Klein, C. Mannweiler, J. Schneider, and H. D. Schotten, Access schemes for mobile cloud computing, in *Proceedings of the 11th International Conference on Mobile Data Management*, Kansas City, MO, IEEE, pp. 387–392, 2010.

35. J. Flinn and M. Satyanarayanan, Powerscope: A tool for profiling the energy usage of mobile applications, in *Second IEEE Workshop on Mobile Computing Systems and Applications*, New Orleans, LA, pp. 2–10, 1999.
36. K. Banerjee and E. Agu, Powerspy: Fine-grained software energy profiling for mobile devices, in *International Conference on Wireless Networks, Communications and Mobile Computing*, Maui, HI, pp. 1136–1141, 2005.
37. C. Seo, S. Malek, and N. Medvidovic, An energy consumption framework for distributed Java-based systems, in *Proceedings of the 22nd IEEE/ACM International Conference on Automated Software Engineering*, Atlanta, GA, ACM, pp. 421–424, 2007.
38. E. Koukoumidis, D. Lymberopoulos, K. Strauss, J. Liu, and D. Burger, Pocket cloudlets, *ACM SIGPLAN Notices*, 47(4), 171–184, 2012.
39. P. Ganapati, Apple blocks palm pre iTunes syncing again, *Wired.com*, 2009.
40. W. Li, H. Yang, and P. He, The research and application of embedded mobile database, in *International Conference on Information Technology and Computer Science*, Kiev, Ukraine, pp. 597–602, 2009.
41. J. H. Wu and S. C. Wang, What drives mobile commerce?: An empirical evaluation of the revised technology acceptance model, *Information & Management*, 42(5), 719–729, 2005.
42. A. Nugraha, S. H. Supangkat, and D. Nugroho, Goesmart: Social media education in cloud computing, in *International Conference on Cloud Computing and Social Networking*, Bandung, Indonesia, pp. 1–6, 2012.
43. S. M. Butt, *Cloud Centric Real Time Mobile Learning System for Computer Science*, GRIN Verlag GmbH, Munich, Germany, 2014.
44. C. O. Rolim, F. L. Koch, C. B. Westphall, J. Werner, A. Fracalossi, and G. S. Salvador. A cloud computing solution for patient's data collection in health care institutions, in *Second International Conference on eHealth, Telemedicine, and Social Medicine*, St. Maarten, The Netherlands, IEEE, pp. 95–99, 2010.
45. D. De and A. Mukherjee, Femtocell based economic health monitoring scheme using mobile cloud computing, in *Fourth International Advance Computing Conference*, IEEE, pp. 385–390, 2014.
46. D. De and A. Mukherjee, Femto-cloud based secure and economic distributed diagnosis and home health care system, *Journal of Medical Imaging and Health Informatics*, 5(3), 435–447, 2015.
47. A. Ahmad, M. M. Hassan, and A. Aziz, A multi-token authorization strategy for secure mobile cloud computing, in *Second IEEE International Conference on Mobile Cloud Computing, Services, and Engineering*, Oxford, United Kingdom, pp. 136–141, 2014.
48. A. Rafiq and B. Ahsan, Secure and dynamic model for book searching on cloud computing as mobile augmented reality, *International Journal of Modern Education and Computer Science*, 6(1), 72, 2014.
49. E. Ahmed, A. Akhunzada, M. Whaiduzzaman, A. Gani, S. H. A. Hamid, and R. Buyya, Network-centric performance analysis of runtime application migration in mobile cloud computing, *Simulation Modelling Practice and Theory*, 50, 42–56, 2014.
50. V. Bharath and K. Priyadarsini, Energy efficient data traffic management in mobile cloud computing, *Energy*, 2(4), 59–64, 2014.

4

Offloading in Mobile Cloud Computing

ABSTRACT   Mobile devices suffer from poor battery life, limited resource, and limited storage capacity. To deal with these constraints, offloading is performed. Offloading refers to a mechanism where data storage and computations are done inside the remote cloud instead of the mobile device. Consequently, the battery life of the device is increased as well as the difficulties of storage and resource limitations are removed. In this chapter, we have discussed on offloading with its applications toward energy-efficiency.

KEY WORDS: *offloading, energy, cloud, mobile cloud computing.*

4.1 Introduction

In this era of cloud computing, people leverage cloud services from diverse aspects and enjoy various benefits of cloud computing. Cloud functionalities can be exploited in many ways: infrastructure-as-a-service (IaaS), such as Amazon EC2; platform-as-a-service (PaaS), such as Google App Engine build and deliver web applications; software-as-a-service (SaaS), such as e-mail services (e.g., Hotmail); and web applications (e.g., Google Docs). The increasing commercial adoption of cloud computing is attributable to its advantages over conventional computing, which include reduced cost, easy maintenance, and automatic scaling. Despite the combined advantages of cloud computing, the full potential of mobile cloud computing is far from being fully exploited. When it comes to mobile handheld apparatus, computing, storage resources of mobile, and serious power constraints due to limited battery lifetime are the major contributors leading to a bottleneck. Offloading as distributed computing can be used to solve this problem.

Offloading means the transfer of data from a computer or digital device to another digital device [1]. Offloading is a solution to augment these capabilities of mobile systems by migrating computation to more resourceful computers, such as servers. This is different from the traditional client–server architecture, whereas thin client always migrates computation to a server. Computation offloading is also different from the migration model used in multiprocessor systems and grid computing, where a process may be migrated for load balancing. The key difference is that computation offloading migrates programs to servers outside the users' immediate computing environment; process migration for grid computing typically occurs from one computer to another within the same computing environment, that is, the grid [2]. For mobile devices, cyber foraging is proposed by Kumar et al. [2]. It is described as a mechanism to augment the computational and storage capabilities of mobile devices through task distribution.

Energy-efficient resource allocation is cost effective as well as environment friendly. For power optimization, a game theoretic approach is proposed for resource allocation by Ge et al. [3]. For the past two decades, there have been many attempts to enable mobile devices to use remote execution for the purpose of improving energy efficiency and application performance [3,4]. These approaches reduce application execution time on mobile devices, thus decreasing the energy consumption of both CPU and memory. These attempts could be classified into two approaches. The first approach involves fine-grained, energy-aware offloading of mobile code to the infrastructure [5]. This approach relies on programmers to modify the program to handle partitioning, state migration, and adaptation to various changes in network conditions. Application can offload only part of the methods, which benefits from remote execution for the purpose of energy saving. This approach is also known as *partial offloading*. For instance, a media streaming application contains a decoder component and a video player component, with the former being CPU-intensive mainly consuming energy for the CPU and memory. As such, this CPU-intensive component can be offloaded without offloading any of the screen-intensive portion. The second approach is the coarse-grained task offloading scheme in which the full process/program or full virtual machine is migrated to the infrastructure, and then programmers do not have to modify the application source code to take advantage of computation offloading. This approach is referred to as *full offloading*, which reduces the burden placed on programmers.

4.2 Offloading Decision

Since offloading migrates computation to a more resourceful computer, it involves decision making regarding whether and what computation to migrate. A vast body of research exists on offloading decisions for (1) improving performance and (2) saving energy. Figure 4.1 shows the computation performed if offloading is used or not.

4.2.1 Improving Performance

Offloading becomes an attractive solution for meeting response time requirements on mobile systems as applications become increasingly complex; for example, a navigating robot may need to recognize an object before it collides with the object; if the robot's processor is too slow, the computation may need to be offloaded. Another application is context-aware computing, where multiple streams of data from different sources such as

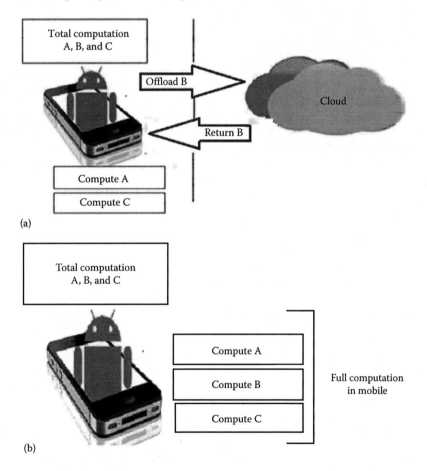

FIGURE 4.1
(a) Computation process performed using offloading to cloud. (b) Computation process performed without using offloading.

TABLE 4.1

Parameters Used

Parameter	Definition
s_m	Speed of the mobile system
s_s	Speed of the server
W	Amount of computation for the second part
d_i	Amount of data
B	Bandwidth
p_m	Power consumed by the mobile system per unit time
p_c	Power required to send data from the mobile system over the network
p_i	Power consumed per unit time for executing instruction inside the server

GPS, maps, accelerometers, and temperature sensors need to be analyzed together in order to obtain real-time information about a user's context. In many of these scenarios, the limited computing speeds of mobile systems can be enhanced by offloading.

The condition for offloading to improve performance can be formulated as follows [3]:

Without loss of generality, the program can be divided into two parts: one part that must run on the mobile system and the other part that may be offloaded. This is a case of *partial offloading*. The first part may include user interface and the code that handles peripherals.

The parameters used in performance measurement and energy calculation are presented in Table 4.1.

The time to execute the second part on the mobile system is calculated as follows:

$$T = \frac{w}{s_m} \tag{4.1}$$

If the second part is offloaded to a server having bandwidth B, then sending d_i amount of data takes d_i/B seconds for sending the input data d_i ignoring the initial setup time for the network. Offloading can improve performance when execution, including computation and communication, can be performed faster at the server. The time to offload and execute the second part is given by

Total time = Communication time + Computation time

$$T' = \frac{d_i}{B} + \frac{w}{s_s} \tag{4.2}$$

where
(d_i/B) is the communication time
(w/s_s) is the computation time at the server side

Offloading will improve performance if the following condition holds:

$$T > T' \Rightarrow \frac{w}{s_m} > \left(\frac{d_i}{B} + \frac{w}{s_s}\right) \Rightarrow w\left(\left(\frac{1}{s_m}\right) - \left(\frac{1}{s_s}\right)\right) > \frac{d_i}{B} \tag{4.3}$$

This inequality holds for:

- Large w: the program requires heavy computation.
- Large s_s: the server is fast.
- Small d_i: a small amount of data is exchanged.
- Large B: the bandwidth is high.

This inequality shows limited effects of the server's speed. If the computation and data exchange are both high even though the server is infinitely fast, offloading cannot improve performance. Hence, only tasks that require heavy computation (large w) with light data exchange (small d_i) should be considered. This requires analyzing programs to identify such tasks. Moreover, if $[w((1/s_m) - (1/s_s)) - (d_i/B)]$ is defined as the performance gain of offloading, the server's speed has diminishing return: doubling s_s will not double the gain.

If the complete data or computation is offloaded, it is the case of *full offloading*.

4.2.2 Saving Energy

Energy is a primary constraint for mobile systems [6]. Recently, developing smartphone systems having a large number of functionalities consume more power and shorten the battery life. Offloading can extend the battery life by transferring the energy-intensive parts of the computation to the servers.

The power consumed to perform the task within the mobile system is given by

$$E = p_m * \frac{w}{s_m} \tag{4.4}$$

The total power consumed considering the transmission and computation is determined by

$$E' = p_c \left(\frac{d_i}{B} \right) + p_i \left(\frac{w}{s_s} \right) \tag{4.5}$$

where
(d_i/B) is the communication time
(w/s_s) is the computation time at the server side

Offloading saves power if the following condition holds:

$$E > E' \Rightarrow p_m * \left(\frac{w}{s_m} \right) > p_c \left(\frac{d_i}{B} \right) + p_i \left(\frac{w}{s_s} \right) \Rightarrow w \left(\left(\frac{p_m}{s_m} \right) - \left(\frac{p_i}{s_s} \right) \right) > p_c \left(\frac{d_i}{B} \right) \tag{4.6}$$

Offloading will save power if heavy computation (large w) and light communication (small d_i) are considered. Figure 4.2 shows offloading is beneficial when large amounts of computation C are needed with relatively small amounts of communication D.

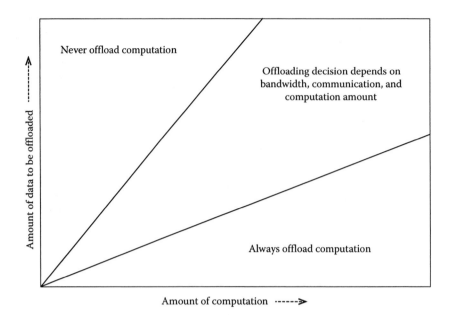

FIGURE 4.2
Decision of offloading based on the amount of data and computation.

Figure 4.2 also shows the different areas requiring a decision on whether to offload data or not and supports the fact that data should be offloaded only if the amount of computation is high [2]. Thus, it is observed that in the area where computation is high, *full offloading* should be used and where moderate computation is required, *partial offloading* can be done.

4.3 Types of Offloading

Offloading can be categorized into two different groups depending on two different criteria, as shown in Figure 4.3.

4.3.1 Depending on Material Being Offloaded

Data offloading: In this scheme, data are migrated from one congested network to another network [7].

Computational offloading: In this scheme, an expensive computational process is migrated from the mobile device to the server with cloud to improve performance and battery life [7].

4.3.2 Depending on Approaches to Time Reduction

4.3.2.1 Fine-Grained Offloading or Partial Offloading

An energy-aware offloading approach sends mobile code to infrastructure [1]. This is intuitively proposed to transmit as little data as possible by partitioning a program and

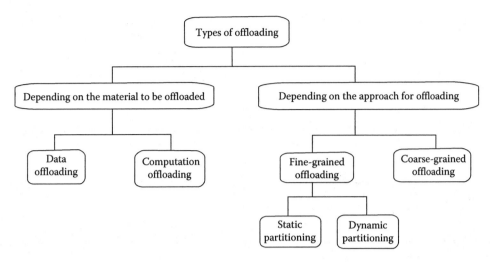

FIGURE 4.3
Types of offloading.

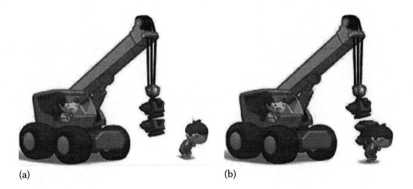

(a)   (b)

FIGURE 4.4
(a) Full offloading and (b) partial offloading.

offloading only the power-hungry parts of the application, as shown in Figure 4.4. This can be further divided into the following:

- *Static partitioning offloading scheme*: The communication cost depends on the size of the transmitted data and network bandwidth. The computation cost depends on the number of instructions [1]. The proposed client–server architecture of offloading is divided into several parts: monitors, offloading engine, module, CPU utilization, and wireless bandwidth of mobile device. The partitioning program is the transform of application into directed graph, the vertex set is java class, and the edge set denotes the interaction among classes. It is divided into $(k + 1)$ partitions: one unoffloadable and k disjoint partitions offloaded to the cloud [1]. The communication cost and the computation time can be obtained before the execution.

- *Dynamic partitioning offloading scheme*: This scheme models the failure recovery time and total execution time of pervasive applications running under the control of the offloading system [1]. Assuming the application is performed with the presence of offloading and failure recoveries, when any failure occurs, the code should be offloaded. The state machine is an approach that only re-offloads the failed subtasks and thus reduces the execution time.

4.3.2.2 Coarse-Grained Offloading or Full Offloading

In this case, as shown in Figure 4.4, the full program is migrated to the infrastructure, and then the programmer cannot modify the source code [1]. This approach does not require estimating the computation time prior to the execution. Instead, the program is initially executed on the portable client with a timeout. If the computation is not completed after the timeout, it is offloaded to the server. The timeout is first set to be the minimum computation time that can benefit from offloading. This method has the following advantage: The instances with short computation time are executed at the client side, and the instances with large computation time are offloaded to the server. It requires no burden on the application programmers and brings little calculation overhead. This method also has the following limitation: The timeout wastes energy for the computation instances that are eventually offloaded, in which case the total application offloading may result in unnecessary energy consumption in transmission.

4.4 Topologies of Offloading

A connection is shown between the mobile devices and the cloud of self-reliant multi-cloud offloading in Figure 4.5. This is known as star topology. If the mobile device communicates with different servers, then a lot of time and energy are spent for communication with multiple servers [8]. Figure 4.4 also presents ring topology where a connection exists between the mobile devices and the cloud of self-reliant multi-cloud offloading. Here, two nodes are communicated with each other and dependent on other nodes. When failure occurs in the middle of the nodes, then the program cannot be successfully executed [8].

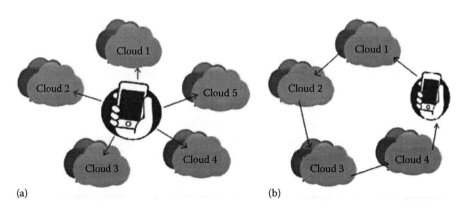

(a)   (b)

FIGURE 4.5
(a) MCC offloading star topology and (b) MCC offloading ring topology.

4.5 Offloading in Cloud Computing and in Mobile Cloud Computing: Similarities and Differences

The word "mobile computing" essentially means computing related to mobile devices. To state more clearly, mobile computing deals with data and computation related to devices that are not fixed and can change the network as and when required. On the contrary, cloud computing, as shown in Figure 4.6, is the computation of a whole set of devices that are mobile and, at the same time, also immobile devices. So, it can be said that cloud computing is a super set of mobile cloud computing.

Using the same analogy, it can also be deduced that offloading in cloud computing is nothing but a super set of the offloading in mobile cloud computing. All the architectures and algorithms for the two cases remain the same; the only difference that comes into the picture is the mobility of the computing devices that are being considered.

Moreover, as the devices can move in the mobile cloud computing environment, the network they access also may change with their movement, as shown in Figure 4.7. So, in the case of offloading in a mobile cloud computing paradigm, the change in the network is an essential factor, as with the change in network, the server that the device accesses changes and the need for a constant calculation of energy optimization techniques also arises.

4.6 Adaptive Computation Offloading from Mobile Devices

Offloading of computationally intensive applications from the mobile platform to the remote cloud can solve the problems of limited battery lifetime and limited processing power of mobile devices. Nowadays, most of the mobile applications are developed by using the standard Android development pattern. Android is already established as the most prominent mobile phone platform. Mobile Augmentation Cloud Services (MACS) is

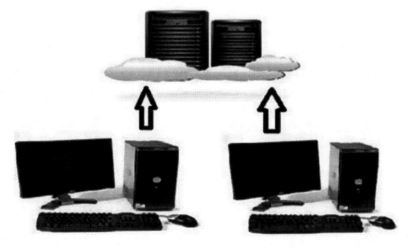

FIGURE 4.6
Offloading to cloud from the connected immobile devices.

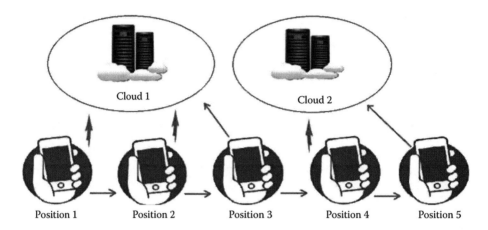

FIGURE 4.7
Offloading in MCC deals with moving devices.

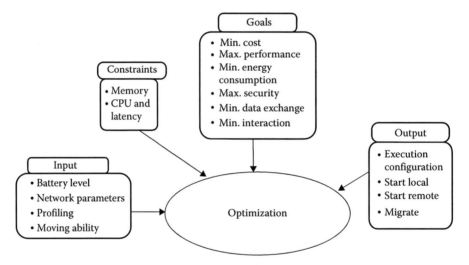

FIGURE 4.8
Optimization problem in mobile augmentation cloud services.

a middleware that enables adaptive extension or optimization, as shown in Figure 4.8, of Android application execution from mobile client into the cloud.

4.6.1 Mobile Augmentation Cloud Services

MACS middleware [9] is capable of making Android applications offloadable and supports seamless adaptive computation offloading, as shown in Figure 4.9. The modules of the program are divided into two groups: one runs locally and the other runs on the cloud side. The decision for partitioning is approached as an optimization problem according to the input on the conditions of cloud side and mobile devices, such as available memory, CPU load, and remaining battery power of devices, bandwidth between the cloud and mobile devices. Finally, based on the solution to the optimization problem, this middleware offloads parts to the remote clouds and returns the corresponding results back. The goal

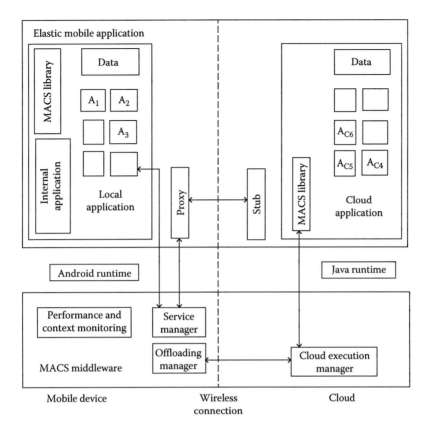

FIGURE 4.9
Mobile augmentation cloud services architecture.

of MACS is to enable transparent computation offloading for mobile applications. A MACS application consists of an application core, such as Android activities, GUI, access to device sensors that cannot be offloaded. It also consists of multiple services (S_i) that encapsulate separate application functionalities, usually resource-demanding components, which can be offloaded, SR_i. Services communicate with the application through an interface defined by the developer in the Android interface definition language. MACS allows dynamic application partitioning at runtime; it monitors the execution of the services and the environment parameters. Whenever the situation changes, this middleware can adapt the offloading and partitioning. At the cloud side, the MACS middleware handles the offload requests from the clients, installs and then initializes the offloaded services, and invokes the methods. MACS middleware monitors the resources on the mobile execution environment and available clouds. It then forms the optimization problem whose solution will decide whether the service that contains the called function should be offloaded or not. When the service is determined to be offloaded to the remote cloud, MACS tries to execute the service remotely.

4.6.2 Adaptive Computation Offloading

Adaptive computation offloading is the partial computing as stated earlier. The whole task is divided into smaller units, and only the power-hungry parts are sent to the cloud for offloading, depending on certain conditions. Suppose there are n modules that can be offloaded, $A_1, A_2, ..., A_n$. For a specific module i, let its memory cost be mem_i and code size be $code_i$. Let us consider k number of related modules that can be offloaded. For each of them, we denote the transfer size $tr_1, tr_2, ..., tr_k$, send size $send_1, send_2, ..., send_k$, and receive size $rec_1, rec_2, ..., rec_k$, where $send_k + rec_k = tr_k$. We introduce x_i for module i, which indicates whether the module i is executed locally ($x_i = 0$) or remotely ($x_i = 1$). The solution $x_1, x_2, ..., x_n$ represents the required offloading partitioning of the application. The cost function is represented as [9] follows:

$$\min_{x \in 0,1}(c_{transfer} * w_{tr} + c_{memory} * w_{mem} + c_{CPU} * w_{CPU}) \qquad (4.7)$$

where

$$c_{transfer} = \sum_{i=1}^{n} \sum_{j=1}^{k} tr_j * (x_j \text{ XOR } x_i) \qquad (4.8)$$

$$c_{memory} = \sum_{i=1}^{n} mem_i * (1 - x_i) \qquad (4.9)$$

$$c_{CPU} = \sum_{i=1}^{n} code_i * \alpha * (1 - x_i) \qquad (4.10)$$

In the cost function, the first part depicts the transfer cost for remote execution of services, including the transfer cost of its related services. The latter part of Equation 4.8 includes the dependency relationship between modules, c_{memory} denotes the memory cost of the mobile device, c_{CPU} denotes the CPU cost on the mobile device, and α is the convert factor mapping the relationship between code size and CPU instructions. Here, w_{mem}, w_{CPU}, and w_{tr} are the weights of each cost representing the lowest memory cost, lowest CPU load, and lowest interaction latency, respectively. Three constraints considered in this case are discussed as follows:

Constraint 1—*Minimized memory usage*: The memory cost of resident service should not be more than the available memory of the mobile device, that is,

$$\sum_{i=1}^{n} mem_i * (1 - x_i) \leq avail_{mem} * f_1 \qquad (4.11)$$

where f_1 is the factor to determine the memory threshold to be used.

Constraint 2—*Minimized energy usage*: For the offloaded services, the energy consumption of offloading should not be greater than the case where offloading is not used and local execution occurs, that is,

$$E_{local} - E_{offload} > 0 \qquad (4.12)$$

Constraint 3—Minimized execution time: This is enabled when the user prefers very fast execution, that is,

$$t_{local} - t_{offload} > 0 \tag{4.13}$$

where
The local execution time t_{local} is the ratio of CPU instructions to local CPU frequency
The remote execution time $t_{offload}$ consists of the time consumed by CPU, file transmission, and the overhead of the middleware

According to these three constraints, the partitioning problem can be transformed into an optimization problem. The solution of $x_1, x_2, \ldots, x_n$ is the optimized partitioning strategy. By using integer linear programming on the mobile device, MACS gets a global optimization result. Whenever the parameters in the model change, such as available memory or network bandwidth, the partitioning is adapted by solving the new optimization problem. MACS also does profiling for each offloaded module or service to dynamically change its execution plan and adjust the partitioning.

Offloading can lower the CPU load on a mobile device significantly. It also saves lots of energy, which indicates that the battery time can be increased compared to the local execution. With the increase in needed computation, it is better to push those computations that utilize considerable resources to the remote cloud. The more computation is needed, offloading has more advantage. The overhead of this framework is small and acceptable. This framework supports offloading of multiple Android services. If there are multiple services in one application and all of those services can be offloaded to the remote cloud, MACS resource monitor natively supports this situation and can make the corresponding allocation determination, so that some of the services should be offloaded to cloud and the rest of the services can run locally.

4.7 Cloud Path Selection for Offloading

Data traffic over mobile networks has observed an exponential increase during the past few years, mainly because of the very high popularity of tablets, smartphones, and laptops. As the data traffic on the mobile networks has increased substantially, it is clear that there is an immediate need to offload data traffic for the best performance of voice and data services. As a result, different methods, including such technologies as Wi-Fi, IP flow mobility, and femtocells, have been proposed to manage the data traffic. Along with the development of cloud computing, offloading has become a more attractive way to extend the battery life and reduce the execution time on mobile devices. But a large number of clouds appear in the sky with different charging and conditions, and offloading the same program to different clouds may result in different amounts of computing within the same duration due different cloud speeds, and may cost different communication time due to bandwidth and cloud availability. Therefore, an optimal cloud path selection method is needed when choosing the best cloud.

The selection process is shown in Figure 4.10 [10]. It can be a hard task since a variety of data needs to be analyzed and many factors need to be considered. AHP and fuzzy TOPSIS are ideal ways to do multiple criteria decision making [10].

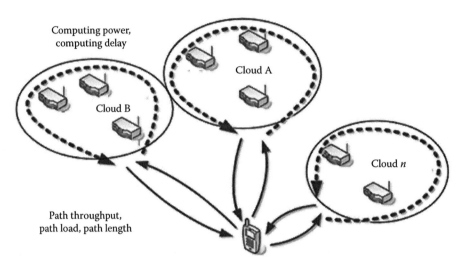

Computing power,
computing delay

Cloud A

Cloud B

Cloud *n*

Path throughput,
path load, path length

FIGURE 4.10
Cloud path selection.

4.7.1 Cloud Path Selection Methods

There are different ways to determine the optimal cloud path pair when more than one such path exists. They can be summarized as follows:

- *Random*: Select the cloud path pair randomly.
- *Bandwidth dependent*: Choose the cloud path pair with the highest bandwidth.
- *Link failure rate dependent*: Select the cloud path pair with the lowest link failure rate.
- *Speedup factor dependent*: Select the cloud path pair with the highest speedup factor.
- *Cost dependent*: Select the cloud path pair with the lowest cost.

4.7.2 Cloud Path Selection Issues

The cloud path selection method must be chosen taking into consideration the understated issues:

- *Bandwidth*: It depends on the wireless link between the mobile devices and cloud. When the wireless connection is excellent, a large amount of application execution and data could be offloaded to the cloud, but when it is poor, only a small amount of application execution and data can be offloaded during limited time.
- *Price*: It denotes the cost for the same amount of computing and its variation in different cloud services. For the mobile cloud service providers, how to build an economic service provisioning scheme is critical particularly when the mobile cloud resource is restricted.
- *Speed*: It presents how fast a server can compute on cloud. Relatively, it can be measured though speedup factor *F*, which compares the execution speed of cloud to that of mobile device.

- *Security*: First of all, shifting all data and computing resources to the cloud is dangerous; for example, tracking individuals through location-based navigation data offloaded to the cloud. Besides, security and privacy settings depend on the cloud providers as the data are stored and managed inside the cloud.

- *Availability*: It is related to the link's failure and cloud's unavailability during the whole offloading process. Failures may occur due to the mobility of mobile devices and unstable connectivity of wireless links, which render a less predictable performance of a program running under the control of offloading systems.

4.8 Mobile Data Offloading Using Opportunistic Communication

Smartphones and their various applications are very popular nowadays. Due to the increasing popularity of different applications for smartphones, 3G cellular networks have become overloaded. Offloading of mobile data using opportunistic communications is one of the most promising solutions to this problem. The main advantage of this technique is that it does not require any monetary outlay. Using opportunistic communications, we can promote information broadcast in mobile social networks (MoSoNets) [11] to reduce the amount of mobile data traffic. The application of MoSoNets has already expanded from online social networking sites to different powerful mobile applications and mobile social software. Currently, a high percentage of mobile data traffic is generated using these mobile social applications. The major side effect of these applications along with other mobile applications is that the 3G cellular networks are overloaded already.

4.8.1 System Model

In MoSoNets [11], the target-set selection problem is the first step toward offloading of mobile data for delivery of information. It may be decided by the service providers to deliver the information to only a few selected customers or target users to minimize the mobile data traffic and thus reduce the operation cost. Then, the information can be further generated by the target users among all other subscribers via social participation, when their smart mobile phones are within the range of communication and opportunistic communication is possible.

First, the procedure of choosing the initial target set with only k users has to be known, so that the mobile data traffic can be reduced. This objective can be translated for maximizing the number of users who can receive the delivered information using opportunistic communications. When the number of users is large, the mobile data traffic will be less. Next, the regularity of human mobility has to be exploited and the identified needs of the target set have to be applied using mobility history for future information delivery. Then ultimately the feasibility of opportunistic communications for mobile smartphones will be evaluated.

MoSoNets, as shown in Figure 4.11, can integrate all friends from all major social networking sites as well as family members and colleagues. Using MoSoNets, it is possible to

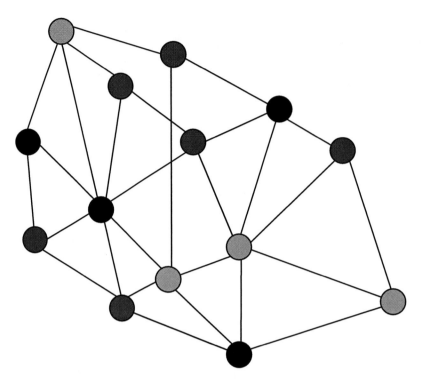

FIGURE 4.11
Social graph of mobile users.

provide face-to-face interactions among people who know each other. In MoSoNets, there are two types of connections:

1. *Local connections*: These are realized by short-range communications through Wi-Fi or Bluetooth networks.

2. *Remote connections*: These connections are realized by long-range communications through cellular networks such as EDGE, HSPA, etc. In real life, this type of communication occurs between friends only. Here, users have to pay for data transmissions. According to the social relationship of mobile users, a social graph can be drawn, as shown in Figure 4.11. Here, users who are directly connected by an edge are called friend uses, and three types of communities are denoted by three different colors: light gray, gray, and black. Users in the same community form a clique. Different communities are connected by different connections.

MoSoNets are a combination of traditional social networks with opportunistic networks. In MoSoNets, both types of communications can be exploited to facilitate information dissemination. Here, information can be forwarded actively by all friend users. Similarly, information can be pulled directly from each other by all mobile users who are in contact.

4.8.2 Target-Set Selection

First, it needs to be proved that the information broadcast function is sub-modular. Then only the greedy algorithm can be applied to find out the target set. For any subset S of users, the information dissemination function $g(S)$ provides the final number of users who are infected when the initial target set is S. Function $g(\cdot)$ is sub-modular if it satisfies the diminishing returns rule, that is, the marginal gain of adding a user, say u, into the target set S is always equal to or greater than that of adding the same user into a superset S' of S:

$$g(S \cup \{u\}) - g(S) \geq g(S' \cup \{u\}) - g(S') \tag{4.14}$$

Generally, it is very hard to compute exactly the underlying information dissemination function $g(\cdot)$ and to get a closed form expression of it. However, the value of $g(\cdot)$ can be estimated by using Monte Carlo sampling.

4.8.3 Greedy, Heuristic, and Random Algorithms

For a target-set selection problem, there are three algorithms: greedy, heuristic, and random algorithms.

For a greedy algorithm, initially the target set is blank. First, the information broadcast function $g(\{u\})$ will be evaluated for every user u, and then the most active user will be selected into the target set. Then, this process will be repeated, and in each round, the next user will be selected from the rest with maximum increase of $g(\cdot)$ into the target set, until k users are selected. This target-set selection is a type of NP-hard problem. The limitation of this greedy algorithm is that it needs the information of user mobility during the broadcasting process, which sometimes may not be available at the beginning.

The heuristic algorithm uses the history of user mobility to find out the target set, and then this target set is used for the delivery of information in the future.

Finally for the random algorithm, k target users are chosen by the service providers randomly from all subscribers. This algorithm may be very simple, but it is still very much effective in the offloading process.

As a comparison among these three algorithms, for all cases, the greedy algorithm performs best followed by the heuristic algorithm. The greedy algorithm may not be practical, but it is the basis for the heuristic algorithm by which the regularity of human mobility is exploited. Compared to these two algorithms, a unique advantage of the random algorithm is that service providers need not collect contact information from users.

4.9 Three-Tier Architecture of Mobile Cloud Computing

The general case of architecture, as shown in Figure 4.12, is intended to support computation offloading for mobile nodes where both a middle tier consisting of nearby resource-limited cloudlets and a remote tier consisting of resourceful distant cloud servers is available. For such architecture, both managed and unmanaged usage scenarios can be envisioned. The managed usage scenario typically corresponds to the case in which a wireless service provider deploys its own cloudlet infrastructure at its own APs to be used by its mobile subscribers [9]. Hence, the WSP can be expected to centrally regulate the access to the cloudlet with the goal of offering a good service experience to its subscribers and of fulfilling its own utility goals. The unmanaged usage scenario corresponds to the case where cloudlet-augmented Wi-Fi hot spots are deployed by public authorities or private entrepreneurs at

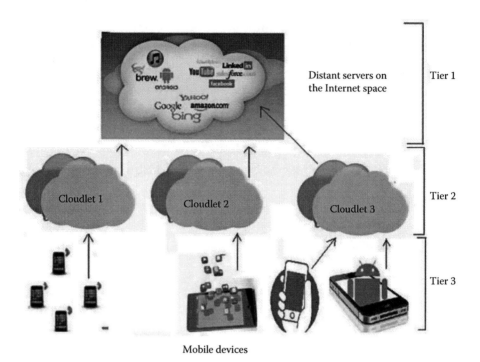

Distant servers on
the Internet space — Tier 1

Tier 2

Tier 3

Mobile devices

FIGURE 4.12
Three-tier architecture of mobile cloud computing.

facilities such as airports, train stations, public buildings, and cafes for the benefit of their citizens or customers [12]. This scenario is an extension of the current plan in which free-access Wi-Fi hot spots are deployed just as an additional service, or as a way to attract more customers on a simple best effort basis and without any attempt of regulating their use.

4.10 Requirements of Data Offloading

Energy optimization is one of the important reasons for offloading [13]. Due to the recent improvement in wireless technology and distributed computing, high-speed wireless connection, powerful computing, and storage capability are possible. Energy efficiency is a very important issue in this recent development. Through a power estimation model by Alex Shye, we can say power consumption varies drastically depending on the workload, the screen, and the CPU being the largest consumers. Mobile cloud computing through offloading saves processing energy, which is one of the most significant ways to minimize power consumption. Remote execution is another technique to reduce energy consumption; here, the client sends the task to the server or servers where the task is being executed and it is sent back to the client. If the energy required for sending the parameter and receiving the result is less than the energy required to execute the task locally, then only some energy will be saved. On the other hand, if the offloading time and the server execution time are less than the local execution time, then an improved performance of the mobile application will be noticed.

The solution proposed for various mobile computational environment changes considers three common environmental changes: client-side power level, connection status, and bandwidth. The highest priority is given to the client-side power levels as they are irreversible. Among many solutions, the Memory Arithmetic Unit and Interface (MAUI) is one of the best [13]. Based on the input, the solver decides whether to execute the method locally or remotely. Profiling information helps to predict if the future invocation needs to be offloaded. The decision made by the MAUI solver is implemented by proxy, handling both data transfer and control.

A running application decides the nature of power consumption. Better utilization of the resource on the cloud needs a trade-off between the transmission energy cost and local execution energy cost. Some models reduce transmission cost while bringing partition overhead by offloading some parts of the program to the server. Some try to reduce the workload of the programmer by transplanting the entire OS or application runtime to the cloud.

4.11 Performance Analysis of Offloading Techniques

4.11.1 Analysis of Energy Consumption in Offloading for Different Data Amounts

The energy consumed when offloading 10 and 100 KB of data with Wi-Fi using the platform-specific model is shown in Figure 4.13. The energy consumption of code offloading grows almost linearly with the round trip time (RTT). In fact, for offloading 10 KB

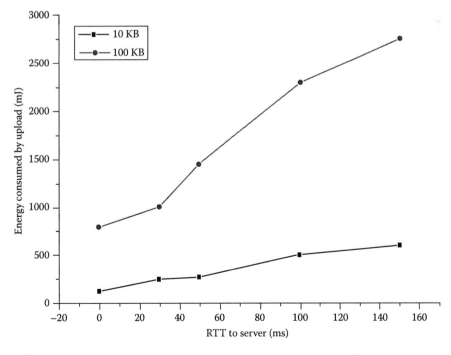

FIGURE 4.13
The energy consumed by offloading 10 and 100 KB of code to the cloud; various round trip times to the server.

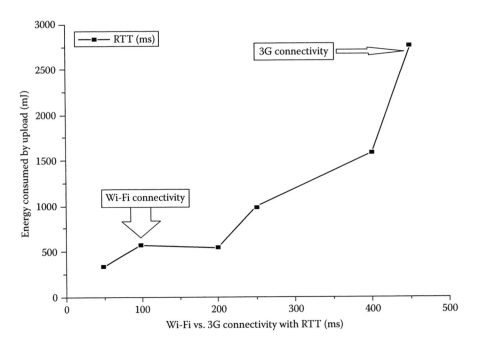

FIGURE 4.14
Wi-Fi vs. 3G connectivity vs. energy consumed by upload.

of code, the energy consumption almost doubles when increasing the RTT from 10 to 25 ms only, and it doubles again when the RTT reaches 100 ms.

4.11.2 Analysis of Energy Consumption in Offloading for Different Connectivity

From this result, two important decisions can be reached: First, cloud providers should strive to minimize the latency to the cloud for mobile users as shorter RTTs can lead to significant energy savings. Second, the benefits of remote execution are most impressive when the remote server is nearby, such as on the same LAN as the Wi-Fi access point, rather than in the cloud. Figure 4.14 shows a comparison of Wi-Fi connectivity and 3G connectivity.

4.12 Multi-Cloud Offloading in Mobile Cloud Computing Environment

A self-reliant multi-cloud offloading system is described in Figure 4.15. It is actually multiple offloading from the single mobile device to a single server one by one. The number of available remote servers is k; A{A1, A2, A3} contains three tasks A1, A2, A3 needed to be offloaded; I = {I0, I1, I2,..., I k} where I0 is the mobile device and the offloading server is {I1, I2, I3,..., IK}. Here, the tasks A1, A2, and A3 are allocated to I1, I2, and I3. The mobile device offloads the computation to each server separately, and after executing the program, the server then sends the processing results back to the mobile device [14].

A program can be offloaded to different servers one by one. Then the time complexity is very high and offloading is very complex, which is the disadvantage of self-reliant mobile

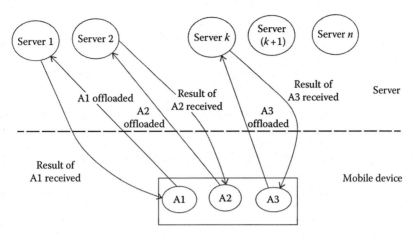

FIGURE 4.15
Self-reliant multi-cloud offloading systems.

cloud computing. This problem can be solved by the parallel offload of different application parts of the program to different servers.

4.12.1 Performance Analysis of Multi-Cloud Offloading Schemes

The non-cooperative transmission when the amount of data (d) is small can still be better energy efficient than multiple input, single output. However, when d is larger, energy consumption is high, as shown in Figure 4.16. The graph underneath shows that when the amount of data sent to the server is small, the energy requirement is less. It indicates that the system is more energy efficient. But as the amount of data is increased, the energy requirement is also increased simultaneously.

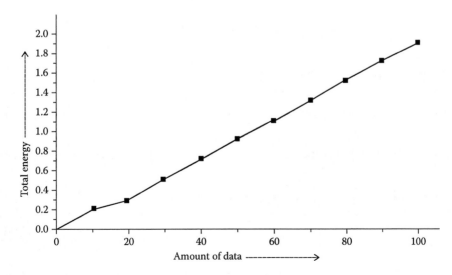

FIGURE 4.16
Amount of data vs. total energy.

4.13 Conclusion

Offloading involves power consumption in executing an application and during communication with the remote cloud. To better utilize the resource on the cloud, there is a trade-off between transmission energy cost and local execution energy cost. Though many works have made an effort to find the balance point, only a subset of those studies focuses on energy efficiency, while in many cases they merely focus on response time and other resource consumption. A large part of the researches uses modeling and simulation. Some of the models only offload parts of the program to the server, which reduces transmission cost, while at the same time bringing extra partition overhead. Other works have focused on transplanting the entire OS or application runtime to the cloud as this ease the burden on external application programmers.

Questions

1. What does offloading mean? Why is offloading necessary?
2. Show how offloading is beneficial for a large amount of computing.
3. What are the types of offloading? Discuss.
4. What is MACS? Show how it complements the necessity of offloading.
5. List the different methods of cloud path selection. Discuss the issues related to it, if any.
6. What is a MoSoNet? Discuss the algorithms for target selection in a MoSoNet.
7. Discuss the trade-offs in offloading.
8. Discuss the topologies in offloading.
9. What are full and partial offloading? Which one is considered better and why?
10. Explain with different examples when full offloading and partial offloading is to be done. Also provide a case in which offloading will not be beneficial.

References

1. X. Ma, Y. Cui, L. Wang, and I. Stojmenovic, Energy optimizations for mobile terminals via computation offloading, in *Second IEEE International Conference on Parallel Distributed and Grid Computing*, Solan, India, pp. 236–241, 2012.
2. K. Kumar, J. Liu, Y. H. Lu, and B. Bhargava, A survey of computation offloading for mobile systems, *Mobile Networks and Applications*, 18(1), 129–140, 2013.
3. Y. Ge, Y. Zhang, Q. Qiu, and H. Y. Lu, A game theoretic resource allocation for over all energy minimization in mobile cloud computing system, in *ACM/IEEE International Symposium on Low Power Electronics and Design*, Redondo Beach, CA, pp. 279–284, July, 2012.
4. L. Jiao, R. Friedman, X. Fu, S. Secci, Z. Smoreda, and H. Tschofenig, Cloud-based computation offloading for mobile devices: State of the art, challenges and opportunities, in *IEEE Future Network and Mobile Summit*, Lisboa, Portugal, pp. 1–11, 2013.

5. K. Kumar and H. Y. Lu, Cloud computing for mobile users: Can offloading computation save energy? *Computer*, 43(4), 51–56, 2010.

6. H. Wu, Q. Wang, and K. Wolter, Tradeoff between performance improvement and energy saving in mobile cloud offloading systems, in *IEEE International Conference on Communications Workshops*, Budapest, Hungary, pp. 728–732, 2013.

7. H. Pan, Cellular and wireless offloading, in *Making the Most of Mobile*, Telekom Innovation Laboratories, Cambridge, United Kingdom, September 24, 2012.

8. B. Gao, L. He, L. Liu, K. Li, and S. A. Jarvis, From mobiles to clouds: Developing energy-aware offloading strategies for workflows, in *13th International Conference on Grid Computing*, Beijing, China, pp. 139–146, September 2012.

9. D. Kovachev, T. Yu, and R. Klamma, Adaptive computation offloading from mobile devices into the cloud, in *Tenth International Symposium on Parallel and Distributed Processing with Applications*, Leganes, Spain, pp. 784–791, 2012.

10. H. Wu, Q. Wang, and K. Wolter, Methods of cloud-path selection for offloading in mobile cloud computing systems, in *CloudCom*, Taipei, Taiwan, pp. 434–448, 2012.

11. B. Han, P. Hui, A. V. Kumar, V. M. Marathe, J. Shao, and A. Srinivasan, Mobile data offloading through opportunistic communications and social participation, *IEEE Transactions on Mobile Computing*, 11(5), 821–834, 2012.

12. S. X. Wang, H. Shen, and D. Wetherall, Accelerating the mobile web with selective offloading, in *Proceedings of the Second ACM SIGCOMM Workshop on Mobile Cloud Computing*, Hong Kong, China, pp. 45–50, August 2013.

13. E. Lagerspetz and S. Tarkoma, Mobile search and the cloud: The benefits of offloading, in *IEEE International Conference on Pervasive Computing and Communications Workshops*, Seattle, WA, pp. 117–122, 2011.

14. H. Qi and A. Gani, Research on mobile cloud computing: Review, trend and perspectives, in *Second International Conference on Digital Information and Communication Technology and Its Applications*, Bangkok, Thailand, pp. 195–202, 2012.

5

Green Mobile Cloud Computing

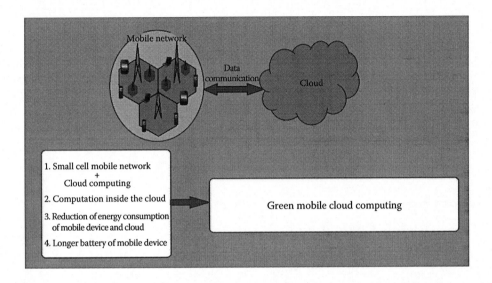

ABSTRACT Green mobile cloud computing is an emerging research area today. Green mobile network means an energy-efficient mobile network that consumes low power. Small cell network provides a green mobile network. By offloading computation inside the cloud, the power consumption by a mobile device can be reduced. But this can cause more power consumption and expense inside the cloud. Thus, there should be a tradeoff between an energy-efficient mobile network and a green cloud environment. In this chapter, we discuss several existing approaches for green mobile networks and green cloud computing. Based on comparative studies, we discuss how green mobile cloud computing can be achieved by merging green cellular networks with the cloud environment.

KEY WORDS: *macrocell, femtocell, cloud, power, energy.*

5.1 Introduction

Energy efficiency is gaining importance for future mobile networks day by day. The increased usage of mobile devices, together with rising energy costs and the need to lessen greenhouse gas emissions, calls for energy-efficient technologies that reduce the overall energy utilization of computation, storage, and communications [1]. A fraction of energy savings in mobile networks could lead to significant financial savings and would protect the environment with the reduced production of the harmful greenhouse gases [1].

Cloud computing (CC) has been widely accepted as the coming generation's computing framework [1]. CC allows its users to make use of the platforms, software, and infrastructures provided by the cloud providers at low cost. With the enormous growth of the usage of mobile applications and the gaining popularity of cloud computing, CC is integrated to the mobile environment, leading to the development of mobile cloud computing (MCC) [2].

MCC has received substantial attention as it is a promising technology that allows the processing and storage of data outside the mobile devices, that is, in the cloud, thus resulting in considerable energy saving for the devices. It provides virtually endless resources that are obtainable as per the requirement and charged accordingly. This provides economic advantages both for the cloud providers and the users. MCC is characterized by lower operating cost, high scalability, and better accessibility [3]. The notions of user mobility and wireless access pattern give rise to certain challenges, such as mobility management, quality of service (QoS), energy management, and security and privacy issues, to MCC [2]. The most critical among them is the energy efficiency of mobile devices, which is discussed in this chapter. We also review the literature on the same.

5.2 Green Mobile Computing

Energy efficiency of a mobile network is a major issue, which has cost effectiveness and environmental benefits. Research in the field of mobile communication is focused currently on the advancement of greener mobile networks, that is, networks that consume low energy. Mobile devices have become an intrinsic part of human life. The effectiveness of upcoming technologies lies in their capability to offer more and more innovative applications. Mobile devices are expected to meet the goals of effectiveness and convenience. Development of green mobile network is a demanding concern. Mobile communication techniques seek to maximize performance metrics, such as throughput, QoS, and reliability, and minimize energy consumption [3]. Except during peak times, mobile devices are not operated at their full capacity, and consequently the energy level they require is much less than the maximum energy supplied. The rapid growth of energy consumption by the mobile devices poses a serious threat to biodiversity; for instance, the greenhouse effect has become more alarming with high emission of CO_2. Nearly half of the operational expenditure of the mobile networks is for the consumed energy.

Developing green, that is, energy-efficient, techniques for mobile networks is essential for the reduction of energy consumption of the entire mobile network. A mobile network consists of the following five components, as shown in Figure 5.1:

- Data centers
- Macrocell base station (MBS)
- Femtocell base station (FBS)
- Mobile devices
- Mobile application and services

For the development of a green mobile network, energy-efficient techniques are needed to be implemented at each of the five components.

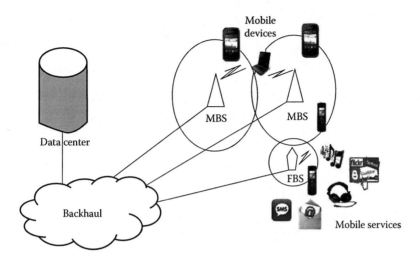

FIGURE 5.1
Mobile network consisting of data center, MBS, FBS, mobile devices, and mobile services.

5.2.1 Green Data Center

In the present scenario, the number of data centers in the backhaul is constantly increasing with the growing surge of data storage and computations online. In the interim, the energy consumption of the data center is increasing tremendously. In order to reduce the energy consumption for ecological and economic benefits, energy-efficient, that is, green, techniques must be implemented. The techniques involve switching the on/off device and developing hardware and software in accordance with the traffic load and user demands [4]. It is observed that many links of the core network lie idle, and thus it is better to shut them down for energy saving. A possible approach toward energy saving is to switch off inactive network devices, links, and switches, aiming at an energy-efficient data center [5].

5.2.2 Green Macrocell Base Station

It has been pointed out that nearly 60% of the energy consumption of a mobile network is for operating the MBSs [3]. Moreover, as the MBSs are densely deployed for guaranteeing effective coverage, the cost of the required resources also increases. Therefore, green techniques that reduce the energy consumption of this sector of mobile network are absolutely necessary. The common techniques include turning the base station on/off depending on the changing traffic load [6,7]. It has been is proposed that some of the macrocell base stations can remain switched off while others remain switched on and take care of the coverage at the time of low traffic [6,7]. In Niu et al. [8], a cell zooming strategy is implemented that effectively adjusts the cell size according to the traffic load, user requirements, and channel conditions. Cell breathing is another popular congestion control scheme, which also adjusts cell size depending on the traffic load [9].

5.2.3 Green Femtocell Base Station

FBSs are low-cost, low-power, self-deployed base stations that are used effectively for the reduction of energy consumption of mobile networks. As femtocells are deployed by users, they are cost effective [3,10]. Figure 5.2 shows the deployment of a femtocell in the core network.

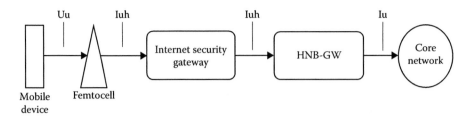

FIGURE 5.2
Deployment of femtocell in the core network.

Mobile users and the femtocell are connected via a Uu interface [10]. A femtocell is connected to the core network by the HNB-GW (home node base station gateway) [10]. A security gateway (SeGW) provides a proper security mechanism between the femtocell and the HNB-GW over the Internet [10]. A femtocell connects with the HNB-GW through the Iuh interface [10]. HNB-GW is connected to the core network via the Iu interface [10]. Though each femtocell is a low-power device, when a number of femtocells operate together, the energy consumption is quite large. Thus, reduction of energy consumption is necessary at this sector too. Yeh et al. [11] observed that energy control is of utmost importance between coverage and performance for greening the femtocell. Adaptive coverage based on outdoor and indoor user mobility is proposed by Claussen et al. [12] to minimize energy consumption. Ashraf et al. [13] propose an algorithm to improve the energy efficiency of femtocell base stations as per user activity detection. According to the proposed algorithm, the femtocell switches off its transmission and the associated processing in the absence of active users. A low-energy sniffer capability is included in the femtocell base station that detects any active call from a mobile to the underlying macrocell [13]. Once an active call is detected from a registered mobile device by the sniffer, the femtocell can reactivate the air interface and its pilot energy transmission, which means that it enters the active mode. On switching to the active mode, the mobile reports the femtocell pilot to the macrocell to which it is connected, and the mobile is handed over from the macrocell to the femtocell. The user activity detection algorithm is presented in Table 5.1 [13].

In a two-tier macrocell–femtocell network, many femtocell base stations are deployed in the coverage area of the macrocells. Chen and Wu [14] defined an energy-efficient handover protocol in two-tier OFDMA (orthogonal frequency division multiple access) macrocell–femtocell network, which is based on the idea of switching off hardware parts and processing capabilities of the base station at the right time and the right place.

TABLE 5.1

User Activity Detection Algorithm

1. **Start**
2. Femtocell measures the received energy on macrocell uplink band.
3. **If** no active mobile device is detected close to the femtocell.
4.   **Step 2** is repeated until an active mobile device is detected close to the femtocell.
5. **Else** Femtocell activates processing and starts pilot transmission.
6. **End If**
7. Mobile is handed over from the macrocell to the femtocell.
8. Femtocell serves the mobile device until the call terminates.
9. Femtocell goes back to the idle mode and disables pilot and processing.
10. **End**

TABLE 5.2

Green Handover Algorithm

---

1. **Start**
2. Femtocell measures the received energy on macrocell uplink band.
3. Femtocell measures the dwell time, expected time, and the free spectrum band for data transmission.
4. **If** the available spectrum bandwidth for the femtocell is less than or equal to the bandwidth required for serving the user request,
5.   **Step 2** is executed until the available spectrum bandwidth for the femtocell is greater than the bandwidth required for serving the user request.
6. **Else Step 8** is executed.
7. **End If**
8. **If** dwell time is greater than expected time.
9.   **Step 2** is executed until the dwell time is less than the expected time.
10. **Else** Femtocell activates processing and starts pilot transmission.
11. **End If**
12. Mobile is handed over from the macrocell to the femtocell.
13. Femtocell serves the mobile device until the call terminates.
14. Femtocell goes back to the idle mode and disables pilot and processing and repeats. **Step 2**
15. **End**

A femtocell is in active mode if it has multiple connected mobile devices with it. When a femtocell is not in active mode, it can switch off its energy amplifier, RF transmitter, RF receiver, and other hardware performing functions such as data encryption and hardware authentication. It has been proved that the energy consumption of an active femtocell is 10.2 W and that of an idle femtocell is 6 W [13,14]. Thus, when a femtocell is switched off, there is a saving of its processing energy by 4 W. An energy-efficient handover scheme is proposed in Table 5.2 [14].

With the technique of switching off transmission in the absence of user traffic, the energy consumption of a femtocell base station can be effectively reduced. Comparative analyses of the user activity detection algorithm and the green handover algorithm based on different parameters are shown in Figures 5.3 through 5.8.

Figure 5.3 is the graphical representation of energy consumption with respect to data size ranging from 10 to 14.5 MB [14]. Generally, energy consumption decreases as the data size increases, as depicted in Figure 5.3. The energy utilization includes the total energy consumed by the femtocell base stations (i.e., FBSs), the macrocell base stations (i.e., MBSs), and user equipment (i.e., UEs existing in the network) [14]. The result shows that the green handover protocol achieves substantial savings in energy [13,14] because with the green protocol, frequent handover is minimized and the FBS stays in the idle mode [14].

It is observed that the Ashraf protocol [13] has more handover events than the green handover protocol [14]. The energy consumption of the Ashraf protocol is constant at 42 W [13,14], whereas that of the green handover protocol is less than 42 W [13,14]. Figure 5.4 is the graphical representation of energy consumption with respect to velocity of the user, ranging from 30 to 80 km/h. The energy consumption of the green handover protocol is less than that of the Ashraf protocol, as depicted in Figure 5.4 [13]. It is observed that the energy consumption drops as the velocity increases. It is also seen that the energy consumption is about 46 W if the velocity is 30 km/h, and it is about 41.28 W if the velocity is 80 km/h. Hence, the probability of keeping the FBS idle under the high-speed environment is high [14].

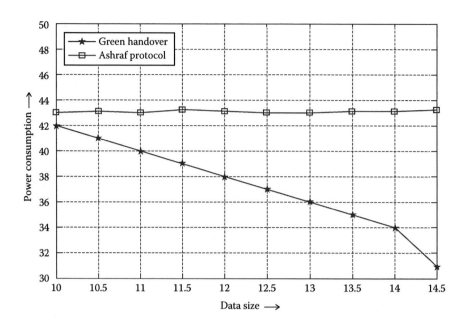

FIGURE 5.3
Data size versus energy consumption.

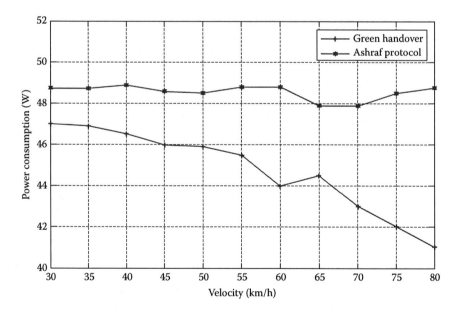

FIGURE 5.4
Velocity versus energy consumption.

Figure 5.5 is a graphical representation of the average handover latency (HL) [14], which is defined as the period of time required by user equipment (UE) to change its association from the current FBS/MBS to another one [13,14]. With a reduced number of handovers, HL also decreases. Figure 5.5 shows the variation of HL with data size ranging from 10 to 14.5 MB. It is observed from the figure that HL drops as the data size increases [14]. In case of green

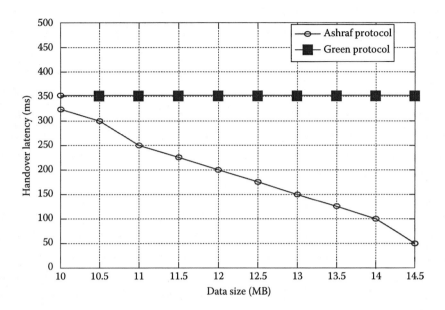

FIGURE 5.5
Data size versus handover latency.

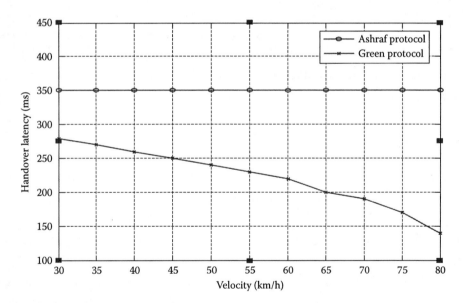

FIGURE 5.6
Velocity versus handover latency.

handover, HL is 325 ms for a data size of 10 MB and 74 ms for a data size of 15 MB, while the HL of the Ashraf protocol [13] is fixed at 360 ms [14], as seen from Figure 5.5. This is because a less frequent number of handover events occur.

Figure 5.6 is graphical representation of the average HL as a function of the velocity of the UE ranging from 30 to 80 km/h [14]. It is observed that HL decreases as the velocity increases. In case of a green handover, the HL is 275 ms if the velocity is 30 km/h and

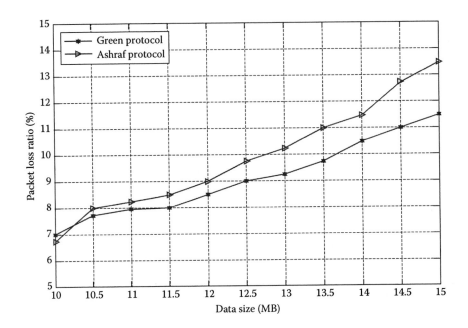

FIGURE 5.7
Data size versus packet loss ratio.

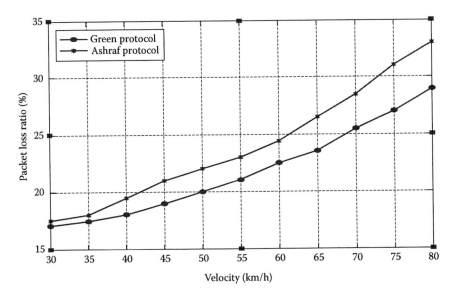

FIGURE 5.8
Velocity versus Packet loss ratio.

125 ms if the velocity is 80 km/h, while in the Ashraf protocol HL is fixed at 360 ms. According to Figure 5.6, the HL of the green protocol is better than that of the Ashraf protocol [13,14]. Figure 5.7 shows the packet loss ratio as a function of the data size ranging from 10 to 14.5 MB [14]. The packet loss ratio (PLR) is defined as the total number of lost packets to the total number of transmitted packet by the UE. Packet loss occurs when

TABLE 5.3

Comparison between User Activity Detection Algorithm and Green Handover Algorithm

Comparison Parameters	User Activity Detection Algorithm	Green Handover Algorithm
Energy consumption	Due to frequent handovers from macrocell to femtocell, the energy consumption of this algorithm is quite high.	Due to less frequent handovers from macrocell to femtocell, the energy consumption of this algorithm is quite low compared to the user activity detection algorithm.
Handover latency	This algorithm involves frequent handover, which increases handover latency significantly.	This algorithm causes fewer handovers than the user activity detection algorithm and hence handover latency decreases significantly.
Packet loss ratio	With frequent handovers, the probability of packet loss increases. Thus, in this case the packet loss ratio is quite high.	In this, case packet loss ratio is reduced as compared to the user activity detection algorithm.

handover occurs [14]. PLR increases as the data size increases. It is observed that the PLR using the green protocol and Ashraf protocol is about 7% if the data size is 10 MB. The PLR in the green protocol and the Ashraf protocol is 11.67% and 11.5%, respectively, if the data size is 15 MB. The PLR of green protocol is less than that of the Ashraf protocol under various data sizes [14].

Figure 5.8 is a graphical representation of the PLR as a function of the velocity of the UE, ranging from 30 to 80 km/h. PLR increases with the increase in velocity.

The PLR of the green protocol is less than that of the Ashraf protocol under various velocities, as shown in Figure 5.8. It is considered that the PLR is 29% if the velocity is 80 km/h in the green handover protocol, and it is 33% if the velocity is 80 km/h in the Ashraf protocol [13,14]. Consequently, the PLR of the green handover protocol is less than that of the Ashraf protocol. The results of the graphical comparisons are presented in Table 5.3.

5.2.4 Green Mobile Device

The fast development of mobile devices and the increasing demand for high computational energy is a matter of concern for green mobile networks. Green techniques are implemented in mobile devices considering the energy demand, resource needs, traffic pattern, and user behavior. A dynamic resource allocation scheme based on user requirement prevents unnecessary resource and energy wastage designed for mobile devices [15,16]. The implementation of these techniques results in green mobile devices and thus green mobile networks.

5.2.5 Green Mobile Application and Services

The key goal of mobile networks is to offer satisfactory services to mobile users. Different techniques for green applications are discussed in Wang et al. [3]. Lu et al. [17] proposed an approach for reducing energy consumption of mobile applications and services. This approach uses data compression to reduce the size of the data being transmitted by eliminating redundant data. This results in processing a reduced volume of data, and therefore the process consumes lesser energy. Energy saving in VoIP service is proposed in

Baset et al. [18]. The proposed technique includes the use of persistent TCP connection, which results in fewer keep-alive message exchanges between devices in the network, thus resulting in less energy consumption. This technique efficiently results in green mobile applications and services. Using this procedure and implementing green techniques at the five components of mobile networks, the green mobile network is achievable.

5.3 Green Mobile Network

In this section, we study the energy-efficient technologies proposed for the development of green mobile networks. It was proposed that reduction of energy consumption at the base station can be significantly improved with proper designing of base station hardware [13,19,20]. Proper resource allocation techniques such as efficient use of the RF amplifier or switching off the transceiver equipment or base station can result in significant energy saving. In a mobile network, nearly two-thirds of the calls and 90% of the data services originate indoors. But research shows that maximum household and businesses suffer from poor indoor coverage. Good indoor coverage would both gain customer loyalty and generate more revenue for the service providers [10]. However, utilization of a macrocell for attaining this sort of better indoor coverage comes with certain disadvantages. Hence, indoor solutions such as DASs (distributed antenna systems) [10] and picocells are gaining popularity in areas like large business centers, offices, and malls [10]. These internal systems are installed by the operators. These indoor solutions provide satisfactory internal coverage, offload traffic from the heavily loaded macrocells, enhance the quality of service and facilitate high-data-rate services. Though the aforementioned solutions are more efficient than outdoor macrocells in terms of expenditure, still to provide indoor coverage for voice and high-speed data services, such solutions are still too pricey to be used in some scenarios such as SOHO (small office and home office) and home users. However, the development of femtocells provides low-cost indoor solutions for such situations. The self-deployment feature of femtocell eliminates the deployment cost incurred in case of other indoor coverage techniques such as picocells. The use of femtocell can make mobile networks greener than they are today.

5.3.1 Congestion Control for Energy-Efficient Mobile Network

Mukherjee and De [9] proposed a green network deployment strategy with spectrum and energy efficiency in addition to congestion control. The cell size of each location area of the proposed network is determined according to the traffic in each cluster of the location area, where each location area consists of a number of clusters. A macrocell-based network is considered, with each of the clusters covered by the MBS. The number of users visiting the macrocells, the traffic load provided to the macrocell, and the probability of call blocking in each macrocell are forecasted according to the mobility information of the user maintained in dynamic location area list in the profile of the individual user. Based on the number of users and the call-blocking probability, congestion is controlled. Deployment of microcells, picocells, macrocells, and femtocells based on traffic load of each cluster in each location area achieves low power consumption by the base stations [9].

5.3.2 Energy Efficiency of Femtocell, Microcell, and Picocell Network over Macrocell-Based Mobile Network

In this section, we theoretically prove that FBS consumes low energy compared to the others [20]. The parameters used for the calculations are given in Table 5.4.

The received power by an MT at a distance d from the BS is given by Mukherjee et al. [20]

$$P_r(d) = \frac{P_t G_t G_r \lambda^2}{(4\pi)^2 d^2 L} \tag{5.1}$$

The received power by an MT from a BS at the border region of the cell having radius R is given by [20]

$$P_r = \frac{P_t G_t G_r \lambda^2}{(4\pi)^2 R^2 L} \tag{5.2}$$

Maximum path loss is given as [20]

$$PL_{dB} = -10 \log_{10}\left(\frac{G_t G_r \lambda^2}{(4\pi)^2 R^2 L}\right) \tag{5.3}$$

TABLE 5.4

Parameters Used in Energy Calculation

Parameters	Definition
P_t	Transmission power of a base station (BS) (W)
P_{tm}	Transmission power of an MBS (W)
P_{tmi}	Transmission power of a microcell base station (MiBS) (W)
P_{tp}	Transmission power of a picocell base station (PBS) (W)
P_{tf}	Transmission power of an FBS (W)
G_t	BS antenna gain
G_r	Mobile terminal (MT) antenna gain
λ	Wavelength of light
L	System loss not related to path loss
S	Minimum received energy density (W/m^2)
G_{tf}	Antenna gain of FBS
D	Normalized radiation pattern in the direction (θ, Φ); D is equal to unity in the direction of maximum radiation
R_f	Radius of femtocell
R_p	Radius of picocell
R_{mi}	Radius of microcell
R_m	Radius of macrocell
A_f	Area of femtocell
A_p	Area of picocell
A_{mi}	Area of microcell
A_m	Area of macrocell

Thus, the transmission power of a BS is given by [20]

$$P_{tdB} = P_{rdB} + PL_{dB} = 10\log_{10}\frac{P_t G_t G_r \lambda^2}{(4\pi)^2 R^2 L} - 10\log_{10}\left(\frac{G_t G_r \lambda^2}{(4\pi)^2 R^2 L}\right) = 10\log_{10}(P_t) \tag{5.4}$$

Hence, the transmission power (in dB) of an MBS is given by [20]

$$P_{tdBm} = P_{rdBm} + PL_{dBm} = 10\log_{10}(P_{tm}) \tag{5.5}$$

Hence, the transmission power (in dB) of an MiBS is given by [20]

$$P_{tdBmi} = P_{rdBmi} + PL_{dBmi} = 10\log_{10}(P_{tmi}) \tag{5.6}$$

Hence, the transmission power (in dB) of a PBS is given by [20]

$$P_{tdBp} = P_{rdBp} + PL_{dBp} = 10\log_{10}(P_{tp}) \tag{5.7}$$

The transmitted power by an FBS is calculated as [20]

$$P_{tf} = \frac{S 4\pi R_f^2}{D G_{tf}} \tag{5.8}$$

The area of a femtocell having a radius R_f is determined as [20]

$$A_f = \frac{3\sqrt{3}}{2} R_f^2 \tag{5.9}$$

The minimum received power (in watts) by an MT in a femtocell is given [20] by

$$P_{rf} = S\left(\frac{3\sqrt{3}}{2R_f^2}\right) \tag{5.10}$$

The transmission power of FBS is given by [20]

$$P_{tdBf} = P_{rdBf} + PL_{dBf} = 10\log_{10} P_{tf} \tag{5.11}$$

where
$P_{rdBf} = 10\log_{10}(P_{rf})$
PL_{dBf} is the path loss for FBS

It is observed that transmission power of a BS is directly proportional to its coverage area. As $R_f < R_p < R_{mi} < R_m$, then $A_f < A_p < A_{mi} < A_m$. This implies that $P_{tf} < P_{tp} < P_{tmi} < P_{tm}$ [20]. From Equation 5.2, it is observed that the received power is directly proportional to the transmitted power. Thus $P_{rm} > P_{rmi} > P_{rp} > P_{rf}$, where P_{rm}, P_{rmi}, P_{rp}, and P_{rf} are the received power (W) by an MT under the coverage of a macrocell, microcell, picocell, and femtocell, respectively. Hence, it is concluded that the transmitted power by the FBS and the received power by the MT in the femtocell network are less than that of other cell categories. For this reason, by deploying femtocells, a greener network can be obtained, as discussed in

Mukherjee et al. [20]. An MT will receive a signal from its nearest base station, which can be a macro-, micro-, pico-, or femtocell base station. As the path loss of an MT is directly proportional to the distance between itself and the BS, the mobile terminal will receive a signal from its nearest BS in order to minimize the path loss. Finally, five network classes are represented in Mukherjee et al. [20], where it has been successfully proved that the total power consumption of a femtocell-based network is much less than that of a macrocell-based network. The five cases are as follows:

1. *Class A network*: In a class A network, femtocells are used in urban areas rather than macrocells, as shown in Figure 5.9. Mukherjee et al. [20] observed that the power consumption of a network containing only femtocells is much less than that of a network containing only macrocells. This is so because the power consumption of the FBS is much lower than that of the MBS.

2. *Class B network*: In a class B network, femtocells, macrocells, and portable femtocells are used in urban, suburban, and rural areas, respectively, depending on the user traffic, as shown in Figure 5.10. Mukherjee et al. [20] observed that a class B network results in less power consumption than where the entire area would have been covered by macrocells.

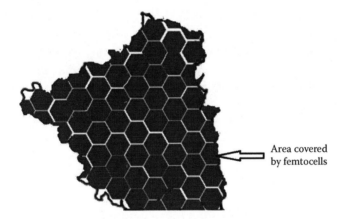

FIGURE 5.9
Femtocell covering the entire area.

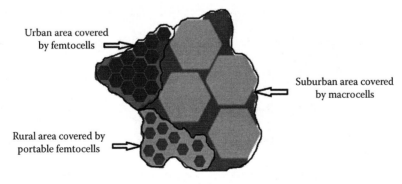

FIGURE 5.10
Allotment of femtocells, macrocells, and portable femtocells based on user traffic.

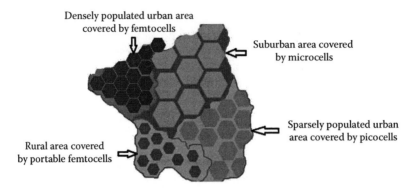

FIGURE 5.11

Allocation of femtocells, picocells, microcells, and portable femtocells in densely populated, sparsely populated, suburban, and rural areas based on user traffic.

3. *Class C network*: In a class C network, femtocells, picocells, microcells, and portable femtocells are allocated in densely populated, sparsely populated, urban, suburban, and rural areas, respectively, as shown in Figure 5.11. It is observed that by using a class C network, the power consumption can be reduced as compared to when using only macrocells [20].

4. *Class D network*: In this network, macrocells, microcells, picocells, and femtocells are deployed at the border region of an area according to the coverage needed, as shown in Figure 5.12. It is observed that using microcells, picocells, and femtocells at the border region instead of covering total area by macrocells can minimize total power transmission by the base station [20].

5. *Class E network*: In this network, femtocells are allocated at the boundary of a macrocell, as shown in Figure 5.13. It is observed that in the scenario considering macrocell at the center and femtocells at the boundary region of each macrocell in an area, the total transmitted power by the BSs can be reduced compared to that when only macrocells are used to cover the total area [20].

Figure 5.14 shows the five cases discussed graphically. From Figure 5.14, it is clear that in all the five cases the power consumption of the network is less than the macrocell-only-based network.

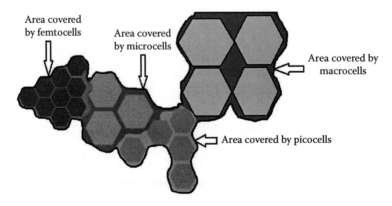

FIGURE 5.12

Border region covered by microcells, picocells, and femtocells, and the rest of the region covered by macrocells.

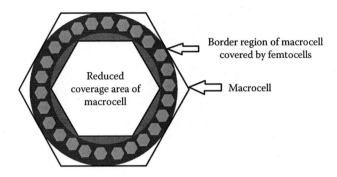

FIGURE 5.13
Femtocells located at the boundary of a macrocell.

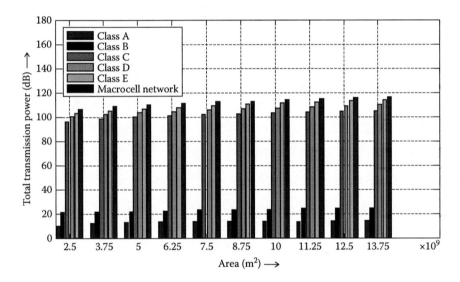

FIGURE 5.14
Comparison of power transmission for a fully macrocell network and five femtocell-based network classes.

From Figure 5.14, it is observed that the approximate power transmission in each of class A, B, C, D, and E networks are less than that of a macrocell network.

5.4 Green Cloud Computing

In the past few years, the use of mobile devices has increased tremendously. Smartphones have replaced the traditional mobile phones. Smartphones are characterized by a number of functionalities other than voice calling and text messaging. Various Android and Windows applications are popular with smartphone users [1]. Executing these applications on mobile devices is associated with many challenges such as high energy consumption, high CPU cycle, storage capacity, low bandwidth connection to the Internet, and so on [1]. These problems are solved with cloud computing. The introduction of cloud computing

TABLE 5.5

Comparison between Cloud Data Centers and Traditional Data Centers

Comparison Parameters	Cloud Data Center	Traditional Data Center
Scalability	Highly scalable	Not scalable
Elasticity	High elasticity	Low elasticity
Workload structure	Simple workload structure	Complex workload structure
Nature of hardware	Homogeneous hardware structure	Heterogeneous hardware structure
Ease of organization	Easy to organize data in the data centers	Difficult to organize data in the data centers

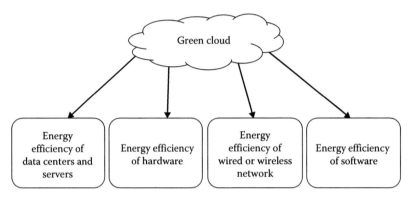

FIGURE 5.15
Energy consumption in the cloud.

has led to the simplification of delivering services and resources present in various data centers over the Internet [2]. Cloud computing gives support to computation, storage, and communication. Cloud computing is capable of storing bulk data in the cloud data centers and performing various computations and processes [21,22]. A comparison between the cloud data centers and the traditional data centers is presented in Table 5.5.

Though very efficient, the high-performing capabilities of CC eventually lead to increase in energy consumption [21]. This would definitely lead to release of greenhouse gases, which would be detrimental to the environment. However, there are techniques that can lead to energy saving, thus resulting in green cloud computing. For minimization of energy consumption in this field, energy saving is aimed at the followings regions of the cloud infrastructure [21,22], as shown in Figure 5.15:

1. Cloud data centers and servers
2. Cloud hardware
3. Cloud software
4. Wired or wireless network

5.4.1 Green Cloud Servers

The servers on the cloud consume energy due to a variety of reasons, such as enormous workload present on hardware, duplication of disk arrays and machines for the ease of usability and high bandwidth, large memories for storing the large quantity of data, various interconnection switches, and sudden changes in the load [21]. With server virtualization/consolidation, a very large number of users can share a single server, which not only increases utilization but also reduces the total number of servers required [22]. Users need not know of the tasks being performed by other users. Moreover, users utilize the server as though they are the only ones on the server [22]. During periods of low demand, some of the servers enter into the sleep mode, which reduces the energy consumption [22].

5.4.2 Green Cloud Data Center

Rapid growth in demand for computational power has resulted in the development of a large number of data centers. The cloud data centers store bulk data, but consume an enormous amount of energy, thus incurring high operational cost and generating carbon dioxide [21–23]. Thus these data centers have an adverse impact on the ecology and economy. The important technology for the energy-efficient framework of the data centers is virtualization, which is shown in Figure 5.16.

Virtualization eases software modification and portability [22,23]. Virtual machines (VMs) or hypervisors are capable of shifting from one cloud to another [23]. Virtualization simplifies the process of resource management because of the ease of mapping between virtual and physical resources. Beloglazon and Buyya [23] present a resource management scheme for virtualized cloud data centers, the objective of which is to constantly unite VMs influencing reallocation and switching off idle nodes to minimize energy consumption. In Beloglazon and Buyya [23], a dynamic reallocation of VMs is proposed based on resource availability. The number of physical nodes serving the current workload is minimized, and idle nodes are switched off for reduction of energy consumption. The algorithm for reallocation of VMs is shown in Table 5.6.

It is observed that the reallocation algorithm for virtualized cloud data centers leads to significant energy saving and efficiently forms a part of green cloud computing.

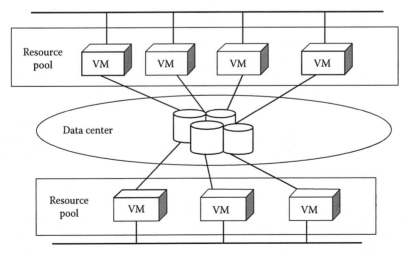

FIGURE 5.16
Virtualization of cloud data center.

TABLE 5.6

Algorithm for Reallocation of Virtual Machines (VMs) for Energy Efficient Cloud Data Centers

---

1. **Start**
2. VMs are arranged in decreasing order of its current utilization.
3. Each VM is allocated to a host that provides the minimum increase in energy consumption due to this allocation.
4. **If** the VMs are to be reallocated from one host to another, any of the two rules specified here is followed:
   a. Minimization of migration: Minimum number of VMs are migrated to reduce migration overhead;
   b. Highest potential growth: The VMs with lowest CPU utilization are migrated in order to increase the corresponding CPU utilization.
5. **End If**
6. **End**

5.4.3 Green Cloud Hardware

Energy is consumed by the CPU and various hardware components such as disks and network devices. In the idle state, about 60% of the energy is consumed by these units. This is a matter of concern for the environment [21,22]. Energy-efficient hardware is promoted by substituting hard disk drives with solid disk drives that consume less energy. The different redundant hardware can be also be made to lie dormant in order to achieve energy efficiency. Energy management techniques can be static energy management or dynamic energy management. Static energy management deals with the circuit, logic, and architecture at the system design stage for providing an energy-efficient design [21]. On the other hand, dynamic energy management aims at attaining energy efficiency by adapting the system behavior according to the requirement [21].

5.4.4 Green Cloud Software

The energy required for the execution of system software, applications, and online services are ever increasing. The cloud provides the platform and infrastructure to execute a bulk of software and applications [21,22], but the result is huge energy consumption. Thus, to combat this, a reduced amount of data replication and task transfer is used [21].

5.4.5 Green Wired or Wireless Network

Energy consumption can be minimized in wireless and wired network. Wireless networks require various energy-saving routing protocols [21]. It is crucial to optimize the number of hops made by the packets because the flow of packets also consumes energy. Load balancing is another technique that distributes the service requests to resources and results in low latency and increases the system throughput [21]. This technique can be implemented in hardware and software [21]. In load balancing, the load balancer listens to network port in order to receive the service requests [21]. When the service request arrives, scheduling algorithms such as round robin arrange them according to the fastest response time to set the order of execution of the workload [21,22]. Whenever the services are migrated from one site to another, the generation of energy should be measured. If the energy is high on one node, the service is transferred to another node of lower temperature and load [22]. In this way, proper load balancing is achieved to effectively reduce energy consumption of wired and wireless networks [22].

5.5 Green Mobile Cloud Computing

Efforts to reduce energy consumption are achievable by decreasing the energy consumption from the conventional sources to offer both economic and ecological benefits [2,24]. Intuitively, major benefits include cost reduction, energy conservation, and environment protection.

5.5.1 Energy Saving in Mobile Cloud Infrastructures

A large portion of the energy consumed in MCC is wasted because of idle cloud servers. Rahman et al. [24] and Li et al. [25] discuss an energy conservation architecture for mobile cloud, which takes into consideration efficient placement of applications on the cloud servers. Inefficient placement and migration of application may result in allocations of server machines that consume comparatively high energy. The architecture proposed by Rahman et al. [24] uses an energy-aware algorithm to create application placement and scheduling schemes in response to the workload arrival, departure, and resizing events. The strategy introduced in Li et al. [25] recognizes two types of servers:

- *Open Box*: represents an active server
- *Close Box*: represents an idle server

This method addresses the following issues:

1. *Rate of workload arrival*: Newly arrived workload replaces existing ones.
2. *Rate of workload departure*: Released resource by a departing workload is rearranged to reduce the number of running nodes.
3. *Rate of workload resizing*: The rate involves either of the following:
   a. *Workload inflation*: This occurs when the number of applications or workload increases. It affects the performance of other workloads.
   b. *Workload deflation*: This occurs when the number of workloads decreases. As a result some resources are released, which causes idling of resources and energy usage.

In this architecture, workloads are coupled tightly to minimize the number of active servers, that is, open boxes. Proper mapping of resources with workload arrival, departure, or resizing is necessary to prevent wastage of resources and energy. The scheme has two primary aims [25]:

1. Reduction of the number of active servers that is open boxes
2. Reduction of the number of migrations.

Energy-saving architecture of MCC consists of several functional components [25] as shown in Figure 5.17:

- Global controller
- VM controller

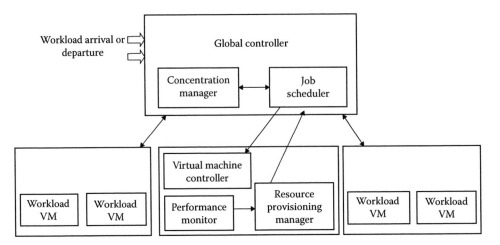

FIGURE 5.17
Architecture for energy saving in mobile cloud infrastructure.

- Concentration manager
- Resource provision manager
- Performance monitor

The functionalities of the different parts forming the architecture of energy-efficient mobile cloud infrastructure [25] are given in Table 5.7.

The provision of transferring the content, or rather migrating it from the main cloud computing data center to the local cloud data centers, results in significant energy saving [2]. This technique is useful to cope with many content-based applications, such as distance learning for academic lectures, conferences, museums, and so on. With this approach, energy consumption is reduced to 63%–70% [24]. MCC data center-based approach, though plays a vital role in green cloud technology, this approach loses its importance when the cost becomes higher for purchasing and maintaining the same sets of equipment for many MCC data center [24].

TABLE 5.7

Functionalities of the Constituent Parts of Energy-Efficient Mobile Cloud Infrastructure

---

1. *Global controller* consists of concentration manager and job scheduler, and they are organized throughout the architecture while the other components are deployed in every node.
    a. *Concentration manager* provides the strategy for application placement based on inclusion and migration of applications and sends the scheme to the job scheduler.
    b. *Job scheduler* decomposes the placement strategy to a set of commands for inclusion and migration of applications and sends the commands to the VM controller.
2. *VM controller* executes or migrates an application depending on the command received from the job scheduler.
3. *The resource provision manager* adjusts the resource allocation based on the information received from the performance monitor and prevents repeated application migration.
4. *Performance monitor* supervises the performance of all the components forming the system.

5.5.2 Issues and Requirements for Green MCC

The issues of energy saving in MCC are as follows [24]:

1. Developing and delivering mobile cloud services with minimum energy consumption of the mobile device and keeping user experience unaffected
2. Reducing energy consumption of cloud data center
3. Designing computation offloading mechanism that will learn computation needs and make decision regarding reduction of energy consumption

The requirements of energy saving in MCC are as follows [24]:

1. Energy-aware infrastructures required for MCC, which facilitates proper resource management
2. Development of proper offloading management algorithm
3. Development of standard solutions for energy-efficient communication
4. Development of proper application migration scheme for cloud data centers

5.6 Green Mobile Devices Using Mobile Cloud Computing

Green computing for mobile cloud and mobile computing services has become a challenging issue because mobile cloud vendors are looking for energy-efficient solutions to achieve cost reduction [24–27]. In this section, we discuss and review the existing energy-saving strategies for mobile devices using MCC.

For efficient reduction of energy consumption of mobile devices with MCC, two strategies are available:

1. Computation offloading
2. Resource management

5.6.1 Computation Offloading

Offloading computation to the cloud from the mobile device results in significant energy saving for the latter [24,26]. However, it is necessary to determine the optimum condition for offloading the computation from the mobile device to the cloud; that is, offloading computation must not be such that the cost of offloading exceeds the energy saved due to offloading. A number of studies have shown that the battery lifetime is one of the fundamental resources for the mobile devices and extending battery life is considered a necessary factor. Battery life is inversely proportional to the energy consumed by the device [27]. The execution of many computation-intensive applications when occurs inside the mobile device results in an enormous amount of energy consumption. If these applications are offloaded to the cloud, it results in significant energy saving of the mobile client [26,27]. Three approaches can be used to save energy of mobile clients [27]:

1. Using smaller transistors that consume less energy due to their smaller size, but as the transistors become smaller, more transistors are required and the energy consumption increases.
2. Reducing the clock speed by half, resulting in doubling the execution time; but energy consumption is reduced to a great extent.
3. Offloading the computation as a whole from the mobile device to some other device capable of executing the application.

Energy efficiency of mobile client utilizing MCC is discussed in Miettinen and Nurminen [26]. The concept of offloading computations to the cloud results in trivial energy saving. It is pointed out that the critical aspect of mobile clients in the context of cloud computing is the tradeoff between the energy consumed for computation and that for communication [26]. The parameters used in the calculation for energy saving with computation offloading to the cloud are given in Table 5.8.

If the energy consumed by the mobile device for performing certain computations locally is E_{local} and for transferring computation data is E_{cloud}, then the necessary condition of computation offloading is given by [26]

$$E_{local} > E_{cloud} \tag{5.12}$$

where

$$E_{cloud} = \frac{D}{D_{eff}} \tag{5.13}$$

and

$$E_{local} = \frac{C}{C_{eff}} \tag{5.14}$$

TABLE 5.8

Parameters Used in Energy Calculation

Parameters	Definition
D	Data to be transferred in bytes
C	Computational requirement in CPU cycles
D_{eff}	Data transfer efficiency; that is, the amount of data that can be transferred with given energy
C_{eff}	Computational efficiency; that is, the computation that can be performed in cycles per joule
I	Number of instructions that is to be executed
S	Speed of cloud server
M	Speed of mobile system
B	Network bandwidth
P_c	Energy required by mobile device (in watts) for performing the computation
P_i	Energy consumed by mobile device (in watts) when idle
P_{tr}	Energy consumed by mobile device (in watts) to transmit and receive data to and from cloud

From Equations 5.12 through 5.14, it is observed that [26]

$$\frac{C}{D} = \frac{C_{eff}}{D_{eff}}$$ (5.15)

where
D is the data to be transferred in bytes
C is the computational requirement in CPU cycles

Similar studies have been performed in Kumar and Lu [27] to verify whether computation offloading can result in energy saving and under which conditions this is attainable. It is observed that

$$T_{server} = \frac{I}{S}$$ (5.16)

$$T_{device} = \frac{I}{M}$$ (5.17)

where
T_{server} is the time taken to complete the task in the server
T_{device} is the time taken to complete the task by the device

If the server and mobile system transfer and receive D bytes of data and B is the network bandwidth, then it takes D/B seconds to transmit and receive the data [27]. Now, based on the energy consumed by the server and the device itself to execute the task as well as the energy consumed for data transmission, it is decided whether to offload work or not. The proposed decision-making algorithm for offloading is presented in Table 5.9.

If the mobile device performs the computation, the energy consumed is determined as [27]

$$E_s = P_c \cdot \frac{I}{M}$$ (5.18)

If the server performs the operation, the energy consumed is calculated as [27]

$$E_c = P_i \cdot \frac{I}{S} + P_{tr} \cdot \frac{D}{B}$$ (5.19)

TABLE 5.9

Decision-Making Algorithm for Offloading

1. **Start**
2. **If** a mobile device has an application to offload,
3.   The system calculates the energy required to perform the application execution within the mobile device.
4.   The system calculates the energy saved if the execution of the application occurs in the cloud server.
5.   **If** the value of energy saved is positive,
6.   The application is offloaded to the cloud.
7.   **Else** the application is executed within the mobile device without offloading to the cloud.
8.   **End If**
9. **End If**
10. **End**

Then, the amount of energy saved is given by [27]

$$E_{save} = E_s - E_c = P_c \cdot \frac{I}{M} - P_i \cdot \frac{I}{S} - P_{tr} \cdot \frac{D}{B} \qquad (5.20)$$

If the server is N times faster, then $S = N \cdot M$. Substituting the value of S in Equation 5.20 gives [26]

$$E_{save} = \left(\frac{I}{M} \right) \cdot \left(P_c - \frac{P_i}{N} \right) - P_{tr} \cdot \frac{D}{B} \qquad (5.21)$$

Energy saving is attained when Equation 5.21 gives a positive result. In Kumar and Lu [27], it is proved that offloading is beneficial when a large amount of computation C is needed with fairly small amounts of communication D, as shown in Figure 5.18.

In the figure, the energy consumption for an application execution inside the mobile device and the cloud are determined using Equations 5.18 and 5.19, respectively. In this simulation, the parameters values are assumed as $M = 400$, $P_i = 0.3$ W, $P_c = 0.9$ W, and $P_{tr} = 1.3$ W, considering an HP iPAQ PDA with a 400 MHz Intel XScale Processor [27]. Graphical representations show that, for large numbers of instructions, offloading to the cloud gives better results than executing them inside the mobile device if the energy consumption in communication is small [27]. It has been observed that offloading necessitates mobile devices to search for accessible wireless interfaces whenever data transfer is needed [27]. However, too many offloads also increase the energy consumption by mobile devices to discover a network. Hence, it becomes important to manage the number of offloads.

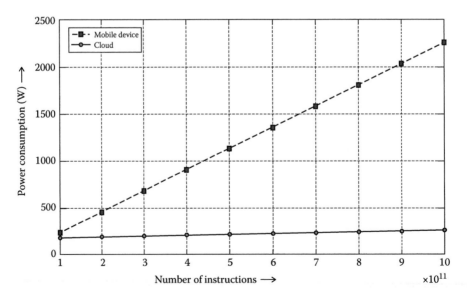

FIGURE 5.18
Number of instructions versus energy consumption for application execution.

5.6.2 Resource Management

Efficient mobile device resource management is a crucial issue for energy efficiency of MCC. Reducing the energy consumption of a mobile device in turn increases its battery life. The use of a resource management architecture for extending the battery life of mobile devices is discussed by Vallina-Rodriguez and Crowcroft [28]. The proposed architecture is an extension of the Android OS. This system is a user-centric resource management system that foresees the resource demands based on periodic monitoring of users, contexts, habits, and actions such as stationary, traveling, and so on. It facilitates mobile devices to access resources from nearby mobile devices through wireless networks. It is proposed that Bluetooth technology and interprocess communication (IPC) can be used to access resources from a nearby mobile device due to its low energy consumption [28]. Figure 5.19 shows the resource management architecture.

This architecture consists of the three parts:

1. Activity manager
2. Access control manager
3. Operating system manager

The various functions of the three parts of the architecture are given in Table 5.10.

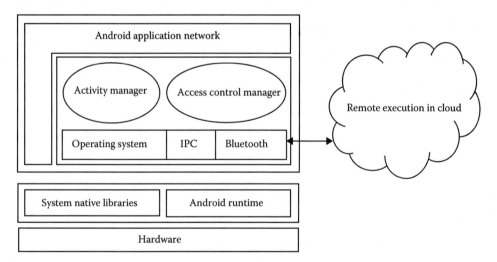

FIGURE 5.19
Architecture for energy-efficient resource management technique for mobile devices.

TABLE 5.10

Functions of the Three Constituent Parts of the Resource Management Architecture for Mobile Devices

---
1. *Activity manager* monitors the user activities and runtime resource usage of Android and its application framework.
2. *Access control manager* manages the access to other devices for resource usage.
3. *Operating system manager* performs the activities of managing local resources, discovering nearby devices, and deciding access patterns based on forecasting algorithms [28].

The resource management scheme works by integrating two features [28]:

1. It determines resource needs that may occur in the future, taking into consideration the user activities.
2. It facilitates access to resources of peer devices using wireless interfaces and IPC.

The scheme gathers knowledge of user activities from the information available in the handset, just as the base station identifier gives the location of the user without consuming any further energy or cost [24,28]. A user activity is defined using a set of positions and times where and when resources are executed, changed, or availed for use. As the scheme allocates resources on the basis of user requirement, it results in significant energy saving as well as flexibility, as the resource allocation is dynamic and there is no unnecessary resource allocation.

5.7 Green Femtocell Using Mobile Cloud Computing

In Section 5.3, we discussed the femtocell and how it helps to reduce the energy consumption of a mobile network. Although an FBS operates with very low energy, when hundreds or thousands of them are activated, the net energy consumption of the network is significantly high. In Mukherjee et al. [29], the concept of MCC with femtocell technology is elaborated to reduce the energy consumption of the FBS. In Figure 5.20, this system model is pictorially depicted where the computations required to be performed by the femtocell are partially offloaded to the cloud. When femtocell (i.e., HNB) has data to process, it sends the data to the cloud via the Internet for processing. The mobile device and femtocell are connected via a Uu interface. The femtocell is connected with the core network by the HNB-GW. A SeGW provides a proper security mechanism between the femtocell and HNB-GW over the Internet [29]. Femtocell connects with the HNB-GW through the Iuh

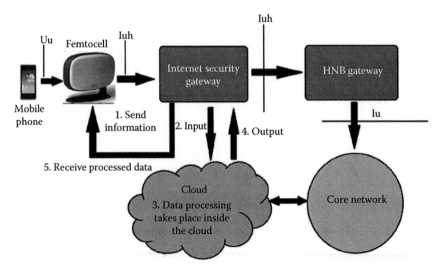

FIGURE 5.20
MCC-based working model of femtocell.

interface. The HNB-GW is connected to the core network via the Iu interface. The core network provides the mobile cloud service.

The femtocell, at the commencement of its functioning, collects information to assess the environment. This information includes the traffic pattern, user mobility, and interference conditions. After collecting the information, the following method is executed:

- Femtocell offloads the processing of the collected information related to network traffic, interference condition, and user mobility and carries out the process of data encryption and hardware authentication to the mobile cloud via the Internet.
- The cloud receives the information through the Internet SeGW.
- The cloud processes the information received and performs data encryption and hardware authentication.
- The cloud sends the result of processing to the femtocell through the Internet SeGW.

Hence, the processing of the collected information, data encryption, and hardware authentication are done in the cloud during the active state of the femtocell [29]. Three cases are considered for specifications of offloading. In case 1, the processing of collected information, data encryption, and hardware authentication are offloaded to the cloud. In case 2, only the information processing is offloaded to the cloud. In case 3, only the data encryption and hardware authentication are offloaded to the cloud. Mukherjee et al. [29] show that by using these three offloading schemes energy consumption by the femtocell as well as total energy consumption including femtocell and cloud can be reduced and battery life of the femtocell can be extended.

5.8 Green Seamless Service Provisioning with Mobile Cloud Computing

MCC efficiently reduces the energy consumption of mobile devices, but the technology comes with the major problem of limited WAN latency. This hampers the quality of service. Mobile users prefer to receive the response without waiting for a long time. Thus seamless service is desired. For this purpose, an energy-efficient, seamless service provisioning method with ad hoc mobile cloud is proposed by Ravi and Peddoju [30]. Mobile ad hoc cloud [24,30] is based on the principle of MANET (mobile ad hoc network). Here, both cloud and mobile ad hoc cloud are used for offloading. When the problem is easy to solve in the mobile ad hoc cloud, it will not be offloaded to the cloud [30]. The application will be executed in mobile ad hoc cloud itself. Hence, needless communication between the mobile device and cloud is minimized. Consequently, energy consumption is reduced, and distributed computation in both environments increases seamless service.

5.8.1 Architecture of a Mobile Ad Hoc Cloud

The architecture of mobile ad hoc cloud is presented by Rahman et al. and Mukherjee et al. [24,29], respectively. The mobile device takes the service from both the mobile ad hoc cloud and the default cloud in the architecture. In such an environment, a portion of the task is carried out locally in a mobile device and the remaining is performed by a nearby device,

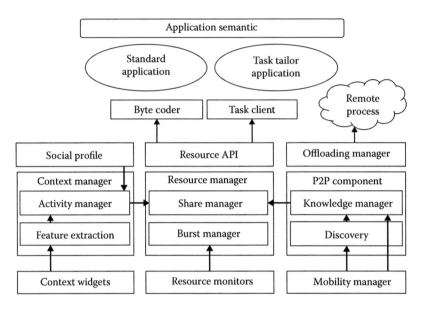

FIGURE 5.21
Mobile ad hoc cloud architecture.

which at that particular time runs the same task. The architecture of the ad hoc mobile cloud is shown in Figure 5.21, which consists of five functional components [24]:

1. Application manager
2. Resource manager
3. Context manager
4. P2P component
5. Offloading manager

The functions of the different parts forming the architecture of a mobile ad hoc cloud [24] are given in Table 5.11.

When a client needs to use a service without any delay, it checks for accessibility of connection with the cloud. If the connection is engaged, it initiates the process of forming a mobile ad hoc cloud, which consists of various domains, each of which consists of varied mobile devices with various computational abilities [30]. A mobile ad hoc cloud is formed with a number of mobile devices residing near the client device. The client commences the process of forming the mobile ad hoc cloud. The devices of one domain form more domains by connecting with other devices that are in their neighborhood. Devices in the domain are interconnected among each other for executing a task in unison. Each device in the mobile ad hoc cloud is connected to the cloud depending upon the availability of connection. The formation of a mobile ad hoc cloud is pictorially depicted in Figure 5.22. In Table 5.12, the process of mobile ad hoc cloud formation is described.

The number of mobile devices that forms a domain depends on the capacity of the wireless network connecting the device to the domain.

TABLE 5.11

Functions of the Constituent Parts of a Mobile Ad Hoc Cloud

1. *Application manager* performs the task of initiating the application and changing the application to incorporate characteristics for offloading such as creation of proxy and RPC support. The application manager also scrutinizes the profiles *that are created by the resource manager to check whether an instance of the virtual cloud needs to be* created or not.

2. *Resource manager* monitors the resource usage on a local device. For each application, a profile is generated to record the number of remote devices needed for developing a virtual cloud. These profiles are monitored by the application manager as mentioned in Step 1.

3. *Context manager* performs synchronization of the contextual information received from the context widgets. It makes the received information accessible to other processes. It has three subcomponents:

    a. *Context widgets*: These handle communication with the information sources.

    b. *Context manager*: This extracts new context from the received information.

    c. *Social manager*: This keeps track of various relationships among the users.

4. *The P2P component* sends a warning to the context manager if a new device is integrated into the environment or if an active device leaves the environment.

5. *The offloading manager* sends out and manages tasks from the current mobile device to the neighboring mobile devices, accepts, and processes tasks from the remote devices.

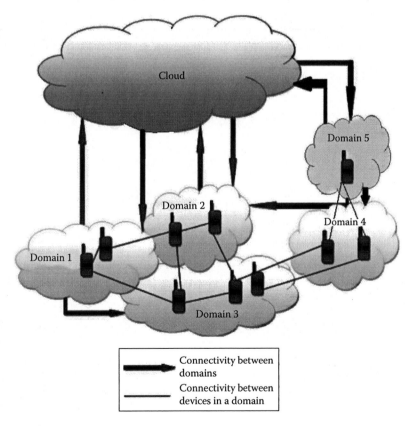

FIGURE 5.22
Formation of mobile ad hoc cloud.

TABLE 5.12

Formation of Mobile Ad Hoc Cloud

1. A mobile device X might be connected with its peer Y by means of one network connection. Y might be connected with another device Z, which is not in the vicinity of X.
2. Devices X and Y form one domain, while Y and Z form another domain.
3. Thus connecting the device between X and Z is Y.
4. If device X requires service of Z, the connecting device Y can be used for communication between the devices providing the service.
5. In this manner, many devices can be connected together to avail the service. The client device can know about the devices of the other domains with the help of the discovery module.

5.8.2 Functioning of Mobile Ad Hoc Cloud

The proposed framework [30] has basically three functions: service discovery, offloading decision, and seamless service. The services provided by the system and the interaction between modules in client and peer devices for the purpose of providing service to the users are depicted in Figure 5.23. The service discovery module helps the client device to find different devices accessible in the mobile ad hoc cloud [30]. This module is connected between the client and peer device so as to discover various services offered by the peer device. The seamless service modules of the client and the peer interact to decide on continuing the service based on the accessibility of connections. Applications and data can be offloaded either to the mobile ad hoc cloud or to the default cloud for execution. Offloading decisions are influenced by the bandwidth, device, and application characteristics [30]. The offloading module interacts with the service discovery and seamless service modules to take the appropriate offloading decision.

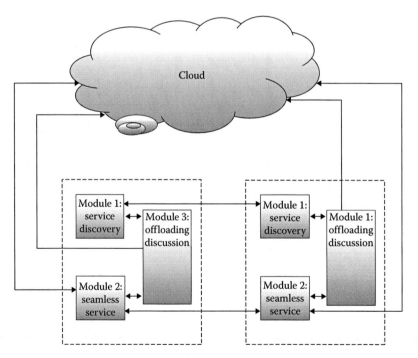

FIGURE 5.23

Interaction between modules forming the architecture of mobile ad hoc cloud.

TABLE 5.13

Service Discovery Table

Unique ID	Service offered	Computation capability	Network characteristics	Distance	Available energy

The three basic functions of a mobile ad hoc cloud are considered to be modules, as described in the following sections [30].

5.8.2.1 Module 1: Service Discovery

Every device in a domain has various services to offer. The services provided by a device are advertised using a discovery message. Through service discovery, a device in a domain can know about the availability of the service in another domain. When a device enters a mobile ad hoc cloud, the discovery message is advertised. The message is received by all other devices, and they update their service discovery table. The structure of the discovery table is shown in Table 5.13.

The parameters of the table are as follows [30]:

Unique ID: It identifies a particular service in mobile ad hoc cloud.

Service offered: It specifies the service that is ready to be offered by the device.

Computational ability: It specifies the processing capability of the device such as the type of processor, memory capacity, and so on.

Network characteristics: It specifies the medium through which it is connected to another network and bandwidth details.

Distance: It specifies the hop count to reach the device when it is advertising. The initial value is set to 0. This value is updated by every device once the message is received. If the received hop count is zero, the device updates the distance by 1. It helps to know how many hops are needed to reach the device.

Available energy: It specifies the energy required to provide service to the other devices in order to assign weights to the devices while deciding whether to offload the contents.

Every device circulates the service discovery table to the devices with which it is connected. On receiving the table, the recipient device compares it with its own table and updates the information accordingly. This table is used by the device to make an appropriate decision on selecting the device for availing the service.

5.8.2.2 Module 2: Seamless Services

To provide seamless service, it is required to keep track of the signal strength of devices in a mobile ad hoc cloud. The received signal strength indication (RSSI) is a measure to do so, and a threshold RSSI is maintained.

The intermediate mobile devices help the client to obtain services from other devices. When the intermediate devices find that the signal strength has fallen below the threshold, it breaks the service and passes it to a nearby device with a similar configuration. If such a device is not found, the interim result is passed to the client for further action. Figure 5.24 shows when to avail service from the cloud or from a peer device. This threshold policy is executed on the client device that requests for service [30].

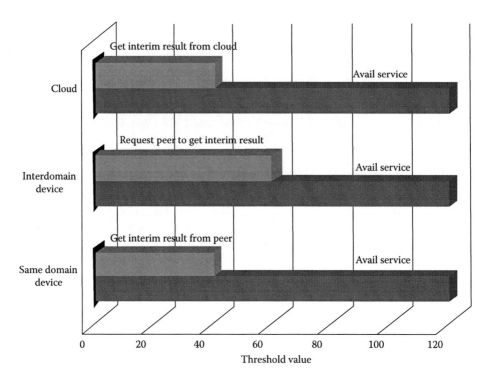

FIGURE 5.24
Diagram showing threshold values for availing service from cloud, interdomain device, or device belonging to the same domain.

5.8.2.3 Module 3: Offloading Decisions

Offloading applications help to execute them in parallel or in a distributed manner. The total computation cost of an application is the sum of the cost of offloading data into the cloud (C_{cloud}), offloading modules to various peers (C_{peers}), and calculating a part of the application within the device itself (C_{mobile}) [30]. In order to provide energy conservation of the mobile device, this total computation cost of using peers, cloud, and the mobile device must be less than the computation cost of all the modules if computed within the mobile device itself. The total computational cost is given by Ravi and Peddoju [30]

$$Total_c = (C_{cloud}) + \sum_{i=0}^{n} C_{peer} + C_{mobile} \tag{5.22}$$

Using Equation 5.22, the total cost for computation is determined.

5.8.3 Issues and Challenges of Mobile Ad Hoc Cloud

The mobile ad hoc cloud has certain issues and challenges associated with it, which are listed as follows [30]:

Trust management: For the purpose of security from the intruders, it is necessary to provide trust management among the peer devices.

Load balancing in mobile ad hoc cloud: Efficient load balancing must be performed so that no device is overloaded at any instant of time when others remain idle.

Elasticity in mobile ad hoc cloud: Dynamic decision making must be provided to facilitate changes as and when required.

5.9 Green Location Sensing within Mobile Cloud Computing Environment

Location-based services and applications have the advantage of both user mobility and cloud resources in the MCC atmosphere. But these applications are characterized by high energy requirement. In this section we discuss some energy-efficient location-based applications in the MCC environment [31]. Recent location sensing technologies make use of GPS, Wi-Fi, and GSM. It is observed that GPS offers continuous service for 9 h, while Wi-Fi and GSM provide 40 and 60 h, respectively. However, GPS is preferred over the others because of its accuracy in spite of its high energy utilization. In this section, we first discuss the business model of MCC and the related commercial products of location-based application (LBA), and then some location sensing approaches used in the MCC environment. LBA is one of the most emblematic applications in the software as a service (SaaS) layer of the MCC architecture. It gains a user's present location and presents a varied range of user-position-related services such as social network, health care, mobile commerce, and so on. Certain energy-efficient, location-based services in the MCC environment are discussed in Ma et al. [31].

EnTracked (energy-efficient robust position tracking for mobile devices) [31,32] uses an accelerometer to detect the position. It locates mobile devices in a robust and energy-efficient way. An EnTracked server sends the request for location determination to the EnTracked client. The EnTracked client uses GPS to determine the location and sends the response to the server, which forwards the response to the requesting application. It was further extended as EnTracked$_T$ [33] (energy-efficient trajectory tracking for mobile devices), which performs trajectory tracking rather than position tracking as in EnTracked. As a substitute of the discussed method, the CAPS (cell-ID aided positioning system) [34], which is based on the steadiness of traversed routes and consistent cell-ID shifting spots, is used. CAPS employs cell-ID sequence matching to determine user location based on the history of cell-ID and GPS positions [34]. It keeps a record of the user's route, that is, the visited cell-IDs, for future usage. EnLoc (energy-efficient localization for mobile phones) [35] is explored for the purpose of recording actual mobility of individuals with the help of the logical mobility tree (LMT), as shown in Figures 5.25 and 5.26. The vertices of the LMT are also referred to as uncertainty points [31]. The basic idea is to sample the activity at a few uncertainty points and allow EnLoc [35] to forecast the remaining. The scheme mentioned earlier highly relies on, and limited by, the spatiotemporal consistency in user mobility [31]. It cannot handle users' deviation from habits. Thus, EnLoc [35] uses mobility of large populations as an indicator of the individual mobility [31].

EnLoc [35] generates a probability map for a given area from the statistical behavior of large populations [30,34]. Then, an individual's mobility in that area can be predicted [31]. For example, think about a person close to traffic junction of street X: since the person had never been to street X previously, it is tricky to foresee how he or she will act at the junction. Nevertheless, if nearly everyone is taking a right turn to Street Y, the person's

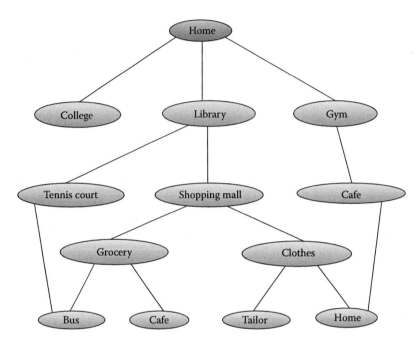

FIGURE 5.25
Spatial logical mobility tree (LMT).

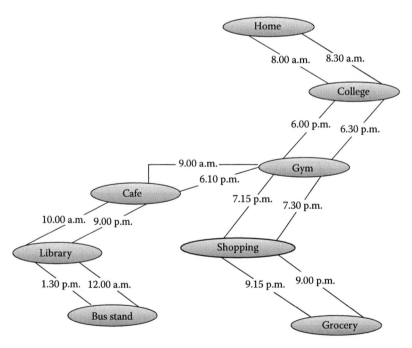

FIGURE 5.26
Spatiotemporal LMT.

TABLE 5.14

Comparative Analyses of Location Management Strategies

Comparison Parameters	EnTracked [32]	EnTracked$_T$ [33]	CAPS [34]	EnLoc [35]
Target	Position tracking	Trajectory tracking	Trajectory tracking	Position tracking
Sensors	GPS, accelerometer	GPS, compass, accelerometer	GSM with GPS	GPS, compass, Wi-Fi, accelerometer
Scheme	Dynamically determines the current position of the user by tracking pedestrian users carrying GPS-enabled devices	Dynamic calculation with reduced energy-sensitive sensors and determines the entire route traversed by the user	Matches cell ID sequence with previous sequences to determine the entire route of the user	Uses logical mobility tree to determine the location of the user

progress can be determined accordingly. A comparative analysis of the discussed methods is presented in Table 5.14.

All the procedures discussed until now are associated with sensing, were organized in real-world situations, and have proven to be energy efficient [31–35].

5.10 Conclusion

MCC is a popular technology that efficiently reduces the energy consumption of mobile devices to develop green mobile networks. In this chapter, we discussed green mobile computing, green cloud computing, and green mobile cloud computing. The architecture of energy-efficient resource management for mobile devices and the issues and requirements of energy saving on mobile cloud computing were studied. The use of MCC to reduce energy consumption of mobile devices was discussed. Graphical representations showed that computation offloading to the mobile cloud is useful when a large amount of computation is to be performed as compared to the amount of communication needed. The architecture of ad hoc cloud for the provisioning of seamless service was discussed along with its functions. A few green location management applications for mobile cloud environment were also described. It was observed that the employment of green MCC can result in efficient green mobile networks and that this technology can be applied to the femtocell networks. The use of MCC in a femtocell-based network can reduce the energy consumption of each femtocell present therein, which can result in a greener femtocell-based mobile network.

Questions

1. What is green mobile computing? Explain its components.
2. Discuss the energy efficiency of femtocell-based network over only macrocell-based network.

3. Define green mobile cloud computing.

4. What is femtocell? How does femtocell help make a mobile network "green"?

5. Explain the MCC-based working model of femtocell.

6. Discuss the green handover algorithm. Also compare handover latency with data size and velocity for green protocol.

7. What are the issues and requirements for green MCC?

8. How do location-based services and applications benefit within the MCC environment when it becomes a "green" network?

References

1. A. Berl, E. Gelenbe, M. D. Girolamo, G. Giuliani, H. D. Meer, M. Q. Dang, and K. Pentikousis, Energy-efficient cloud computing, *The Computer Journal*, 53(7), 1045–1051, 2010.

2. H. T. Dinh, C. Lee, D. Niyato, and P. Wang, A survey of mobile cloud computing: Architecture, applications, and approaches, *Wireless Communications and Mobile Computing*, 13(18), 1587–1611, 2013.

3. X. Wang, A. V. Vasilakos, M. Chen, Y. Liu, and T. T. Kwon, A survey of green mobile networks: Opportunities and challenges, *Mobile Networks and Applications*, 17(1), 4–20, 2012.

4. W. Fisher, M. Suchara, and J. Rexford, Greening backbone networks: Reducing energy consumption by shutting off cables in bundled links, in *Proceedings of the First ACM SIGCOMM Workshop on Green Networking*, New Delhi, India, ACM, pp. 29–34, 2010.

5. B. Hellen, S. Seetharaman, P. Mahadevan, Y. Yiakoumis, P. Sharma, S. Banerjee, and N. McKeown, ElasticTree: Saving energy in data center networks, *National Spatial Data Infrastructure*, 3, 19–21, 2010.

6. A. M. Marsan, L. Chiaraviglio, D. Ciullo, and M. Meo, Optimal energy savings in cellular access networks, in *Communications Workshops*, Dresden, Germany, IEEE, pp. 1–5, 2009.

7. S. Zhou, J. Gong, Z. Yang, Z. Niu, and P. Yang, Green mobile access network with dynamic base station energy saving, *ACM MobiCom*, 9(262), 10–12, 2009.

8. Z. Niu, Y. Wu, J. Gong, and Z. Yang, Cell zooming for cost-efficient green cellular networks, *IEEE Communications Magazine*, 48(11), 74–79, 2010.

9. A. Mukherjee and D. De, Congestion detection, prevention and avoidance strategies for an intelligent, energy and spectrum efficient green mobile network, *Journal of Computational Intelligence and Electronic Systems*, 2(1), 1–19, 2013.

10. J. Zhang, G. Roche, and L. De, *Femtocells: Technologies and Deployment*, John Wiley & Sons Ltd., Chichester, U.K., 2010.

11. S. Yeh, S. Talwar, S. Lee, and H. Kim, WiMAX femtocells: A perspective on network architecture, capacity, and coverage, *IEEE Communications Magazine*, 46(10), 58–65, 2008.

12. H. Claussen, L. Ho, and L. G. Samuel, Self-optimization of coverage for femtocell deployments, in *Wireless Telecommunications Symposium*, Pomona, CA, IEEE, pp. 278–285, 2008.

13. I. Ashraf, L. T. W. Ho, and H. Claussen, Improving energy efficiency of femtocell base stations via user activity detection, in *Proceedings of IEEE Wireless Communications and Networking Conference*, Sydney, Australia, pp. 1–5, 2010.

14. Y. S. Chen and C. Y. Wu, A green handover protocol in two-tier OFDMA macrocell–femtocell networks, *Mathematical and Computer Modelling*, 57(11), 2814–2831, 2013.

15. N. V. Rodriguez, P. Hui, J. Crowcroft, and A. Rice, Exhausting battery statistics: Understanding the energy demands on mobile handsets, in *Proceedings of the Second ACM SIGCOMM Workshop on Networking, Systems, and Applications on Mobile Handhelds*, New Delhi, India, ACM, pp. 9–14, 2010.

16. F. R. Dogar, P. Steenkiste, and K. Papagiannaki, Catnap: Exploiting high bandwidth wireless interfaces to save energy for mobile devices, in *Proceedings of the Eighth International Conference on Mobile Systems, Applications, and Services*, San Francisco, CA, ACM, pp. 107–122, 2010.
17. X. Lu, E. Erkip, Y. Wang, and D. Goodman, Energy efficient multimedia communication over wireless channels, *IEEE Journal on Selected Areas in Communications*, 21(10), 1738–1751, 2003.
18. S. A. Baset, J. Reich, J. Janak, P. Kasparek, V. Misra, D. Rubenstein, and H. Schulzrinnne, How green is IP-telephony, in *Proceedings of the First ACM SIGCOMM Workshop on Green Networking*, New Delhi, India, ACM, pp. 77–84, 2010.
19. J. He, P. Loskot, T. O'Farrell, V. Friderikos, S. Armour, and J. Thompson, Energy efficient architectures and techniques for Green Radio access networks, in *Fifth International ICST Conference on Communications and Networking*, Beijing, China, IEEE, pp. 1–6, 2010.
20. A. Mukherjee, S. Bhattacherjee, S. Pal, and D. De, Femtocell based green energy consumption methods for mobile network, *Computer Networks*, Elsevier, 57(1), 162–178, 2012.
21. M. Kaur, G. Kaur, and P. Singh, A radical energy efficient framework for green cloud, *International Journal of Emerging Trends and Technology in Computer Science*, 2(1), 171–175, 2013.
22. J. Baliga, R. W. Ayre, K. Hinton, and R. S. Tucker, Green cloud computing: Balancing energy in processing, storage, and transport, *Proceedings of the IEEE*, 99(1), 149–167, 2011.
23. A. Beloglazon and R. Buyya, Energy efficient allocation of virtual machines in cloud data centers, in *Tenth IEEE/ACM International Conference on Cluster, Cloud and Grid Computing*, Melbourne, Australia, pp. 577–578, 2010.
24. M. Rahman, J. Gao, and W. Tsai, Energy saving in mobile cloud computing, in *Proceedings of IEEE International Conference on Cloud Engineering*, Redwood City, CA, pp. 285–291, 2013.
25. B. Li, J. Li, J. Huai, T. Wo, Q. Li, and L. Zhong, EnaCloud: An energy-saving application live placement approach for cloud computing environments, in *IEEE International Conference on Cloud Computing*, Bangalore, India, pp. 17–24, 2009.
26. P. A. Miettinen and J. K. Nurminen, Energy efficiency of mobile clients in cloud computing, in *Proceedings of the Second USENIX Conference on Hot Topics in Cloud Computing*, USENIX Association, New York, p. 4, 2010.
27. K. Kumar and Y. Lu, Cloud computing for mobile users: Can offloading computation save energy? *Computer*, 43(4), 51–56, 2010.
28. N. Vallina-Rodriguez and J. Crowcroft, ErdOS: Achieving energy savings in mobile OS, in *Proceedings of the Sixth International Workshop on MobiArch*, Bethesda, MD, ACM, pp. 37–42, 2011.
29. A. Mukherjee, P. Gupta, and D. De, Mobile cloud computing based energy efficient offloading strategies for femtocell network, *Applications and Innovations in Mobile Computing*, 2014, 28–35, 2014.
30. A. Ravi and S. K. Peddoju, Energy efficient seamless service provisioning in mobile cloud computing, in *IEEE Seventh International Symposium on Service Oriented System Engineering*, Redwood City, CA, pp. 463–471, 2013.
31. X. Ma, Y. Cui, and I. Stojmenovic, Energy efficiency on location based applications in mobile cloud computing: A survey, *Procedia Computer Science*, 10, 577–584, 2012.
32. M. B. Kjærgaard, J. Langdal, T. Godsk, and T. Toftkjær, EnTracked: Energy-efficient robust position tracking for mobile devices, in *Proceedings of the Seventh International Conference on Mobile Systems, Applications, and Services*, Krakow, Poland, ACM, pp. 221–234, 2009.
33. M. B. Kjærgaard, S. Bhattacharya, H. Blunck, and P. Nurmi, Energy-efficient trajectory tracking for mobile devices, in *Proceedings of the Ninth International Conference on Mobile Systems, Applications, and Services*, Washington, DC, ACM, pp. 307–320, 2011.
34. J. Paek, K. Kim, J. P. Singh, and R. Govinda, Energy-efficient positioning for smartphones using cell-id sequence matching, in *Proceedings of the Ninth International Conference on Mobile Systems, Applications, and Services*, Washington, DC, ACM, pp. 293–306, 2011.
35. I. Constandache, S. Gaonkar, M. Sayler, R. Roy Choudhury, and L. Co, EnLoc: Energy-efficient localization for mobile phones, in *INFOCOM*, Rio de Janeiro, Brazil, IEEE, pp. 2716–2720, 2009.

6

Resource Allocation in Mobile Cloud Computing

ABSTRACT   Rapid resource allocation and release is a challenging era of mobile cloud computing. This chapter discusses the various resource allocation schemes of mobile cloud computing including energy aware resource management. Different task scheduling methods are also presented. Challenges to be faced in the field of resource allocation are explored.

KEY WORDS: *mobile cloud computing, resource, energy, scheduling.*

6.1 Introduction

The emergence of two different but important fields, mobile computing and cloud computing, has given birth to a new concept of mobile cloud computing (MCC). Mobile devices have the limitation of storage and processing power. In MCC, a mobile device is augmented by offloading its task and data into a resourceful cloud. A cloud is a rich collection of resources such as memory, storage, processing power, network, server, database, and applications. A cloud user employs these resources in a "pay as you use" or "elastic" manner. When the user sends a service request to the cloud, the cloud provider allocates the desired resource to the user. So, it is very important for the cloud provider to use a sound resource-allocation strategy to maintain the quality of service (QoS) of the cloud.

Several resource-allocation methods [1–3], strategies, algorithms, and middleware have already been developed. In this chapter, various frameworks and issues regarding resource allocation for MCC will be discussed.

6.2 Significance of Resource Allocation in Mobile Cloud Computing

There is a huge significance of resource allocation [4] in MCC. Only an appropriate resource allocation can augment mobile devices and fulfill users' demand. Thus, the cloud maintains its QoS. Without a resource-allocation strategy, many diverse situations may occur. An optimal resource-allocation strategy can avoid the following situations:

- *Resource contention*: This situation arises when two applications try to access the same resource at the same time.
- *Scarcity of resources*: This situation arises when there are limited resources.
- *Resource fragmentation*: This situation arises when resources are isolated. Despite being abundant, resources cannot be allocated to the needed application.
- *Over-provisioning*: This situation arises when the application receives more resources than it demanded.
- *Under-provisioning*: This situation occurs when the application is assigned with fewer resources than its demands.

Hence, resource allocation plays a significant role in MCC.

6.3 Resource-Allocation Strategies in Mobile Cloud Computing

Many resource-allocation strategies, shown in Figure 6.1, have been already developed for MCC.

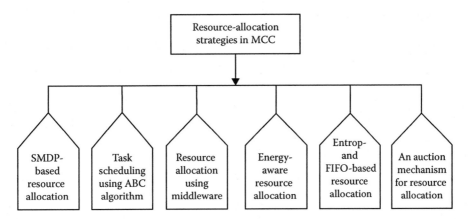

FIGURE 6.1
Resource-allocation strategies.

6.3.1 Semi-Markov Decision Process (SMDP)-Based Resource Allocation in MCC

The cloud comprises several domains according to resources or locations. The steps to be followed for a dynamic selection of adjacent cloud domains are as follows:

Step 1: Receive a service request from the client.

Step 2: Decide whether a home domain can accommodate it or not.

Step 3: If the home domain can accommodate it, then accept the service request and allocate resources to it.

Step 4: If the home domain cannot accommodate it, then send an interdomain transfer request to the adjacent domain.

Step 5: Collect the decision from the adjacent interdomain. If at least one adjacent domain can accept it, then select the domain that can accommodate it and transfer the service to the selected adjacent domain. If the adjacent domain cannot accept it, then reject the service request.

When a service request comes to the cloud, the cloud controller transfers the request to the home cloud domain according to the user's location. Then, single or multiple virtual machines (VMs) are allocated to the users according to their requests. If the home domain is insufficient to fulfill the demand of the user request, then the request is transferred to other resourceful cloud domains. Managing cloud resources in different domains is a critical task. In Liang et al. [5], a service decision-making system, using SMDP (semi-Markov decision process) for interdomain service transferring computation loads to balance among multiple cloud domains, was proposed. It maximizes the rewards by minimizing the number of service rejections that degrade the users' satisfaction level significantly for both the cloud system and the users compared to the greedy approach [5].

According to the decision-making process algorithm, when a new mobile cloud service request arrives, the cloud controller checks for a few things such as expected system gains and expected system expenses, including the cost of occupying VMs during the computation period, the communication cost between the cloud and the mobile devices, and the power consumption of the mobile devices. If resources are available and all the conditions are satisfied, then the home domain accepts the service request. Otherwise, it sends the interdomain transfer request to an adjacent cloud domain. If the adjacent domain accepts the request, then it will allocate the resources; otherwise, the service request will be simply rejected and the user has to run the task on the device itself.

6.3.1.1 System Model of Cloud

An MCC is a system composed of different cloud domains. Each domain consists of different resources such as storage, processing unit, and memory. Suppose one cloud domain has N VM resources and one service occupies n VMs. Here, $n \in \{1, 2, 3 ... N\}$ where $n \leq N$. A new service request and an interdomain service request to the cloud domain both follow Poisson distribution with means γ_{new} and γ_{int}. The computation rate of one VM is σ. So the computation time of n VMs is $1/n\sigma$.

6.3.1.2 System States of Cloud

The number of served service requests allocated to n VMs is $serv_n$. So, the total number of occupied VMs in a single cloud domain will be $\sum_{n}^{N} (serv_n * n)$. The arrivals of a new service request and an interdomain service request are A_{new} and A_{int}. After the completion of a service and the release of VMs, the VMs in the cloud domain need to be updated. Departure of a service is D_n. So, any event ev will be described as $ev \in E\{A_{new}, A_{int}, D_1, D_2, ... D_n\}$. Finally, the system state St will be composed of current services with a number of occupied VMs and events. Events may be arrival or departure denoted by

$$St = \{\overline{serv_n}, ev\} \tag{6.1}$$

where $\overline{serv_n} = \{serv_1, serv_2, ... serv_N\}$.

6.3.1.3 Actions of Cloud

There may be three kinds of actions according to a received request: accept, transfer, and reject, denoted by

$$ac(serv) = n \tag{6.2}$$

where
$n \in \{1, 2, ... N\}$
$ac(n) = 0$
$ac(n) = -1$

When a service is completed and it departs from the cloud domain, only the available VM list will be updated.

6.3.1.4 Dropping Probability

Dropping probability is the probability of rejecting incoming requests, which may be a new request or an interdomain request. So, there are two kinds of dropping probability.

When both the home cloud domain and the adjacent cloud domain cannot handle the requests, they reject the request and the task is forced to run on the client device.

The dropping probability for a new service request is denoted by

$$P_{new0} = \frac{\sum_{a(\overline{serv}, A_{new})=0} \tau_{(\overline{serv}, A_{new})}}{\sum_{r=-2, r=-1} \left(\sum_{ac(\overline{serv}, A_{new})=r} \tau_{\overline{serv}, A_{new}} \right)}$$
(6.3)

The dropping probability for an interdomain service transfer request is denoted by

$$P_{int0} = \frac{\sum_{a(\overline{serv}, A_{int})=0} \tau_{(\overline{serv}, A_{int})}}{\sum_{r=-2, r=-1} \left(\sum_{ac(\overline{serv}, A_{int})=r} \tau_{\overline{serv}, A_{int}} \right)}$$
(6.4)

where $r \in \{1, 2, ..., N\}$ but $r \neq n$ and $\tau_{\overline{serv}, A_{new}}$, $\tau_{\overline{serv}, A_{int}}$ are the probabilities of state according to a new service request and an interdomain service request.

6.3.1.5 Comparison of SMDP with Greedy Approach

A comparison between SMDP and the greedy scheme is presented in Figures 6.2 and 6.3.

The greedy algorithm always looks for the next best choice. It never backtracks to a past choice and never looks ahead if the choice has a negative impact. In the greedy approach, the cloud system allocates as many VMs to any service request. Whereas in SMDP, VMs are allocated after checking the expected system gains and expanses. So, when the number

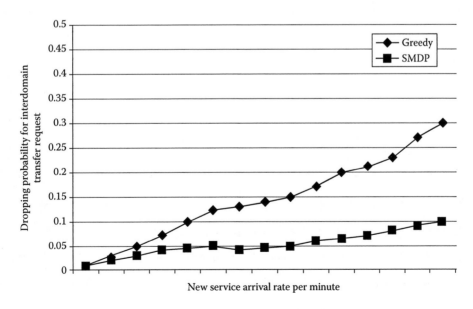

FIGURE 6.2
Comparison of SMDP and greedy approach on dropping probability for interdomain transfer request.

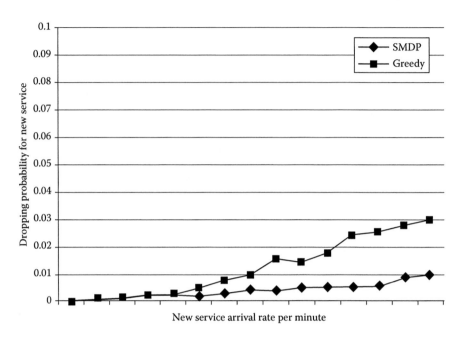

FIGURE 6.3
Comparison of SMDP and greedy approach on dropping probability for new service request.

of new service requests and interdomain service transfer requests increases, the dropping probability increases in the greedy approach. In SMPD, the service rejection is decreased by 20% than in the greedy approach.

6.3.2 Task Scheduling Using Activity-Based Costing Algorithm

Task scheduling in an MCC environment is a very important issue. Here, the tasks are the offloaded jobs from mobile devices. Task scheduling is the process where offloaded tasks are mapped into the available resources in the cloud according to the requirements and characteristics of the tasks. Sing and Ahmed [6] have used the activity-based costing (ABC) algorithm to make task scheduling optimal.

6.3.2.1 Activity-Based Costing Algorithm

The ABC algorithm has a parent–child relationship structure, almost like a tree structure. The tree structure of the activity-based algorithm is shown in Figure 6.4.

Every task has a specific arrival time. According to the arrival time of the tasks, they are stored in one parent queue. After checking the requested resource and data for the task, it is stored in two different queues: available queue and partially available queue. Tasks may have dependency or may be independent. According to the dependency of the tasks, they are stored in different queues. Then, the tasks are stored in different category-based queues if they need data resources from other data centers. After the category queue, there are again three queues based on the priority of the tasks: high, medium, and low. Priorities are measured based on a few factors such as completing time, resources, space needed, and profit.

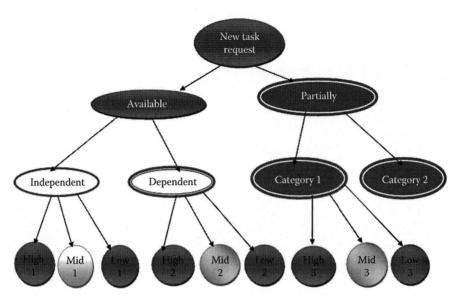

FIGURE 6.4
Activity-based algorithm for task scheduling.

The priority of the tasks is calculated as [6]

$$K_i = \frac{\sum_{j=0}^{n} (T_{i,j} + S_{i,j} + C_{i,j})}{P_i} \tag{6.5}$$

where
K_i is the priority of the ith task
$T_{i,j}$ is the time required to complete the jth activity of the ith task
$S_{i,j}$ is the space needed to operate the jth activity of the ith task
$C_{i,j}$ is the cost of the jth activity in terms of resources of the ith task
P_i is the profit from completing the ith task
n is the total number of activities of the ith task

After measuring the priorities of the different tasks, they are assigned to the high, medium, and lower priority queues. When the high-priority tasks get resources and are executed, then tasks from the medium queues are shifted to the high-priority queue.

Algorithm 6.1 Activity-Based Costing

1. **Begin**
2. **for** all task **do**
3.   Store in a queue **parent**
4. **end for**
5. **for** every ith task **do**
6.   Check **if** all the resource available or not

7.   **if** yes
8.   **move** to queue **available**
9.   **else**
10.   **move** to the queue partially available
11. **end for**
12. **for** all tasks at queue available **do**
13.   Check if independent
14.   **if** yes
15.   **move** to queue **independent**
16.   **else**
17.   **move** to the queue **dependent**
18. **end for**
19. **for** all task at queue
20.   Calculate priority K_i
21. **end for**
22. **for** every K_i **do**
23.   Put task in appropriate priority queue high, mid and low
24. **end for**
25. **compare** priorities, select task with highest priority for execution
26. **while** system is running task **do**
27.   Check if new task is available
28.   **if** yes
29.   Calculate priority and place at appropriate queue
30.   **else** continue
31.   **end if**
32.   Scan queues to modify priorities
33.   **if** queues are not empty
34.   Select new task for highest priorities
35.   **else**
36.   **wait** for new task to arrive
37.   **end if**
38. **end while**
39. **End**

6.3.2.2 Performance of ABC Algorithm

The ABC algorithm is efficient in both execution cost and task completion time. Sing and Ahmed [6] have done a comparison with the greedy approach of resource allocation and have proved that the ABC algorithm performs better when the number of cloudlets [7] increases.

The comparison of these two approaches in terms of execution cost is shown in Figure 6.5 and experimental data are given in Table 6.1.

The comparison of the ABC algorithm and greedy approach in terms of task completion times is shown in Figure 6.6, and experimental data are given in Table 6.2.

6.3.3 Resource Allocation Using Middleware

In MCC, the resource-hungry part of the mobile device is offloaded to the cloud for a better execution environment. These types of applications are called cloud-assistive mobile application or CAM-app. Ferber et al. [8] have proposed a middleware for the allocation of

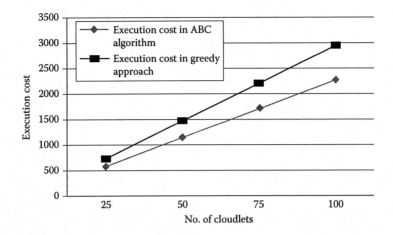

FIGURE 6.5
Comparison of ABC and greedy approach for execution cost.

TABLE 6.1

Execution Cost in ABC and Greedy Approach

Number of Cloudlets	Execution Cost in ABC Algorithm	Execution Cost in Greedy Approach
25	565.91	735.68
50	1131.82	1471.36
75	1697.73	2207.05
100	2263.6	2942.73

Source: Sing, L.S.V. and Ahmed, J., *Int. J. Res. Dev. Technol. Manage. Sci.*, 21, 1, 2014.

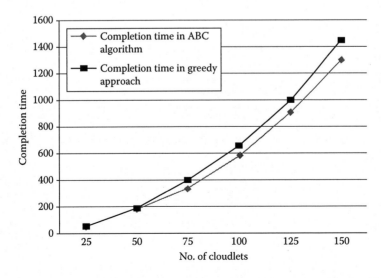

FIGURE 6.6
Comparison of ABC and greedy approach for task completion time.

TABLE 6.2

Task Completion Time in ABC and Greedy Approach

Number of Cloudlets	Completion Time in ABC Algorithm	Completion Time in Greedy Approach
25	52.97	58.97
50	164.66	181.62
75	334.79	399.90
100	584.68	654.03
125	910.04	997.99
150	1298.50	1439.75

Source: Sing, L.S.V. and Ahmed, J., *Int. J. Res. Dev. Technol. Manage. Sci.*, 21, 1, 2014.

resources for Java-based CAM-apps to ensure users' demand of resources from the cloud. It implements the entire server-side feature to enable CAM-apps. The main features are as follows:

- Hosting the remote service
- Resource management and allocation
- Accounting and billing

6.3.3.1 Architecture of Middleware

To achieve a remote service, a remote object is used. When a request comes, the instance of the remote object is created. A season key and wrapper class are used in the middleware. The season key reflects the life cycle of the remote object: create, use, and destroy. It also provides fault tolerance to network failure. Thus, users can switch to another network interface or access point when the current link is down. It also supports REST, RMI, and CORBA. The middleware consists of three modules (Figure 6.7):

1. A fixed master server
2. A fixed server for accounting/billing
3. An elastic set of cloud resource

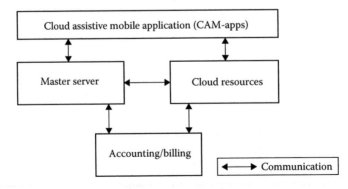

FIGURE 6.7
Architecture of the middleware.

The cloud resource is a complete virtual machine that contains the instances of remote services as remote objects (*robj*) and runs them. The client requests a remote object from the master server and specifies the desired service quality according to the service-level agreement (SLA). During the delivery of the service, the client is authenticated first and a season ID is generated, and communication emanates from VMs to the client using the season ID in a point-to-point manner. All the communications are stored in logs with SSL certification, which will be used in billing later. The season ID is used to describe the life cycle of an allocation time remote object as create, use, and destroy.

6.3.3.2 Resource-Allocation Strategies

There are two different resource-allocation strategies using the middleware.

6.3.3.2.1 Allocating VMs Exclusively

According to the first strategy [8], the middleware never allocates more than one remote service to a single VM. Only one service will be allocated to a single VM. Thus, a remote service can utilize the full capacity of the VM. In the worst case, clients need to wait a certain amount of time until a new VM is started to serve the request. The waiting time for a VM to be free or start is the main disadvantage of this approach.

Algorithm 6.2 Allocate VMs Exclusively

Required: robj{*place a remote object on a vm*}

1. **if** $|VM_{free}| < 1$ **then**
2. $vm_n = createnewVM()$
3. $VM_s = VM_s \cup \{vm_n\}$
4.   **while** $readyvm_n$ **do**
5. $wait(1s)$
6.   **end while**
7. $F_{assign}(vm_n, robj)$
8. **else**
9. $F_{assign}(vm_n, robj)$ with $vm_x \in VM_{free}$
10. $VM_{free} = VM_{free} - vm_x$
11.   **end if**

where $VM_{free} = vm \in VM_{free}$, $F_{assign}(vm_n, robj)$ is assigning a remote object to a specific vm.

6.3.3.2.2 Allocating VMs Immediately

According to the second strategy [8], the middleware allocates a VM instantly on a remote service request from the client. Here, clients have no need to wait for a VM to be free or start. If all VMs are preoccupied, then the least occupied VM will be allocated to the request. The main disadvantage of this approach is that it may overload the VM as there is no upper bound for the allocation of request.

Algorithm 6.3 Allocate VMs Immediately

Required: *robj{place a remote object on a vm}*

1. **if** $|VM_{free}| < 1$ **then**
2. $F_{assign}(vm_n, robj)$ with $vm_n = \mathrm{argmin}(load(vm_n))$
3. **else**
4. $F_{assign}(vm_n, robj)$ with $vm_x \in VM_{free}$
5. $VM_{free} = VM_{free} - vm_x$
6. **end if**

where $VM_{free} = vm \in VM_{free}$, $F_{assign}(vm_n, robj)$ is assigning a remote object to a specific vm.

6.3.4 Energy-Aware Resource Allocation

Cloud computing delivers an infrastructure, platform, and software as services, which are made available to consumers on a pay-as-you-go model. Cloud offers a significant benefit to IT companies for basic hardware and software infrastructures. Data centers have energy usage, which is a challenging and complex issue because large servers and disks are needed to process them fast enough as data are rapidly increased. The goals of energy-aware resource allocation [9] are minimization of energy consumption, efficient processing, and utilization of computer infrastructure. The main technique applied to minimize power consumption is allocating the workload to the minimum of physical nodes and switching off idle nodes.

6.3.4.1 Green Cloud Architecture

Clouds aim to engineer the architecture as networks of virtual services (application logic, hardware, user interface, and database) and to design the next-generation data centers so that the anyone can access and deploy several applications in the world on demand at costs depending on QoS requirements. To develop the green cloud architecture, four entities [9] are required: consumers/brokers, green service allocator, VMs, and physical machines. The higher-level view of green cloud architecture is shown in Figure 6.8.

Consumers or brokers: Cloud consumers or their brokers submit service requests to the cloud. Cloud consumers deploying a web application and the number of users can access it.

Green service allocator: It acts as an interface between the cloud infrastructure and consumers. It has some components such as green negotiator, service analyzer, consumer profiler pricing, service scheduler, VM manager, and accounting.

Green negotiator: It conveys with the consumers/brokers to settle SLAs with specified prices and forfeits (for the violation of SLAs) between the cloud provider and consumer.

Service analyzer: For submitted requests, the service analyzer deduces and analyzes the service requirements before accepting it. Hence, it requires dynamic information and recent load from the VM manager and energy monitor, respectively.

Consumer profiler: It specifically gathers characteristics of consumers so that important consumers can be treated with special privileges and prioritized by other consumers.

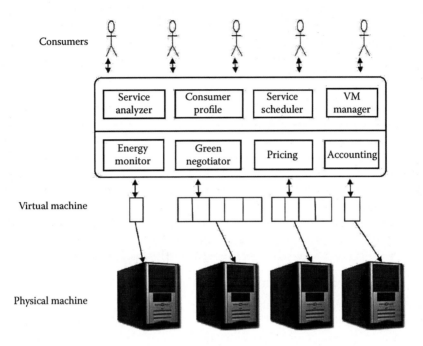

FIGURE 6.8
The higher level view of green cloud architecture.

Pricing: It decides charges and service requests to manage the supply and demand of resources and it facilitates service allocations, prioritizing them effectively.

Energy monitor: It observes energy consumption by VMs and physical machines and provides this information to the VM manager to make energy-efficient resource allocation decisions.

Service scheduler: It assigns the requests to VMs and controls resource privileges for the allocated VMs. If a customer has requested the auto-scaling functionality, it also decides when to add or remove VMs to meet the demand.

VM manager: It tracks the convenience of VMs and their supply usage. It is in charge of provisioning new VMs as well as reallocating VMs across physical machines to adapt the placement.

Accounting: It monitors the actual usage of resources by VMs and accounts for the resource usage costs. Historical data of resource usage can be used to improve resource allocation decisions.

Virtual machines: Multiple VMs dynamically have start time and stop time on a single physical machine according to incoming requests, hence providing the flexibility of configuring various panels of resources on the same physical machine to different requirements of service requests. Multiple VMs can concurrently run applications based on diverse OS environments on a single physical machine.

Physical machines: The physical computing servers provide the hardware infrastructure for forming virtualized resources to meet service demands.

6.3.4.2 Energy-Aware Data Center Resource Allocation

Beloglazov et al. [9] have done VM placement and VM selection based on a few issues to provide energy-efficient resource allocation. These issues include excessive power cycling of a server, which reduces the cloud reliability; turning off resources in a dynamic environment, which effects the QoS; and ensuring SLA every time, which may decrease the accurate performance of applications.

VM allocation: VM allocation [9] is divided into two parts: (1) admitting new requests for VMs and placing the VMs on hosts and (2) optimizing the current VM allocation. The current VM allocation is optimized in two steps. In the first step, the VM selects the need to be migrated, and in the second step, VMs are placed on the host using the modified best-fit decreasing algorithm. Here, VMs are sorted in the decreasing order according to their current CPU utilization and allocated to a host with least increase in power consumption.

Algorithm 6.4 Modified Best-Fit Decreasing Algorithm

1. **Input**: *hostList, vmList* **Output**: allocations of VMs
2. *vmList.sortDecreasingUtilization()*
3. **for each** VM in *vmList* do
4.   min*Power* ← *MAX*
5.   *allocatedHost* ← *NULL*
6.   **for each** host in *hostList* do
7.   **if** host has enough resource for VM **then**
8.   *Power* ← *estimatePower(host,VM)*
9.   **if** *Power* < min*Power***then**
10.   *allocatedHost* ← *host*
11.   min*Power* ← *Power*
12.   **if** *allocatedHost* ≠ *NULL* **then**
13.   Allocate VM to *allocatedHost*
14.   **return** allocation

VM selection and minimization of migration: For VM selection, three double-threshold policies are used by Beloglazov et al. [9]. Here, the upper and lower thresholds are used for the hosts, and the total utilization of the CPU is kept under these thresholds. If a host goes under the threshold value, then all VMs will be removed from the host and the host will be in idle power consumption mode. If the CPU utilization of one host exceeds the upper threshold, then some VMs will be migrated from the host to reduce power consumption.

If a host crosses any of the thresholds, then a minimum number of VMs should be migrated from the host. For the minimum number of VM migration, an algorithm has been proposed by Beloglazov et al. [9]. The first list of VMs is sorted according to decreasing CPU utilization. Then, it looks for VMs with two conditions: the VM should have a utilization higher than the difference between the host's overall utilization and the upper utilization threshold. If the VM is migrated from the host, the difference between the upper threshold and the new utilization is the minimum across the values provided by all VMs. If any VM satisfies these two conditions, then the particular VM will be removed from the host.

Algorithm 6.5 Minimization of Migration

1. **Input:** *hostList* **Output:** *migrationList*
2. **for** each host in *hostList* **do**
3.   *vmList* ← *h.getmvList*()
4.   *vmList.sortDecreasingUtilization*()
5.   *hUtil* ← *h.getUtil*()
6.   *bestFitUtil* ← *MAX*
7.   While *hUtil* > *thresh_up* **do**
8.   **for each** VM in *vmList* **do**
9.   **if** *vm.getUtil*() > *hUtil* – *thresh_up* **then**
10.   *t* ← *vm.getUtil*() – *hUtil* + *thresh_up*
11.   **if** *t* < *bestFitUtil* **then**
12.   *bestFitUtil* ← *t*
13.   *bestFitvm* ← *vm*
14.   **else**
15.   **if** *bestFitUtil* = *MAX* **then**
16.   *bestFitvm* ← *vm*
17.   **break**
18.   *hUtil* ← *hUtil* – *bestFitVm*() ()
19.   *migrationList.add*(*h.getVmList*())
20.   *vmList.remove*(*bestFitvm*)
21.   **if** *hUtil* < *thresh_low* **then**
22.   *migrationList.add*(*h.getVmList*())
23.   *vmList.remove*(*h.getVmList*())
24.   **return** *migrationList*

6.3.5 Resource Allocation in MCC Using Entropy-Based FIFO Method

At present, research on the utilization of mobile devices as resources in the MCC environment is increasing due to the enhancement of the processing power (use of quad-core, octa-core chip) of mobile devices. The users of mobile devices (smartphones, tablets, laptops) have increased tremendously during the past 2–3 years, and as time progresses, they will continue to increase. This increasing trend has motivated researchers to investigate the use of mobile devices in a cloud computing environment.

In this method [10], resources are allocated on the basis of the entropy of the mobile device and in the FIFO order. Entropy is calculated as [10] follows:

$$E(M_i) = -\left(\frac{t}{M}\right)\log 2\left(\frac{t}{M}\right) \tag{6.6}$$

where
   t is the time for which the mobile device has sent a request for a resource to the cloud
   M is the total number of mobile devices in the cloud

A predefined threshold value is set. For each mobile device, the entropy is first calculated using Equation 6.6 and then compared with the predefined threshold value. If the entropy is greater than or equal to the threshold value, the resource is allocated to the device; otherwise, the request is denied. If more than one mobile device satisfies the condition, the resource is allocated to the devices in the FIFO order.

For example, consider 10 mobile devices M1, M2, ..., M10. They try to access three resources R1, R2, and R3. If multiple mobile devices try to access the same resources, then there will be resource contention. To avoid this situation, we calculate the entropy for each device (Tables 6.3 and 6.4).

Suppose M1, M5, and M9 are requesting for resource R1; M2, M4, M6, and M10 are requesting for resource R2; and M3, M7, and M8 are requesting for resource R3. Let the predefined threshold value be 0.5.

Now the entropy value is calculated for each mobile device and results are given in Table 6.5.

From Table 6.5, we can see that M1 satisfies the condition; that is, the entropy of M1 is equal to the predefined threshold value, so resource R1 is allocated to M1 and the requests of other devices are denied. For resource R2, more than one device (M2, M4) satisfies the condition. Since M2 has requested for the resource first, the resource is allocated to M2 and the remaining devices will be served later. Resource R3 is allocated to device M3 since M3 satisfies the threshold condition.

TABLE 6.3

Mobile Devices with Requesting Resources

M	M1	M2	M3	M4	M5	M6	M7	M8	M9	M10
R	R1	R2	R3	R2	R1	R2	R3	R3	R1	R2

Notes: M denotes the mobile device and R denotes the resource. M1, M5, M9 requesting for resource R1; M2, M4, M6, M10 requesting for resource R2; M3, M7, M8 requesting for resources R3.

TABLE 6.4

Mobile Device with Time

M	M1	M2	M3	M4	M5	M6	M7	M8	M9	M10
t	3	2	5	4	6	7	6	9	8	9

Note: Here M denotes the mobile devices and t denotes the time period.

TABLE 6.5

Mobile Device with Entropy Value

Resources	Mobile Devices	Entropy	Time
R1	M1	0.5	3
	M5	0.4	6
	M9	0.3	8
R2	M2	0.5	2
	M4	0.5	4
	M6	0.4	7
	M10	0.1	9
R3	M3	0.5	5
	M7	0.4	6
	M8	0.1	9

Algorithm 6.6 Entropy-Based FIFO Method

1. A threshold value is set according to which resource is allocated.
2. Store the threshold value to **TRS**.
3. **for** each mobile device requesting for the resource **do**
4.   Calculate Entropy
   $$E\ (M_i) = -\ (t/M)\ \log2\ (t/M)$$
                   [*t* denotes the time period for which the mobile device is requesting for resource and *M* denotes the total number of mobile devices in the cloud.]
5.   Store the entropy value to **ENT**
6.   **if** *ENT>=TRS*
7.   Move the request to the queue
8.   **else**
9.   Resource is not allocated to the device
10.   **end if**
11.   **end for**

6.3.6 Auction Mechanism for Resource Allocation in MCC

In the auction mechanism, the resource is treated as a commodity, the cloud service provider is treated as a seller, and the resource user is treated as a buyer. The auction mechanism [11] works in the following manner, as shown in Figure 6.9.

First, users send the request for the resource. After getting the request, the cloud service provider sends an acknowledgment to the users, which contains pricing details and the validity period of the resource. Now the users submit their bid for the resource to the cloud service provider. The bid contains the valuation of the resource. The valuation of user *i* for the resource *j* is denoted by V_{ij}. After getting all the bids, the cloud service provider compares the valuation of all users and selects and stores the highest bid value. Now the cloud service provider sends the highest value to all users except the highest bidder and also starts a timer within which the next bid valuation should be submitted. If the timer is expired, the user with the highest bid gets the resource; otherwise, the process continues.

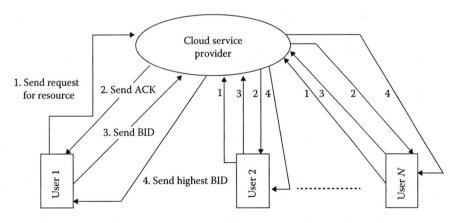

FIGURE 6.9
Auction method.

Algorithm 6.7 Auction-Based Resource Allocation

1. Users send the request R_i for resource to Cloud.
2. Cloud sends the ACK [ACK contain pricing and validity of the resource]
3. Users submit their bids (V_{ij}) to CSP
4. CSP **check** all the value of V_{ij}, select the highest value and store it to H_{bid}.
5. Send H_{bid} to all users (except who send this) and start a timer within which next bid is submitted.
6. **if** timer expires
7.   User with highest bid gets the resource.
8. **else**
9.   **Repeat** step 3 to 5.
10. **end if**

6.4 Research Challenges in Resource Allocation in Mobile Cloud Computing

There are several resource-allocation strategies and approaches in the MCC system that provide higher level of QoS and ensure SLAs. Still, there are some areas that need more focus.

6.4.1 Energy-Aware Memory Management

The processing and memory units are the heart and the soul of a cloud system. CPUs consume the highest amount of energy in a cloud, and memory is the second highest power-consuming unit. Day by day, the uses of cloud are increasing rapidly, and memory and CPUs are also increasing. So, this has resulted in huge power consumption. Multi-core processors are very energy efficient. On the other hand, memory technologies have not shown energy efficiency [9]. Data centers have the same problem in networking and disk storage.

6.4.2 Maintaining Strict Service-Level Agreements (SLAs)

SLAs play a vital role in maintaining the QoS of a cloud system. A strict SLA always tries to avoid performance degradation, which is a very difficult job. Many questions arise here [9]: How to predict performance pick? How to determine which VMs, when, and where should be migrated to prevent performance degradation if multiple system resources are considered? How to develop fast and effective algorithms for the VM placement optimization across multiple resources for large-scale systems? These questions have to be answered.

6.4.3 Merging of Different Resource-Allocation Strategies

Many resource-allocation strategies have been developed. But every time they cannot optimize each and every field. For example, the activity-based resource-allocation strategy [6] performs task scheduling very well, but it does not consider the energy efficiency factor.

The modified best-fit decreasing approach does not consider the priority factor of tasks [9]. So by merging different approaches, a positive tradeoff between performance and energy efficiency of the system can be made.

6.5 Conclusion

Resource allocation is one of the main operational issues in an MCC environment. After the task is offloaded from the mobile device into the cloud, the cloud provider allocates the task with the desired resources. In this chapter, we discussed the significance, strategies, and future challenges of resource allocation with the comparison of different approaches. We wish to modify the difficulties of the present approaches and overcome all the challenges of resource-allocation strategies to maintain proper QoS of the MCC system.

Questions

1. What is resource allocation in MCC?
2. What is the significance of resource allocation in MCC?
3. Describe the different resource-allocation strategies in MCC.
4. Discuss the SMDP-based resource allocation in MCC.
5. SMDP or greedy approach: which one is better and why?
6. How activity-based algorithm is used to make task scheduling optimal?
7. Describe different resource-allocation strategies using the middleware.
8. Explain resource allocation in MCC using the entropy-based FIFO method with a suitable example.
9. Explain the resource-allocation strategy using the auction method.
10. What is a virtual machine? How VM allocation is done in energy-aware data center resource allocation?

References

1. J. S. Park and E. Y. Lee, Entropy-based grouping techniques for resource management in mobile cloud computing, *Ubiquitous Information Technologies and Applications*, 214, 773–780, 2013.
2. Y. Ge, Y. Zhang, Q. Qiu, and Y. H. Lu, A game theoretic resource allocation for overall energy minimization in mobile cloud computing system, in *Proceedings of the ACM/IEEE International Symposium on Low Power Electronics and Design*, Rome, Italy, pp. 279–284, 2012.
3. D. Minarolli and B. Freisleben, Utility–based resource allocations for virtual machines in cloud computing, in *IEEE Symposium on Computers and Communications*, Kerkyra, Greece, pp. 410–417, 2011.
4. R. Patel and S. Patel, Survey on resource allocation strategies in cloud computing, *International Journal of Engineering Research & Technology*, 2(2), 1–5, 2013.
5. H. Liang, L. X. Cai, D. Huang, X. Shen, and D. Peng, An SMDP-based service model for interdomain resource allocation in mobile cloud networks, *IEEE Transactions on Vehicular Technology*, 61(5), 2222–2232, 2012.
6. L. S. V. Sing and J. Ahmed, A greedy algorithm for task scheduling & resource allocation problems in cloud computing, *International Journal of Research & Development in Technology and Management Science*, 21(1), 1–17, 2014.

7. M. Satyanarayanan, P. Bahl, R. Caceres, and N. Davies, The case for VM-based cloudlets in mobile computing, *IEEE Pervasive Computing*, 8(4), 14–23, 2009.
8. M. Ferber, T. Rauber, M. H. C. Torres, and T. Holvoet, Resource allocation for cloud-assisted mobile applications, in *IEEE Fifth International Conference on Cloud Computing*, Honolulu, HI, pp. 400–407, 2012.
9. A. Beloglazov, J. Abawajy, and R. Buyya, Energy-aware resource allocation heuristics for efficient management of data centers for cloud computing, *Future Generation Computer Systems*, 28(5), 755–768, 2012.
10. P. Akki and Y. M. Roopa, Resource allocation using entropy based FIFO method in mobile cloud computing, *International Journal of Engineering Research*, 4(1), 1289–1292, 2013.
11. Y. Zhang, D. Niyato, and P. Wang, An auction mechanism for resource allocation in mobile cloud computing systems, in K. Ren et al. (eds.), *Wireless Algorithms, Systems, and Applications*, Springer, Berlin, Germany, pp. 76–87, 2013.

7

Sensor Mobile Cloud Computing

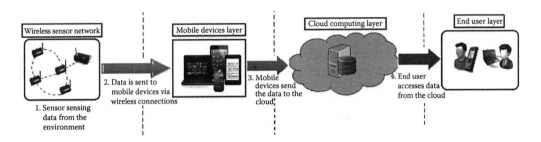

ABSTRACT Sensor mobile cloud computing (SMCC) is an emerging research area today. It is an integration of wireless sensor network with mobile cloud computing (MCC). In this chapter, we study the architecture and applications of SMCC. A life cycle model of this architecture is developed. Different challenges of SMCC are also discussed.

KEY WORDS: *wireless sensor network, mobile cloud computing, sensor cloud, urban sensing.*

7.1 Introduction

Wireless sensor networks (WSNs) have been gaining much attention, from both commercial and technical points of view, because of their potential for providing attractive solutions in areas such as health care, industrial automation, asset management, environmental monitoring, transportation business, and so on. Limited processing power, battery life, and communication speed are the main problems of WSN [1]. Cloud computing provides new opportunities in aggregating sensor data and exploiting the aggregates for greater coverage and relevancy and provides scalable processing power. Cloud computing is becoming increasingly pervasive in our daily lives. Its increasing popularity in distributed computing environment is influencing the trend of using cloud environment for storage and data processing. The rapid growth of sensor network and cloud computing technology has led to the emergence of a new platform called sensor clouds. It integrates WSN with the data center model of cloud computing [2]. The primary goal of a sensor cloud is to facilitate connecting sensors and software objects to build community-centric sensing applications. To explore this sensor, data of all types will drive the need for an increasing capability to do analysis and mining on the cloud [3]. One of the applications of sensor cloud computing is doctors' virtual community, where various sensors and cloud computing technologies are used for monitoring health of patients.

Cloud computing is a real paradigm that provides applications and services that are executed on distributed networks using virtualized resources accessed by the Internet protocol. Cloud provides software as a service (SaaS), where software is deployed in the cloud in such a way that users can access the software through the Internet. This eliminates the need to install the software. Cloud services is a layer of the cloud computing stack, which includes software components running in a distributed fashion across the commercial Internet [4].

To extend the services of a sensor cloud, mobile devices can be integrated with it and this infrastructure is known as sensor mobile cloud computing (SMCC) [5]. In this scheme, the sensor data are sent to the cloud through the mobile devices. Because of the incorporation of mobile devices, communication becomes more real time and pervasive than the basic sensor–cloud communication. One of such SMCC applications is mobile health monitoring, which is used for monitoring patient health remotely.

7.2 Wireless Sensor Network

A sensor network is composed of a large number of sensor nodes that are densely deployed over a geographical area. Sensor network protocols and algorithms must possess self-organizing capabilities. WSN nodes are comprised of four basic components: a low-power sensing device, an embedded processor, a wireless communication subsystem, and a power module. The embedded processor is generally used for collecting and processing the signal taken from the sensors. The wireless communication subsystem is used for data transmission. For this purpose, different communication technologies such as IEEE 802.15.4, Zigbee, and radio frequency identification (RFID) are used. The power source consists of a battery with a limited energy level.

Sensor networks consist of many different types of sensors, such as thermal, radar, seismic, acoustic, magnetic, and visual, that are able to monitor a wide variety of conditions, which include temperature, humidity, vehicular movement, pressure, soil makeup, noise levels, the presence of certain objects, and so on. Sensor nodes can be used for continuous sensing, event detection, event ID, location sensing, and local control of actuators.

7.2.1 Different Deployment Technologies of WSN

In regular deployment, data are routed through a predefined path. This deployment is generally used in home networks, the industrial sector, and so on. Random deployment is where sensor nodes are scattered over a finite area. When the deployment of nodes is not predefined, optimal positioning of the cluster head becomes a critical issue. Random deployment is generally used in rescue operations. The mobility of sensor nodes compensates the deployment shortcomings and can be passively moved around by some external force such as wind, water, vehicles, and so on. This type of sensors is generally used in battle field surveillances and emergency situations such as fire, volcanic eruption, and tsunami [6].

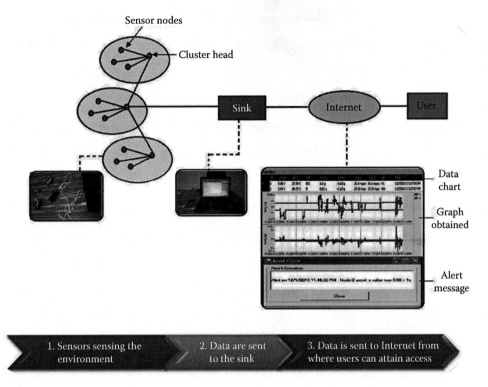

FIGURE 7.1
Basic architecture of sensor network.

7.2.2 Architecture of WSN

In WSN, one of the important methods is clustering to prolong the network lifetime. This method groups the sensor nodes into clusters and elects cluster heads (CHs), which acts as coordinators of the clusters. CHs gather data from sensor nodes and send the aggregated data to the sink or base station. The cluster heads can be selected randomly or based on one or more criteria. Sensor nodes are scattered all over the network. Sensors send data to the cluster head, and finally the cluster head sends the aggregated data to the sink. Then, through the Internet, users or the controllers can access those data from a remote place (Figure 7.1).

7.3 Sensor Cloud

Sensor cloud infrastructure is a broader form of cloud computing. Cloud computing elastically stores and processes data sensed by the sensors, which are scattered through the network. Sensor cloud service architecture integrates cloud computing with WSN to overcome the obstacles faced by WSN and provides new and better facilities [3]. A sensor cloud accumulates and processes information sensed by several sensor networks, facilitates information sharing, and collaborates with the applications on the cloud among users.

Cloud infrastructure manages a set or chain of filters that perform online analysis on sensor data. Physical sensor nodes are used to sense different applications such as transport monitoring, health monitoring, military uses, weather forecasting, and so on. Sensor nodes are programmed with the desired application. On each sensor node, application is sensed by the application program and is sent back to the gateway in the cloud directly through the base station or through multiple hops through other nodes.

Cloud provides on-demand services and storage resources to the clients. It provides access to these resources through the Internet and comes in handy when there is a sudden requirement of resources [2].

7.3.1 Architecture of Sensor Cloud

The main components of the sensor cloud are the client module, the portal module, the provisioning module, resource management, the monitoring module, the virtual sensor group module, and the physical sensor module.

Client module: Client layer is where end users can employ sensor cloud infrastructure by accessing the user interface through web browsers.

Portal module: This module offers interfaces to the client to access the sensor cloud infrastructure.

Provisioning module: This provides automatic provisioning of virtual sensor groups, in accordance with the user's specific need [3].

Resource management: Server, storage, and so on, are the IT resources used for the virtual sensor and the templates for the provisioning module used in sensor cloud architecture.

Monitoring module: This layer monitors the virtual sensor layer and the cloud infrastructure.

Virtual sensor group module: This layer is provided to help the user employ the physical sensor layer dynamically.

Physical sensor module: The sensors are used to sense the environment or a patient's body, as the application requires.

The main components of the layered architecture of sensor cloud are described in the following section.

1. *Login*: The portal is logged into by the end user through a web browser.
2. *Select the templates of virtual sensor group*: The portal wants the list of the templates of virtual sensors groups and virtual sensors from the database. The required template is selected by the end user from the list.
3. *Request the virtual sensor group*: By selecting the templates on the portal, the end user requests the virtual sensor groups. The portal calls the provisioning server providing the input parameters (e.g., the template IDs, user ID, and the virtual group names).
4. *Reserve IT resource*: The IT resources are first reserved by the provisioning server for the virtual sensor group and then, if there is no resource left on the existing virtual servers, a new virtual server is automatically provided with a monitoring agent, and the IT resource is reserved.

5. *Fetch the templates and provision*: The templates of the virtual sensors and the virtual sensor group are obtained by the provisioning server from the database. The virtual sensor groups are provisioned on the virtual server being selected.

6. *Notify the completion*: The end user is notified by the provisioning server of obtaining the virtual sensor group requested by e-mail. It also incorporates the latest records to describe virtual sensor groups (Figure 7.2).

The layered architecture of sensor cloud is described as follows:

Layer 1—Portal server: The portal server provides the web pages to the client browser for provisioning, for requesting or demolishing virtual sensor groups, for controlling those created templates of virtual sensors, for monitoring their virtual sensors, for checking their usage-related charges, and for logging in and logging out [6].

Layer 2—Provisioning server: The provisioning server takes initiatives to arrange the virtual sensor groups in response to requests from the portal server. Workflows are predefined and are controlled by a workflow engine. In a proper order, the workflows are executed. When the provisioning server receives a request from the client, the IT resource pool is checked first and reserved for further processing. The client uses virtual sensor groups through the templates. The server retrieves those templates and then provisions the virtual sensor groups. The job or definition of the virtual sensor groups is updated after provisioning, and agents are provisioned to monitor virtual servers [3].

Monitoring server: The status of virtual sensors is watched by the monitoring server through the agents of the servers and the virtual servers. The received data relating to the status report are stored in a database. Through the web browsers, monitoring information about the virtual sensors is available to the client. The status of the servers is also monitored by the sensor cloud administrators.

7.3.2 Benefits of Sensor Cloud

There are several advantages of merging WSN with cloud computing, thus forming the sensor cloud. The disadvantages of WSN are overcome by adding it to cloud computing [7]. The advantages of sensor cloud are shown in Figure 7.3.

7.3.3 Extension of Sensor Cloud with Mobile

The main reason for the extension of sensor cloud to SMCC is the features it provides [8]:

1. *Mobility*: To handle computing and to establish connection with the Internet to send the data.

2. *Low power consumption*: The mobile device consumes lower battery power than other devices, for example, limited energy availability on portable devices.

3. *Communication capabilities (Pervasive)*: The mobile devices help establish wireless connection to access data from anywhere, anytime, or user profiles from host, and communication between users. Communication through voice, video calls, and massages is possible.

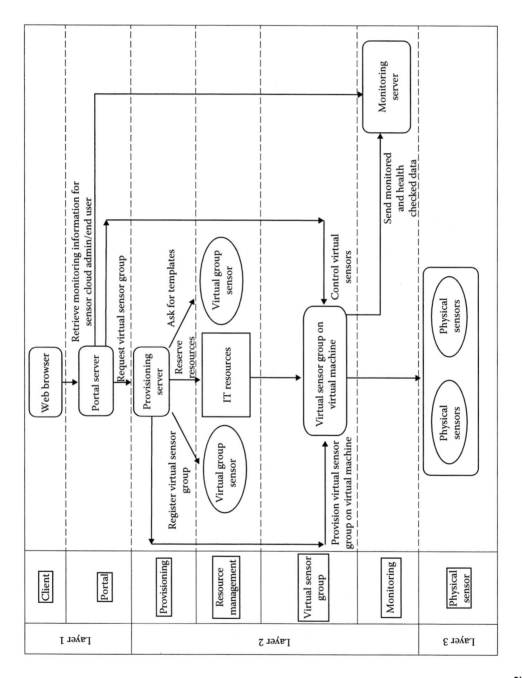

FIGURE 7.2
System architecture of sensor cloud computing.

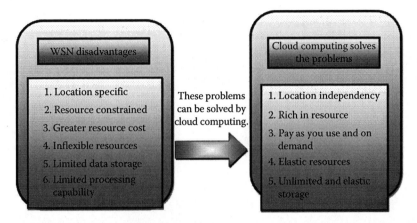

FIGURE 7.3
Incorporation of cloud computing with WSN.

7.4 Sensor Mobile Cloud Computing

SMCC is a new field of mobile cloud computing (MCC). It is used in some applications such as rescue services, healthcare, and so on. Mobile devices are being equipped with various sensors to sense data from the environment or the human body and send the aggregated data to the cloud through the Internet. By introducing a mobile phone between a sensor and the cloud server, data communication overhead can be reduced with the help of intelligent data filtering and fusing techniques. It has been shown that data transmission in a sensor mobile cloud requires less energy than that in a sensor cloud. Therefore, MCC plays an important role in wireless sensor networks.

7.4.1 Architecture of Sensor Mobile Cloud Computing

The sensor mobile cloud architecture is developed to improve the capability of a sensor mobile network. Here, capability means data processing, memory management, data communication, and energy efficiency. Since the capability increases from the sensor to the mobile and from the mobile to the cloud, integration of the sensor, the mobile, and cloud, which is SMCC, increases the capability tremendously [9]. The main components of the architecture are described in the following section, and their diagram is shown in Figure 7.4.

1. *Physical sensors*: Sensors are placed arbitrarily in various locations (e.g., on the human body) for monitoring. Different sensors are used in different applications. A portable electrocardiography (ECG) system uses smart phones attached to the heart and transmits heart rhythm data to the health provider. An asthma sensor has been developed to track the environmental conditions that can cause possible problems to asthma patients.

2. *Mobile phone*: From Figure 7.4, it is clear that the sensor's data are sent to the mobile phone via Bluetooth or Zigbee networks. A mobile phone collects the sensor data, processes the data, and transfers it to the cloud for further processing [9]. Low computational devices such as mobile phones can be used to filter the sensor data.

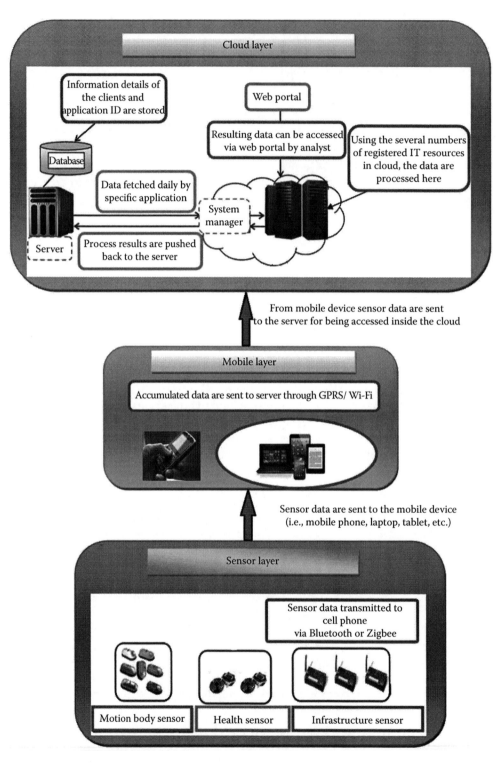

FIGURE 7.4
Block diagram of sensor mobile cloud.

Generally, Wi-Fi is used to establish communication between a mobile and the cloud. If Wi-Fi is not available, 3G and 4G technologies allow sending the collected sensor data to the cloud.

3. *Mobile network*: Mobile phones send sensor data to the WAP (wireless application protocol) server placed in the mobile network. A mobile network that contains a WAP server and a backend database is shown in Figure 7.4.

Mobile devices are connected to the mobile networks via a base station, that is, base transceiver stations (BTSs) or access points that establish connection to the existing mobile network and provide functional interfaces between the networks and mobile devices [10]. HLR is an important database in mobile network, storing the mobile device's identification number (IMEI number) and user details with the corresponding SIM. This way, the particular user can be traced via the WAP server. Authentication, authorization, and accounting are controlled by mobile network operators based on home agent and the subscriber's information stored in databases. The subscriber's requests are delivered to the cloud through the Internet. The cloud provider processes the requests and sends them to the corresponding cloud services, which are developed with the concepts of utility computing, virtualization, and service-oriented architecture.

4. *System Manager*: The system manager manages the cloud, which is connected to the WAP server through the Internet. System manager fetches data from the WAP server to process it in the cloud server, which allots IT resources before starting data processing. The cloud server runs the user application and computes the data collected from sensor nodes. There exist web portals through which the analyst can access results and provide appropriate decisions to the client for particular application.

In SMCC, events are generated from the client side by mobile phones having certain event IDs and subscriber IDs and are sent to the cloud for processing. One of the main components of this architecture is the system manager which can retrieve data for the particular client from the HLR database placed in the mobile network. After identification of the client, the event is sent to the cloud server. Experts are logged in through web portal, and suitable decisions are sent to the client.

7.4.2 Service Life Cycle Model of Sensor Mobile Cloud Computing

In this section, the service life cycle of SMCC is described. The operation is triggered from the client side. A sensor collects data from the client or the environment, and this signal is passed to the cloud via the client's handheld mobile device. Finally, the experts monitor the information and send the response to the client. The work flow of this service model is described in detail in the following:

Physical sensor: The sensor senses data depending on the application.

Client side: An event is triggered by the end user, and the application sensor data are collected and sent to the cloud server through mobile devices.

Reserve IT resources: The cloud server is used to dynamically store the sensor data. Data management and computation are also handled by the cloud.

Expert monitoring: Experts such as doctors or rescue teams monitor the data received from the mobile phones and take action if there is any abnormality.

Response: Expert teams transmit their advice to the subscriber by sending messages to his or her mobile and taking quick actions to help him or her [6].

7.4.3 System Architecture for a Rescue Service Model

In this section, the specific system architecture for rescue service is demonstrated. This architecture is separated into four layers:

1. Multiple-sensed mobile device
2. Emergency cloud
3. Nearby people
4. Rescuer

The architecture is depicted in Figure 7.5.

1. *Multiple-sensed mobile device*: Multiple-sensed mobile devices can collect a wide range of sensing data from the environment or from the behavior of the people. The sensors, which collect important data for the rescue authority to make decisions when an emergency occurs, are mostly visual, audio, motion, location, ambient, and physiological. To detect any emergency event, the sensed signals should be processed automatically. To extract meaningful information, these sensing data are filtered into the mobile. Information is considered to be meaningful if it contains a predefined pattern. According to user-defined criteria, events are categorized and sent to the emergency cloud [6].

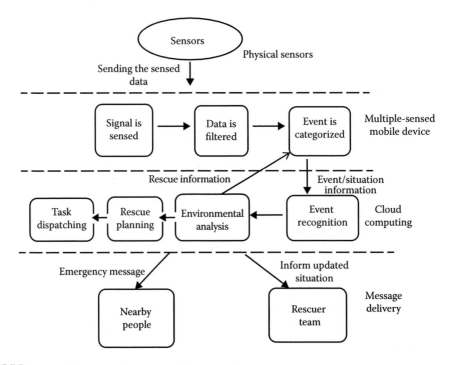

FIGURE 7.5
System architecture of sensor mobile cloud computing for rescue service. (From Chen, Y.J. et al., Sensors-assisted rescue service architecture in mobile cloud computing, *Wireless Communications and Networking Conference*, 2013, pp. 4457–4462.)

2. *Emergency cloud*: Since the information regarding an event from mobile devices is generated from a local perspective and may be incomplete as the sensing is done in an emergency situation, emergency cloud has to further recognize the complete event from a universal perspective. Then the environment within the range of the emergency event is analyzed with the collected data to support rescue planning. For example, emergency events in an urban district and a mountain area may require different rescue methods and policies. In addition, rescue information, such as the shortest path to the shelter that can provide guidance to the people in emergency, is sent to mobile devices. Finally, emergency cloud distributes the rescue tasks to appropriate rescue units according to their work load and locations.

3. *Nearby people*: It is possible that somebody near the victims can provide instant help compared to distant rescue units in an emergency. Thus, the emergency message can also be broadcast to nearby people. Sometimes, prediction of an evacuation process from the disaster area can be broadcast as the emergency message to people in the locality.

4. *Rescuer*: Mobile devices will keep updating information about the situation of people in the disaster area and send such information to the cloud in emergency service. The information about the situation is also analyzed and updated to provide urgent information to the rescue units.

7.4.3.1 Performance Analysis of Rescue Service Model

Performance analysis of the earlier rescue model is described in this section. Power consumption for transmitting a continuous signal is one of the main constraints in SMCC. Figure 7.6 shows the remaining battery life of the mobile device with respect to time. Figure 7.7 shows the delay time between the local server operation and the cloud server operation.

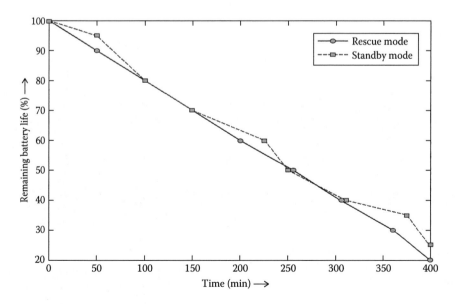

FIGURE 7.6
Time versus remaining battery life.

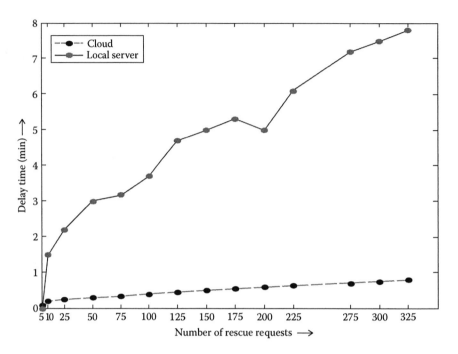

FIGURE 7.7
Number of rescue requests versus cloud and local server delay.

Power consumption: The rescue system works at the background, triggered by the users' smartphones during an unexpected emergency incident. It is necessary to calculate the power consumption by the mobile device. Figure 7.6 shows the power consumption in the rescue mode and in the standby mode. In the rescue mode, after a particular time period, data is sent from the mobile device to the cloud, and the GPS system helps identify the exact position of the person. During this time, the battery keeps losing power with time [6]. When a person needs help, he triggers an event. Data collected from sensors are sent from his mobile device. Rescue program is initiated by the cloud, starting the rescue mode. When the user does not require the application, it defaults to standby mode.

Delay analysis: Figure 7.7 shows the comparison of delays regarding the capacity of local server and the cloud server. Generally, the server placed in a mobile network is denoted "local server," which is equipped with an Intel Core i5 processor running at 2.4 GHz, and 8 GB of RAM. A cloud server can be built with up to 16 CPUs, 128 GB RAM, and 2.5 TB of storage. In the cloud server, the rescue route planning runs a dynamic programming algorithm to compute the fastest route to the disaster spot according to the real-time road status information [6]. But in the local server, the specification is not sufficient to run a dynamic programming algorithm. As a result, the delay in the local server is much higher than that in the cloud server. Here, the delay time is measured from the time of calculation of rescue route planning.

7.5 Internet of Things

Internet of things (IoT) is an upcoming technology that permits interaction between real-world physical elements such as sensors, actuators, personal electronic devices, and so on, over the Internet to facilitate various applications in the fields of e-health, intelligent transportation, and others. IoT is the convergence of different visions—things-oriented, Internet-oriented, and semantic-oriented [11]. Radio frequency identification (RFID) and sensing components are associated with everything used in daily lives, and information is uploaded into the computer, which monitors everything. RFID is the thing that connects the real world to the digital world. The basic idea of IoT is the pervasive utilization of things or objects—such as RFID tags, sensors, actuators, mobile phones, and so on—which, through unique addressing schemes, are able to interact with each other and cooperate with their neighbors to reach common goals. Wireless sensor network, RFID system, and RFID sensor network are used to collect data opportunistically [11]. Many challenges face this upcoming technology, in which technology and social network must be united for unique addressing, storing, and exchange of collected information. A remarkable point of contact for both sensing environments and cloud is IoT, where the underlying physical items can be further abstracted according to thing-like semantics [12]. With emerging technology IoT, a new framework is introduced to converge the utility-driven, cloud-based computing [13]. IoT provides several advantages. They are as follows:

1. It helps people to control household devices to save energy and in turn save money.
2. It can also be used to monitor the health of a person who needs immediate attention.
3. People can control their security systems at home through their mobile phone for their personal safety.
4. IoT can also be used in asset tracking and inventory control, shipping and location, security, individual tracking, and energy conservation.
5. It helps track consumer-based information given by the devices.

7.6 Urban Sensing

A wide range of technical, but also sociopolitical, challenges involve pervasive environmental monitoring, which applies especially to the sensitive context of a city. Urban sensing is an extended form of the existing sensor network. Urban areas comprise many different elements such as buildings, vehicles, citizens, and so on. People-centric urban sensing involves collecting data associated with people, such as their immediate surroundings, their characteristics, and the way they interact with their surroundings [14]. Here, people are involved not only as consumers of collected data but also as collectors of data from natural phenomena or ecological processes. The set of producers and consumers of data overlap in the urban sensing environment. People participate in both roles like a loop sensing and distributing the collected data. Interactions among the elements of the urban sensing scenario, as well as applications achieved through these interactions, are discussed in the following section. Issues that arise and applications that are enabled as a result of these interactions are also discussed. The traditional wireless sensor network is focused on application-specific deployment. Sensors are mobility-driven and data from the sensors are periodically updated into the cloud. Urban sensing uses mobile phone network data, such as the types of activities in different parts of the city, residential and working areas, population distribution, and commuting patterns. Since mobile phone technology is increasingly adopted by the population, every possible micro and macro behavior is available freely. The electronic communication sector is also concerned that the privacy of personal data is only partially addressed.

7.6.1 Opportunistic Sensing

In pervasive or ubiquitous computing, WSN is one of the important elements. The sensor network consists of multimodal sensors that provide opportunistic sensing (OS), which is used to produce more robust event recognition [15]. OS accomplishes automatic target recognition (ATR) and supports large-scale applications. It is adaptive in nature; there is no predefined circumstance for OS. Over the past decade, the focus of wireless sensor networking research has evolved from static networks of specialized devices deployed to sense the environment to networks making use of robotic or other controlled mobility to adapt to the sensing conditions and to a people-centric approach relying on the mobility of people. Hazards (any kind of hazards) are unpredictable, and they can happen at any place at any time, so a rescue team has to continuously monitor a large area in order to identify potential danger and take action. A single type of sensor is not efficient to find victims or to integrate different related information of a situation. A heterogeneous sensor network-based [16], multimodal information integration is required to collect different types of information. This kind of network is not specific or planned for a particular job. OS or networking is often associated with human-centric ubiquitous systems, such as in crowd sourcing and participatory sensing applications, focusing on human activity recognition [17].

The steps of OS are shown in Figure 7.8. In OS, the custodian may not be aware of active applications. Instead, a custodian's device (e.g., cell phone) is utilized whenever its state matches the requirements of an application [18]. OS prefers a standard for information and signal processing in which a network of sensing systems can automatically discover and select sensor platforms based on an operational scenario. In OS, data collection is automatic without user participation [5]. This approach helps the user get rid of the burden to

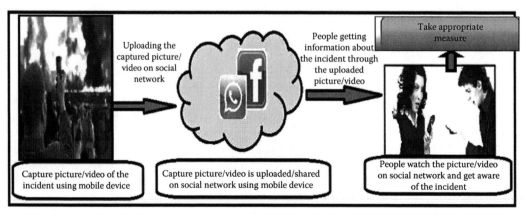

FIGURE 7.8
Steps of opportunistic sensing.

support the system and increases the range of applications and sensing field. The users need not be preoccupied with sensing the data, but system designers should carefully make a system that is only activated if predefined conditions have been satisfied. In the opportunistic approach, data quality is low and suffers from high data-miss rates because the exposure time for sensors may not be sufficient. In the opportunistic approach, missing sensing data is unavoidable in terms of time and space. WSN nodes are homogenous and fixed, whereas users monitor different target areas by moving dynamically during each sensing period in mobile phone sensing.

7.6.2 Participatory Sensing

Participatory sensing (PS) is an improvement on static WSN. Its deployment cost is very low in participatory sensing because it enhances existing sensing and communication infrastructure. Because of the mobility of the cell phone, it can cover different areas [19], and because sensors are already embedded into cell phones, it affords economies of scale. Because of the availability of app stores, it became easy to develop and deploy application in the mobile phone. In PS, people take part directly in the sensing loop, and the applications dramatically improve their daily lives. The sensing tasks by mobile sensors are triggered manually. In PS, the custodian has the opportunity to serve an application request without considering personal interest. People interfere in the sensing system to make decisions about data sharing, and privacy mechanisms should be allowed to impact data fidelity. PS is based on a preplanned system. A person interested in particular applications can install them into his or her mobile, and the sensing operation is triggered by him or her dependent on the condition. In PS, to meet the requests made by applications, people continuously coordinate with the sensing system, and critical decisions are made on sensing the target, location, and data [3]. The system is simplified, as complex operations are solved by the intelligence of the person, and high data quality is assured through users who are actively engaged in data collection. In PS, the burden is on the user and needs a support mechanism to encourage user activity.

For large-scale sensing, these features make it difficult to achieve using PS. PS is also a part of urban sensing and is used to organize less complex trustworthy ad hoc observatory applications that can be implemented in a metropolitan area. The applications are specific in nature, such as a GIS-based noise detection application, transport system, and so on. The steps of PS are shown in Figure 7.9. A public transport operator creates timetables, which contain other statistic information that does not reflect actual traffic conditions [20]. Mobile participatory sensing enhances the basic application with real-time updates that allow the crowd to collect the required data.

7.7 Application

Appropriate and real-time health monitoring of patient is very challenging and important issue nowadays. In a conventional healthcare system, nurses monitor the patient's health, record the data, and forward the data to doctors and other medical staff. All processes are manual. So, the delay or latency is the main drawback of this approach. In case the patient's health deteriorates rapidly, an emergency may occur, which may need real-time health monitoring without delay. To make this possible, an SMCC can be introduced in a conventional healthcare system. As SMCC is a combination of WSN and and CC with

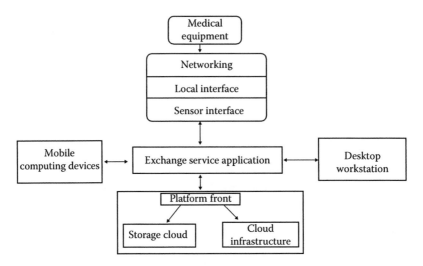

FIGURE 7.9
Steps of participatory sensing.

mobile, we can attach sensors to the patient's body or with various kind medical devices such as x-ray, ECG, or MRI. The sensor node will collect the data and send them immediately to the cloud through mobile devices. Then, doctor and other medical staff can access those data in no time from the Internet through their terminals without any latency. The benefits of the new approach over the existing approach are the system architecture and security issues. Examples are Microsoft's Health Vault and IBM's Smart Health. They provide a cloud solution to health monitoring on a very large scale throughout the world. The traditional scenario of health care is as follows:

1. A nurse or other medical staff collects patient's data and writes the information down on paper.

2. Notes are submitted to a terminal for data entry.

3. The collected data are stored in a database server where they are organized, indexed, and able to be accessed through an interface.

4. Doctors and medical staff are able to access this information through an interface of any application.

The main drawback is the latency between data gathering (1) and information access (4). Real-time monitoring of a patient is not possible here. Besides, this scheme is inaccurate, as there is possibility of wrong input.

7.7.1 A Complete Architecture of Health Service Model

In Figure 7.10, a health service model is described. The main components of this model are the following:

1. *Sensor module*: Physical health sensors are used to measure the blood pressure, ECG, temperature, and other parameters. These data are extracted, transformed, and loaded from the sensor to the attached medical equipment. Software is loaded into the equipment to collect and process data locally. Then, the data are transmitted over the wireless network to "cloud services" [21].

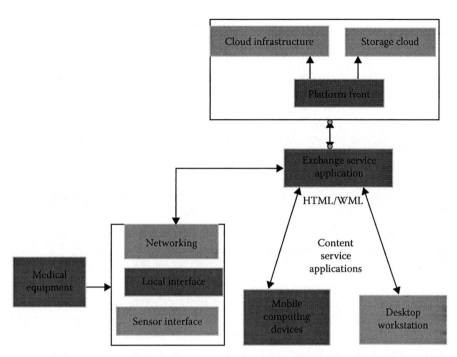

FIGURE 7.10
Detailed architecture of a healthcare system using sensor mobile cloud.

2. *Exchange service application*: This module organizes the patient's overall data related to the disease. It acts as a broker [22] between the local medical equipment and the remote cloud services. There are two main functions of this module:

   a. It is employed as an access point.

   b. It allows medical equipment to preprocess and store sensed data locally (e.g., the data is aggregated, filtered, and analyzed before the transmission to the cloud).

3. *Content service application*: These are the common interfaces used to provide information to doctors and other medical staff.

4. *Utility computing provider*: This module is responsible for providing physical infrastructure to store data, processing, and content delivery services.

Advantage of the scenario:

- This model facilitates continuous real-time data collection.

- Human interruption is eliminated and there is no possibility of erroneous data collection and data transmission to the database.

- Wireless sensors are easily deployed, as there is no fixed architecture of sensor network. Sensors attached to medical devices obviate the necessity of manual data gathering and data entry on the medical system.

- Computer resources available in the cloud are responsible for organizing, indexing, making the data accessible, and distributing the data to doctors and medical staff.

7.8 Challenges of Sensor Mobile Cloud Computing

1. *Wireless communication*: Though in wireless communication, the signals interact with personal digital assistants, the signal paths get blocked, and noise and echoes are introduced, causing more obstacles in wireless communication than in wired communication. As a result, wireless communication is characterized by lower bandwidths, higher error rates, and more frequent spurious disconnections. Hence for retransmissions, communication latency increases [1].

2. *Software development*: Designing software for a networked system for different applications in the mobile phone platform is a challenging process [22] without having the knowledge of the underlying hardware.

3. *Poor resources in mobile devices*: Mobile devices are resource-poor compared with static elements [2]. Computational resources such as disk capacity, processor speed, and memory size are limited due to fixed power consumption and weight limitations.

4. *Finite energy resources*: Battery life of every mobile device is fixed according to the hardware specification. Power consumption should be taken into consideration when developing hardware and software for mobile devices.

5. *Costing and charging issues of cloud computing*: Cloud consumers must be aware of the tradeoffs between integration, computation, and communication [3]. Migration of different applications to the cloud can significantly increase the cost of data communication. Since resources used for computation are likely to be high, the cost per unit of computing drastically increases. An elastic resource pool for intensive customization, performance, and security enhancement is necessary for simultaneous user access, and dealing with complexities increases charges to the customer.

6. *Security issues*: There is no doubt that putting data, running software on a remote machine's hard disk, and using its CPU appear daunting to many. Well-known security issues such as data loss, phishing, botnet [3] (running remotely on a collection of machines) pose serious threats to organization's data and software.

7. *Data availability*: Failure of Internet connectivity is a major risk to running an application in the cloud computing environment [23], especially in unnatural circumstances such as disasters. As applications are accessed by the mobile phones, they are dependent on Internet access for communication. In addition, if vulnerability is identified in a particular service provided by the cloud service provider, the application may have to stop all access to the cloud service provider until they could be secured, so that the vulnerability is resolved.

8. *Low bandwidth*: The main concern for mobile computing is the low bandwidth; wireless networks have much lower bandwidth than wired networks [20].

9. *Bandwidth variation is high*: Mobile devices have problems of greater variation in network bandwidth. Applications can assume high bandwidth and operate when plugged in or can assume low bandwidth or adjust with resources that are available.

10. *Various networks*: The problem with mobile devices in comparison with stationary devices is that the former change networks as they move beyond the range of the network in use. Also, some mobile devices use several networks at the same time.

11. *User interface is small*: It is difficult to open many windows simultaneously on its small screen.

12. *Low power*: The battery is the main reason for the weight of mobile devices. By reducing the size of battery, the need to recharge more frequently increases.

7.9 Conclusion

In this survey, the architecture of sensor mobile cloud was illustrated. The architecture facilitates the sensor data to be processed, categorized, and stored in such a way that it becomes easily accessible anytime, anywhere, and cost effective. Integrating sensors with MCC provides an open, extensible, scalable, interoperable, reconfigurable, and easy-to-use network of sensors for numerous applications. SMCC plays an important role in patient health monitoring by providing remote health data to analysts to handle emergency situations. In SMCC, if the mobile device is far away from the BTS, such as in urban area, the connection is not available properly. Transmitting sensor data from mobile to BTS consumes a lot of energy. So, power management is also a big issue in SMCC. Generally, a macro cell is used for data transmission between the mobile and cloud. The large distance between the mobile phone and the macro cell causes path loss. Femtocell can be used to reduce the path loss from the mobile to the BTS. Using SMCC, the situation of every person in a city can be traced, and gradually the city becomes a smart city.

Questions

1. What are the components of SMCC?
2. What is participatory sensing?
3. What is opportunistic sensing?
4. What are the basic differences between participatory sensing and urban sensing?
5. What is urban sensing?
6. What is a sensor cloud?
7. What is IoT? What are the components of IoT?
8. What are the applications areas of SMCC?
9. What are the open research problems in SMCC?
10. Draw the system architecture for a rescue service model.
11. What is service life cycle model of SMCC?

References

1. J. Yick, B. Mukherjee, and D. Ghosal, Wireless sensor network survey, *Computer Networks*, 52(12), 2292–2330, 2008.
2. W. S. Ansari, A. M. Alamri, M. M. Hassan, and M. Shoaib, A survey on sensor-cloud: Architecture, applications and approaches, *International Journal of Distributed Sensor Networks*, 2013, Article ID 917923, 1–18.
3. M. Yuriyama and T. Kushida, Sensor-cloud infrastructure—Physical sensor management with virtualized sensors on cloud computing, in *13th International Conference on Network-Based Information Systems*, Takayama, Japan, pp. 1–8, 2010.
4. B. P. Rimal, E. Choi, and I. Lumb, A taxonomy and survey of cloud computing systems, in *Fifth International Joint Conference*, Seoul, South Korea, pp. 44–51, 2009.
5. Y. J. Chen, C. Y. Lin, and L. C. Wang, Sensors-assisted rescue service architecture in mobile cloud computing, in *Wireless Communications and Networking Conference*, Shanghai, China, pp. 4457–4462, 2013.
6. S. K. Dash, S. Mohapatra, and P. K. Pattnaik, A survey on application of wireless sensor network using cloud computing, *International Journal of Computer Science and Engineering Technologies*, 1(4), 50–55, 2010.
7. K. Lee, Extending sensor networks into the cloud using Amazon web services, in *IEEE International Conference on Networked Embedded Systems for Enterprise Applications*, Suzhou, China, 2010.
8. R. A. Dhote and S. B. Belsare, The role of cloud computing in mobile, *International Journal of Computer Science and Applications*, 6(2), 262–266, 2013.
9. C. Perera, P. P. Jayaraman, A. Zaslavsky, P. Christen, and D. Georgakopoulos, MOSDEN: An internet of things middleware for resource constrained mobile devices, in *Proceedings of the 47th Hawaii International Conference on System Sciences (HICSS)*, Kona, Hawaii, January 2014.
10. H. T. Dinh, C. Lee, D. Niyato, and P. Wang, A survey of mobile cloud computing: Architecture, applications, and approaches, *Wireless Communications and Mobile Computing*, 13(18), 1587–1611, 2011.
11. L. Atzori, A. Iera, and G. Morabito, The internet of things: A survey, *Computer Networks*, 54(15), 2787–2805, 2010.
12. N. Mitton, S. Papavassiliou, A. Puliafito, and K. S. Trivedi, Combining cloud and sensors in a smart city environment, *EURASIP Journal on Wireless Communications and Networking*, 2012(1), 1–10, 2012.
13. J. Soldatos, M. Serrano, and M. Hauswirth, Convergence of utility computing with the internet-of-things, in *Sixth International Conference on Innovative Mobile and Internet Services in Ubiquitous Computing*, Palermo, Italy, IEEE, pp. 874–879, 2012.
14. A. T. Campbell, S. B. Eisenman, N. D. Lane, E. Miluzzo, and R. A. Peterson, People-centric urban sensing, in *Proceedings of the Second Annual International Workshop on Wireless Internet*, Kuala Lumpur, Malaysia, ACM, p. 18, 2006.
15. Q. Liang, X. Cheng, and D. Chen, Opportunistic sensing in wireless sensor networks: Theory and application, in *Global Telecommunications Conference, IEEE Transactions on Computers*, 63(8), 2002–2010, 2013.
16. H. Scholten and P. Bakker, *Opportunistic Sensing in Wireless Sensor Networks*, Toronto, Canada, 224–229, 2011.
17. R. Tavenard, O. Ambekar, E. J. Pauwels, and M. Waaijers, Opportunistic sensing and learning in sensor networks, in *International Workshop on Content-Based Multimedia Indexing*, Bordeaux, France, IEEE, pp. 46–52, 2007.
18. N. D. Lane, S. B. Eisenman, M. Musolesi, E. Miluzzo, and A. T. Campbell, Urban sensing systems: Opportunistic or participatory? in *Proceedings of the Ninth Workshop on Mobile Computing Systems and Applications*, New York, ACM, pp. 11–16, 2008.

19. S. S. Kanhere, Participatory sensing: Crowd sourcing data from mobile smartphones in urban spaces, in *12th IEEE International Conference on Mobile Data Management*, Lulea, Sweden, IEEE, vol. 2, pp. 3–6, 2011.

20. R. Szabo, K. Farkas, and B. Wiandt, Measurements of a real-time transit feed service architecture for mobile participatory sensing, in *Wireless Days, IFIP*, Valencia, Spain, IEEE, pp. 1–4, 2013.

21. C. O. Rolim, F. L. Koch, C. B. Westphall, J. Werner, A. Fracalossi, and G. S. Salvador, A cloud computing solution for patient's data collection in health-care institutions, in *Second International Conference on eHealth, Telemedicine, and Social Medicine*, St. Maarten, The Netherlands, pp. 95–99, 2010.

22. M. M. Hassan, B. Song, and E. Huh, A framework of sensor-cloud integration opportunities and challenges, in *Proceedings of the Third International Conference on Ubiquitous Information Management and Communication*, Danang, Vietnam, pp. 618–626, 2009.

23. W. Kurschl and W. Beer, Combining cloud computing and wireless sensor networks, in *Proceedings of the 11th International Conference on Information Integration and Web-Based Applications and Services*, New York, ACM, pp. 512–518, 2009.

8

Mobile Social Cloud Computing

ABSTRACT The convergence of mobile, social network, and cloud has turned out to be the platform for cyber digital industry. Digital industry is forming new business designs by big data analysis. The combined power of these four forces explores some of the interesting distinct patterns that define that platform. With the concept of cloud computing, inherent constraints of mobile computing such as resource scarcity, battery life, frequent disconnections, etc., have been addressed, and integrated mobile cloud computing is growing exponentially around the world. Cloud computing also solves the problem of social networking sites that deal with a huge amount of data. With the enlarging pervasive nature of the social networks and cloud computing, users are exploring new methods to interact with, and utilize, these growing paradigms. A social network allows users to split information and build connection for generating dynamic virtual organizations.

Massive use of mobile technologies such as laptop, smartphones, etc., is also drawing attention to the clouds for processing power, storage space, and energy saving, which in turn leads to a new concept known as Mobile Social Cloud.

KEY WORDS: *mobile cloud computing, social cloud, tweet Data, social network, resource sharing, social cloud exchange structure.*

8.1 Introduction

The widespread growth of social networking sites has enormously changed the way we interact and liaise among ourselves. It allows us to interact with an effective platform, to establish an effective community, and to document, represent, and analyze interpersonal relationships. Today, the consumption and acquisitioning of Internet services have rapidly increased, and issues such as security, trust, reliability, and sometimes anonymity have become the major problems of the cloud computing paradigm. The progression of the web in terms of user-generated content, crowd sourcing, and the large number of online social networks has generated a tremendous amount of information describing dynamic interaction among people with their surroundings, in both online and offline mode in the real world. In this respect, the amount of information available from the connected world, belonging to either virtual or physical sources, can describe the aspects of reality in a detailed manner if both sources are combined. The kind of dynamics generated by the integration of the digital and physical worlds will have an impact on different levels involving human socio-environmental dimensions as well as on network organizations. In this regard, many novel research challenges have arisen for dealing with such integration of information.

Social networking is an online service by which many people who are members of the same social networking application can communicate with each other [1–10]. Facebook, Twitter, Orkut, etc., are different types of social networking sites. Sometimes, people may not access social networking sites on their computers or laptops; this increases the use of mobile phones day by day. More than 60% of users access social networks through mobile phones. The sensing, analyzing, and storing of a large amount of social data have become a very important issue. Apart from physical sensors, today social sensor is an emerging terminology [1]. Social sensors are effective human–device combinations that send torrents of data as a result of social interactions and events. Laptops, smartphones, tablets, i-pads, etc., are gaining popularity daily. All these devices are equipped with various kinds of sensor elements such as speed cameras, Internet fridges, transducers, and GPS and proximity detectors. The data generated appear in different formats such as photographs, videos, and short text messages. To gather numerous amount of data, we can use social networking sites such as Facebook, Twitter, and YouTube.

Social networking sites are used to represent realistic types of relationships that allow users to share or distribute information among intended users such as friends and family members. A person can use a smartphone to make videos and upload to YouTube, to tweet on Twitter, or take a picture to post on Facebook. By analyzing all these data, various activities can be done such as event detection, marketing, disaster detection and reporting, mobile social TV, and video streaming. A social sensor also generates a lot of data. But the problem with a huge amount of data is how to analyze, cluster, store, and compute the data. Cloud can be the finest solution to this problem. A seamless integrated cloud platform such as Amazon EC2 and Google App engine can be used to do all these jobs. We call it "social cloud," which is a combination of social networking and cloud. Massive use of mobile technologies such as laptops and smartphones is also drawing attention toward the cloud for processing power, storage space, and energy saving, which in turn leads to a new concept known as "mobile social cloud." The mobile social cloud simplifies the access to different devices and manages the huge volume of data transmitting across different computing networks and storage.

The social cloud furnishes an environment where sharing and provisioning scenarios can be inaugurated based on absolute trust. This trust comes from mutual relationships within social networks [11–20]. The social cloud is the conduit that uses social networks for efficient mutual interactions between the users. Social networking sites serve as an effective resource for multiple users, with different interaction and access patterns that are very difficult to predict. Their websites are basically multi-tiered web applications where each component runs in different virtual machines. Moreover, every plug-in developer has the independence to choose a suitable cloud service provider. As a result, several web applications of different social networking sites are being hosted by different cloud-based data centers. Social cloud storage can be used for developing a cloud framework in a social networking site environment.

This chapter presents a comprehensive survey on mobile social cloud where the applications and problems of mobile social cloud are discussed along with their future scope.

8.2 Mobile Social Cloud Architecture

People have shown significant interest in adopting social networks in mobile computing and cloud computing environments in order to meet their integrated requirements. The mobile social cloud is an emerging concept providing access to social networking sites using cloud, which enables elastic utilization of resources required to store social data in an on-demand fashion. Nowadays, people access social networking sites using mobile phones, even dealing with a huge amount of data. But a conventional computing system is not sufficient to analyze and store high volumes of data, which are referred to as "big data," that can be addressed by cloud only. The architecture of the mobile social cloud, demonstrated in Figure 8.1, is composed of four modules:

1. *User*: A term referring to the social network user who accesses social networking sites through mobile devices.

2. *Mobile devices*: Mobile devices are laptops, smartphones, i-pads, i-phones, etc., with high configuration and mobility.

3. *Social networks*: A social network is actually an implementation of a general network in which each node indicates a user and the edges indicate connection between them. There are a huge number of social networking sites such as Facebook, Twitter, MySpace, etc.

4. *Cloud*: The cloud is a combination of virtualization of a large amount of resources with a distributed computing paradigm incorporated with software as a service,

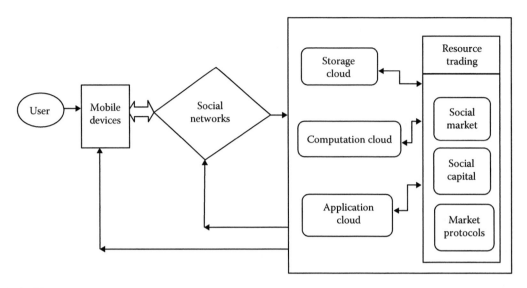

FIGURE 8.1
Architecture of mobile social cloud.

platform as a service, and infrastructure as a service. It can be of different types. The cloud module of this architecture contains three sub-modules:

a. *Storage*: The social data are stored inside the cloud. As social networks generate tremendous amounts of data, it is not feasible to store them without cloud. These data will be used for big data analysis.

b. *Computation*: The computations of data analysis are performed inside the cloud. These data are received from social networks.

c. *Application*: Various social applications such as Facebook and Twitter apps run on the cloud.

Resource trading is the functional module that balances resource sharing, allocation, and trading among the members of a particular social network or community and cloud providers. This module has three functional parts:

1. *Social marketplace*: A social marketplace is a virtual place where users from different social network groups or communities participate in resource trading. They share and utilize resources in return for some benefit, for example, Olx, Quicker, Gumtree, etc.

2. *Social capital*: Social capital generates from the social relationships and interactions between traders and consumers, for example, Amazon.

3. *Market metaphors*: These are real-life economic models for trading such as Posted Price, Auction, Trophy, Spot Price, etc. (e.g., Flipkart).

The Facebook application [2] can be used as an example to describe the social cloud. Its services are mapped to particular users by their identifications or interactions among users. A banking component always handles the transferring of credits among the users and also stores information related to the user's current reservation.

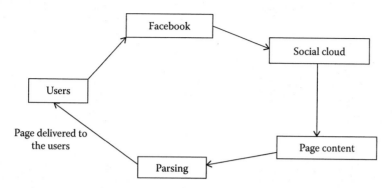

FIGURE 8.2
Facebook application environment.

Facebook handles application programming interfaces (APIs) by REST-like interface. Various Facebook applications can be built to gather data using I-FRAME, FBJS, and FQL languages. FBJS is the Facebook version of Java Scripts, FQL is the Facebook version of SQL, and I-FRAME is the Facebook version of HTML. Previously, instead of I-FRAME, FBML was used. But now Facebook has introduced I-FRAME to build Facebook apps. To integrate all applications with Facebook's look and feel, new Facebook versions are used. These applications are not hosted in Facebook but independently in the cloud. A Facebook URL accessed by users is created. This user-accessed URL is mapped to the remotely hosted callback URL. Through the Facebook URL, users request the page. The Facebook server sends the user-requested URL to the callback-defined URL. The application page is created based on the user's request and finally returns to Facebook. At this point, the Facebook page is parsed, and its important content is added according to the instruction in FBML. The final or ultimate page is returned to the user accordingly, as shown in Figure 8.2.

This is one of the case studies on social cloud. We can use different social networks or social communities instead of Facebook and build applications to fetch a huge amount of data and perform all analyses and store in the cloud. Useful decisions can be made based on the generated results.

8.3 Resource Sharing in Mobile Social Cloud

A cloud itself consists of a huge amount of resources. One of the main goals of the social cloud is to share resources among people through social networking sites as a platform. A social cloud resource represents a physical or virtual entity or capability of limited availability. A resource [3] could be information, storage, computing capacity, software license, personal ability or skills, etc. In a social cloud, one user may share storage in exchange for accessing a specific workflow; for example, a user may keep back up of photos from digital cameras in the hard disk of another member in the social network. To participate in a social cloud, each user must allocate a certain amount of resources to be used by others. This sharing is controlled by a socially oriented marketplace, which adapts common allocation protocols to a social context.

8.3.1 Motivation for Contribution of Resources

In a social cloud [3], social networking users invest in the community by joining the cloud, sharing resources, and utilizing other's resources. Some social incentives present in a social cloud motivate users to participate in, and contribute resources to, their community in different ways. Thus, users become more interested in sharing resources in return for tangible or intangible benefits. The motivation of resource contribution [3] has been studied in a large number of online domains, for example, sharing information and photos on social networks, sharing metadata and tags in online communities, and building collaborative knowledge through online content projects or open source software projects. Motivation is generally categorized as either intrinsic or extrinsic. Extrinsic motivation means motivating users by an external reward so that while they have little interest in the community, they will contribute to that community when the expected benefit exceeds the cost of contribution. Intrinsic motivation represents an internal satisfaction obtained from the task itself rather than the rewards or benefits. This sense of satisfaction may be for completing the task or for the enjoyment and reciprocation of simply working on the task.

8.3.2 Social Capital

Social capital [3] represents an investment in social relationships with expected return. From an individual point of view, social capital is similar to human capital as users of a social network may gain individual returns for specific actions, for example, selling goods or finding a new job. From a group perspective, social capital represents the intrinsic value of the social community, that is, this social community as a whole generates return by the action of its members. The sharing of resources in a social cloud is to invest and generate value from individual actions. The sharing model in a social cloud could be considered for generating both social and physical capital as it reflects the real world. The resource owners invest their resources and produce some individual returns. Thus, investing resource becomes beneficial for both the investors and the community.

8.3.3 Virtualized Resources

Cloud computing depends on virtualized resources. A social cloud provides any resource that users may wish to use, ranging from low-level computation or storage to high-level mash-ups, for example, photo storage. There are two types of requirements: (1) the interface should provide a stateful instance, (2) for discovering those services that need advertisement for incorporating them in the market.

8.3.4 Banking

The social cloud has introduced the credit-based system that recompenses users for contributing resources and charges for using the resources. Every user registers in the cloud for storing credits as well as for their participation in the banking.

8.3.5 Registration

In the registration process, users specify the cloud service they want to render. Then, user instances are created through which banking services can be transparently accessed using the users' IDs. After the registration, users are provided with MDS (monitoring

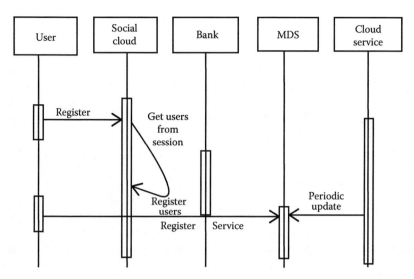

FIGURE 8.3
Registration in social cloud.

and distributed system) endpoint reference and cloud ID. A market service uses the MDS XPath to find good services based on IDs and real capacity. The registration process is shown in Figure 8.3.

8.3.6 Social Market

The social market [3] is the core of the social cloud. It regulates resource sharing among groups that are associated with the separate instances of the market. It allocates resources between peers according to predefined economic or noneconomic protocols. The social market is pictorially depicted in Figure 8.4 where several protocols [3] provide the appropriate allocation of resources for a particular user request.

The choice of protocol depends on the social cloud and the requirements of its members. The protocols are discussed as follows:

1. *Volunteer*: It is an idealistic sharing model in which users contribute resources for no personal gain, but their actions are without accountability.

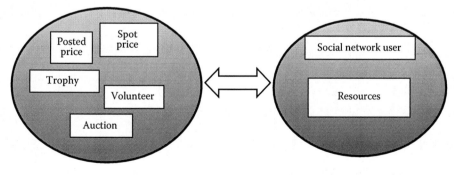

FIGURE 8.4
Social market place.

2. *Trophy*: This is a nonmonetary model in which users are rewarded with intangible credits.

3. *Reciprocation*: In this sharing model, users contributing most to the cloud are proportionally favored when they are requesting resources.

4. *Reputation*: This model is entirely based on the reputation of individuals. Reputation is established through interactions in the community. Higher reputation is favored with higher resources.

5. *Posted price marketplace*: This model provider makes advertisement offers related to a specific service level, and resources are offered at a predefined price. This will help users to create a service level agreement (SLA) with a specified parameter. This service market requires much coordination among different components of the social cloud such as discovering cloud service that first requires checking whether the user is already registered with the bank and has sufficient credits. After that the user selects a particular cloud service, and the social cloud application creates an SLA, which is ultimately sent to the cloud. It is assumed that both the parties have accepted the agreement, which is sent to the bank for transfers of credits among users. Commercial cloud providers use this model frequently. This is shown in Figure 8.5.

6. *Auction/tender*: This is a dynamic multi-participant mechanism designed to establish the market price for a particular resource. Online selling sites often use this model. In auction-based marketing, trades are mainly established by a competitive bidding process between services or users. Here, a list of friends, when discovered, is further passed to a specialized auctioneer for creating or running the auction. In Figure 8.6, reverse auction protocols are used in which cloud services bid (compete) for hosting users' task. The list of friends discovered is used by auctioneers to locate every group with a worthy cloud service. Each service provider needs an agent who acts on behalf to fulfill resource requests. After that

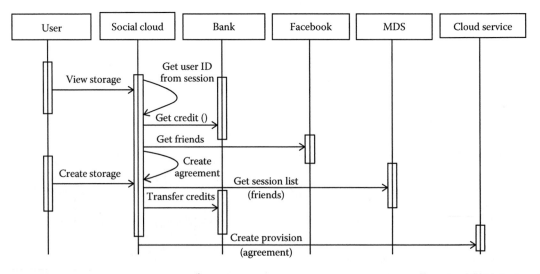

FIGURE 8.5
Posted price in social cloud.

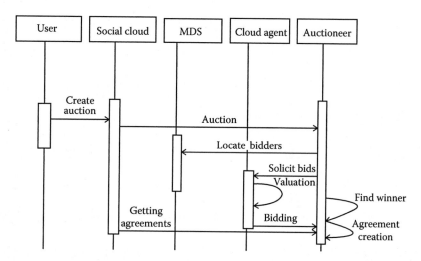

FIGURE 8.6
Auction marketplace in social cloud.

auctioneers determine the winner of the auction and ultimately create an SLA between the winning bidder and the auction initiator. Then, the agreement is sent for instantiation and finally to bank for transfer of credits.

7. *Spot price*: This is a dynamic pricing protocol in which a commodity is offered at a price given at a particular time and location.

8.4 Warehousing and Analyzing Social Data Using Cloud

Warehousing and analyzing social network data are always a big issue due to the high volume of data. These data are messy and noisy. So it becomes time consuming to analyze them. The cloud is the best solution to this problem. Social networks have three types of elements [4]: actors, ties, and relationships. Actors are people, events, objects, organizations, and so forth that are represented by nodes. Nodes are connected by lines that show the relationship between actors. Ties are used to construct the relationships between actors. Ties are divided into strong and weak categories according to the strength of the relationship. So through social network analysis, we can analyze ties, relationships, and actors in social networks.

8.4.1 Architecture of Analysis and Warehousing of Social Network Data

The architecture [4] of the social network analysis using cloud is divided into three major parts:

1. Front-end data collection component
2. Intermediate system analysis component
3. Analysis result producing component

The process of analyzing and warehousing social data is depicted pictorially in Figure 8.7.

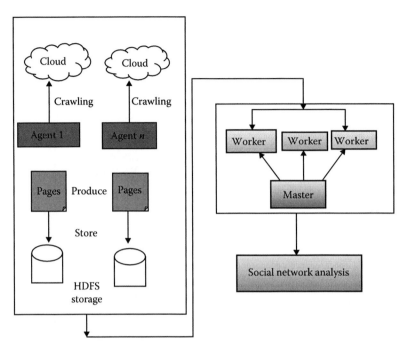

FIGURE 8.7
Process of warehousing and analyzing of social data.

The front-end data collection component gathers social data from various social networking websites such as Facebook and Twitter. Well-designed crawling programs to collect data are attached with this component. Then, it stores these data in a distributed environment, using the Hadoop Distributed File System [4], which is highly reliable. The Hadoop Distributed File System breaks the incoming files into blocks and stores them across the machines in the cluster environment. This prevents hardware failure in the distributed environment. After collecting and storing data, the system performs different levels of data processing according to user requirement. The bulk synchronous parallel (BSP) [4] model is used to process social data according to various algorithms based on the master/worker structure: The master assigns work to workers; the workers perform their work according to the assignment from the master; and finally the master acquires the processed result. This component produces the final result from the analyzed data. Now, users can get results according to the required analysis. They can interact with the help of web interfaces or APIs. This component should provide a platform-independent web interface or API to users, which is treated as the main feature of it. Thus, users can send queries and get analyzed results from various operating systems and platforms.

Two types of methods are used for social network analysis: MapReduce and Hama BSP [4]. MapReduce can be used for computer programs that need to be processed and generate a large amount of data. It has strength in locality, fault tolerance, and parallel processing. It is used to generate the index of Google. But this technique is not feasible while dealing with the processing of a graphical algorithm. In social network analysis, computation and processing of graphs are essential, so the MapReduce [4] approach is not very efficient. The Hama BSP model [4] is used to reduce the drawbacks of the MapReduce model. Pregel technology is based on the concept of BSP developed by Google, and it is used to implement graphical processing. As BSP is not an open source algorithm, Apache is now running a BSP-based project called Hama.

8.4.2 Case Study on Tweet Data Analysis

In a case study on twitter data analysis [1], we have analyzed the mood of people on twitter on the basis of changes in the weather of a particular area. This is something like a fusion of social and sensor data. The data gathered from twitter are social data, and weather data are sensor data. A mood space is created where both social and sensor data are mapped. In the mood space, a mood word is represented as a score according to three dimensions [1]—valance, arousal, and dominance—where value in each dimension ranges from 1 to 9, as shown in Figure 8.8.

Valence is defined by its two poles—negative/bad and positive/good—whereas the arousal dimension spans between the two poles: sleepy/calm for very low arousal and aroused/excited for very high arousal. Dominance is proposed to differentiate subtle emotions such as fear and anger. As valance and arousal are the main dimensions to create the mood space, a 2D plane is considered using valance and arousal matrices (V × A) to generate the mood space. This plane is divided into 12 regions, which are mapped to each mood. For example, a high value of valence as well as arousal indicates someone is happy.

8.4.2.1 Tweet Mapping

Each tweet consists of some words. Each of them has some valance and arousal score [1]. We have computed the overall score of the tweet as well as which data point will be in the mood space. Let us consider a mood space set M, which consists of 12 moods in the mood space. Let T_i be each tweet in the tweet set T, that is, $T_i \in T$. The mood expressed by the tweet using conditional probabilities is $P(M_k|T_i)$ where $M_k \in M$. A tweet word set $\{wrd_i^j\}$ is considered.

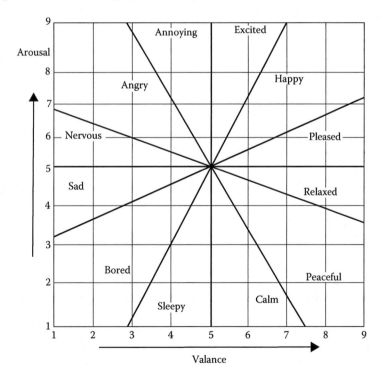

FIGURE 8.8
Graphical representation of mood space.

It is assumed that each word independently contributes to the overall mood of the Twitter message. The words that do not express mood will make zero contribution to the final mood. The mood expressed by the tweet using conditional probabilities is given by

$$P(M_k|T_i) = \frac{P(M_k) \cdot P(T_i|M_k)}{P(T_i)} = \frac{P(M_k) \cdot P\left(wrd_1^i, \ldots, wrd_n^i | M_k\right)}{P(T_i)}$$

$$= C_k \prod_{j=1}^{n} P\left(wrd_1^i | M_k\right) \tag{8.1}$$

where $P(wrd_1^i|M_k)$ signifies the amount of contribution of a particular word to make a mood M_k. It will be derived from "Affective norms of English words (anew): Instruction manual and affective ratings." Along with the term weight, the constant C_k is also determined based on the training set. Depending on the mood M_k for which the term $P(M_k|T_i)$ is largest, the tweet T_i is classified as expressing that mood. For example, consider the following tweet T_0: "Weather it is seasonal, warmish, some rain or sun, greenery or beautiful." The tweet is composed of 12 words, in which 4 of them are listed in the "Affective norms of English words (anew): Instruction manual and affective ratings" set of words. For these four words, valence scores (rain = 5.08; sun = 7.55; green = 6.18; beautiful = 7.60) and arousal scores (rain = 3.65; sun = 5.04; green = 4.28; beautiful = 6.17) are generated by looking up the "anew" list. Finally, applying the aforementioned procedure, the overall tweet valence and arousal scores (6.60, 4.78) are obtained, which form a data point in the mood space and get a "relaxed" mood label.

8.4.2.2 Mood Probabilities

This section describes how the fusion of framework computes a set of mood probabilities [1] according to day, location, and weather. Each tweet T_i carries information about the location L, time stamp t, and weather label W; it also carries the mood information M_i. For each tweet $T_i \in T$, a record is maintained denoted by R: (T_i, L, t, W_j, M_i). Essentially, each tweet is mapped as a point in the 2D mood space. The complete set of twitter data is mapped on the 2D mood space as a distribution of points. For easy querying, the distribution points are summarized on the social metric space.

Once all the tweet records R's are obtained, the mood–weather information can be summarized using p_{ijk} probabilities, where p_{ijk} represents the probability of witnessing mood M_i when the weather is W_j and the day is D_k: {Monday,…, Sunday}, that is, the conditional probability is $P(M_i|W_j, D_k)$. Different models can be considered for computing p_{ijk} ranging from a simple model that summarizes all the events ignoring temporal aspects such as time and weekday. According to the simple model, all the tweet records are grouped corresponding to a particular location L. Different weather labels W_j, mood labels M_i, and day labels D_k associated with each of these tweets are observed and P_{ijk} is computed as follows:

$$P_{ijk} = \frac{(\#\,\text{tweets with } M_i, W_j, D_k)}{\sum_{a=1}^{12} \#(\text{tweets with } M_i, W_j, D_k)} \tag{8.2}$$

It is determined as the fraction of tweets expressing a certain mood M_i for a particular weather label W_j and day D_k.

8.5 Social Compute Cloud: Sharing and Allocating Resources

Nowadays, it is quite easy to share our own resources, services, and data through social networks because we all believe that "apps" are becoming more sophisticated. To justify the statement, a social compute cloud, in which we furnish the cloud infrastructure by "friend" relationship, is presented. Sometimes a group of people needs access to resources that are made available only through connected peers. The social compute cloud is a platform for sharing infrastructure resources within a social network [19]. By using this approach, downloading and installing a middleware, leveraging on personal networks such as Facebook application, and providing resources or consuming resources from friends through a social clearinghouse have become very easy. We predict that resources present in the social cloud are shared as they are idle, underutilized, and made accessible altruistically.

Several challenges of social clouds include technical facilitation for the cloud platform, its interpretation, its inclusion in social network structure, and also its implementation as well as design of socio-economical models, which is required for abetment of exchange and platform infrastructure. Technical facilitation entitles users to supply resources and utilize resources from others. The idea of social cloud is based on a certain height of trust among each other. The social compute cloud architecture requires adequate sandboxing methods [19] as well as security for protecting resources from incompetent and potentially malicious users. Leveraging on social network makes it easy to share and compute resources inside social network. To exploit resource sharing of social cloud users, it must allow access to a social network and trust the platform that contains their social network data. We can say that a social cloud is a type of community cloud because here the resources are consumed, owned, and provided by the members of the social community. This is shown in Figure 8.9.

The models of a social compute cloud are as follows:

Socio-economic model: This model is for allocating resources within the social compute cloud [17]. The concept mainly focuses on sharing (and not selling) of resources.

Platform facilitation: This model mainly focuses on sharing resources. Here, the users are not paying for the services that are offered via the cloud platform. Rather, the platform needs computational resources in order to function.

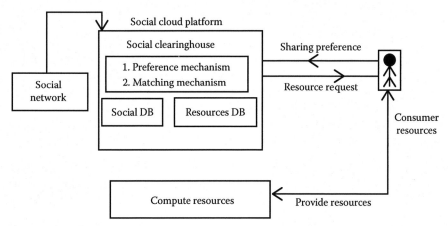

FIGURE 8.9
A social compute cloud and its core components.

8.5.1 Architecture of Social Compute Cloud

The social compute cloud platform requires proper coordination and execution, and its primary functionalities are resource allocation, user management, and so forth.

The social clearinghouse defines how user demands are fulfilled with the perfect supplies. It is the middle point of the system in which all information regarding users as well as shared information, demand, and resource supply are kept. The social clearing home needs two databases:

1. One database to capture the sharing preferences and social graph of the users
2. One database to work as a resource manager for tracking resource availability and resource allocation

The middleware provides basic resource virtualization, resource fabrics, and several mechanisms for consuming and provisioning resources. It also defines different protocols that users mainly need as well as resources to leave or join the system.

The social-technical adapter provides access to the necessary features of social networks. It represents a method of authentication. The preference module mainly provides the necessary functionality to represent and capture the sharing preferences when required. These adapters require special consideration as well as methods, which can be applied to capture preference. Matching mechanism is the socio-economic enactment of the social clearinghouse, which determines where the resources should be allocated by analyzing user's sharing preferences on social networks.

Compute resources are the technical funding of users, what they consume and provide to the social cloud. Here, resources mainly require personal computers, clusters, or servers. For the establishment of social sharing preferences, the social cloud needs to access a user's network. We are motivated to use a social adapter instead of an implementing platform as a social network application. The main misconception is that users always misinterpret between a social network and its application. Another misconception among users is that social networks can access users' data or resources that are kept in the social cloud.

Social graphs can be constructed through matching methods by following authorization. Many platforms are provided APIs for accessing a user's profile and the social graph. In the case of Facebook, Twitter, and Google+, a basic assumption is that the social cloud follows bilateral approval. In a different way, we can say that once a user commences the initiation of the digital tie, the other user must confirm in order to establish the request. But here the users are free to decide whom they want to follow, but the users cannot decide who will follow them. Therefore, another problem of trust among the participants arises.

Until today, there is no perfect methodology for the explanation of these social ties, which are often context dependent. Three methods can be applied either in combination or separately [17]:

1. Users themselves rank their friends.
2. Leverage methods to identify the attribute of social ties, which can be used for artificially constructing preference.
3. Use interaction theories for social networks to establish a social sharing and the interaction models for the social compute cloud.

But these methodologies have some advantages and disadvantages. If we are using user-provided lists, then it will be easy for implementation without the need for special permission. But the fact is that today most Facebook users have more than 200 friends, so the approach could not scale a huge number of friends to join social clouds. In contrast, if we move with computation methods, we can scale a large number of users as social cloud always grows. But the major problem with this approach is that it also requires huge data from social network platform and it is more protruding of users' privacy.

The socio-economic model is used to specify which kind of preference matching can be implemented and used. The first step it follows is capturing users' supply and their demand. This is done by the social clearinghouse. Here, centralized implementation is done, which means we know all the supply and demand of the market. The only disadvantage is that we have to manage an additional overhead for storing and updating the information. Although it has some disadvantages, it is still very useful in finding a solution to matching problems that are either stable or minimizing the overall welfare of users and provides complete fairness between two sides.

8.6 3D Visualization of Social Network Data

In this revolutionized age of technology, 3D data visualization and analysis [5] have become popular in many fields such as process control, decision support systems, and scientific data analysis. As the amount of data is continuously changing, analysis of data becomes a daunting task. The outstanding capabilities of cloud computing can help to solve this 3D data visualization and analysis problem by providing resources. A popular data analysis method called online analytical processing (OLAP) can be used to analyze social data.

8.6.1 Visualization of Social Network Data

The 3D graphical interfaces for knowledge or information navigation are based on a model view controller (MVC) [5] framework implemented in Windows Presentation Foundation. An example is a connection made between cloud computing and the social networking applications. With the help of cloud computing, businesses can search for people who are related to their company. The posted information can be captured, and future conversations about their company can be tracked. The information obtained from these conversations assists to create customer service cases. Thus, cloud computing helps businesses to access the existing social networks, optimize search engine tools, and connect with other businesses.

8.6.2 OLAP Data Analysis and Cube Generation

OLAP is a tool for answering multidimensional analytical problems that encompass data mining and a relational database. It enables users by analyzing multidimensional data interactively from multiple perspectives. An OLAP system generally uses a 2D plot that places multiple dimensions of column or row factors. The user must have some

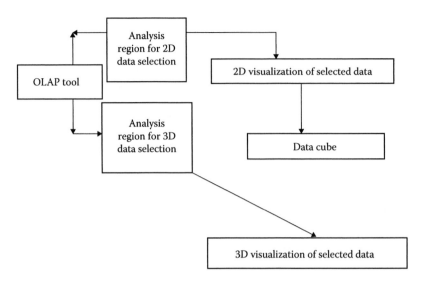

FIGURE 8.10
OLAP data analysis process.

domain knowledge for the selection of appropriate dimensions. But in the presence of new trends, it is very difficult for users to select new dimensions and range. Moreover, most of the existing OLAP systems support 2D or partially limited 3D charts with which users have limited or no interaction. The 3D information represented by an eCube is composed of selected attributes of the given database and its dimensions, as shown in Figure 8.10 [5,8].

The user can visualize the 3D cube by using operations that include rolling-up, drilling-down, slicing, and pivoting operations [17]. Rolling-up consists of combination of data that can be gathered and found in more dimensions. Drilling-down allows users to navigate through the data details. By slicing, users can take out a specific set of data from the OLAP cube. Pivoting allows the user to rotate the 3D cube to see its various faces. Interaction of users with the eCube can happen in a variety of methods, including selection, navigation, and zoom in or zoom out.

8.6.3 MVC-Based Model for Visualization

The system for 3D visualization of large and complex social network data is based on the MVC model. This model represents and manages the targeted information selected by users from huge social networking data. The targeted information is a subset of raw data and statistics obtained from social networking sites. This model uses extensible markup language (XML). The view comprises an interpreter for XML and a 3D graphics renderer. After interpretation of the XML file, the information is shown in a predefined 3D graphical environment. The controller monitors the interaction of the users with 3D visualization space. The flow diagram of MVC is depicted in Figure 8.11.

Based on the type of interaction, the controller governs the model to perform changes. If the state of the model is changed by the controller, the connected views are also changed.

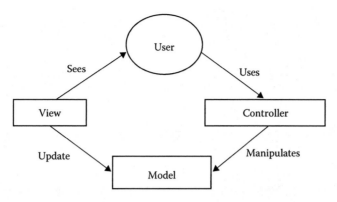

FIGURE 8.11
Model view controller.

8.7 Security in Mobile Social Cloud

As it is the integration of two different fields, that is, cloud computing and mobile networks, MCC has to face many technical challenges. With the innovative blow-up of mobile cloud technology, the demand for social data security is increasing sharply. It is a great responsibility to provide security to social data, which are shared in social networking sites and ultimately stored in the cloud. So, security of the mobile social cloud is the integration of mobile device security, social network security, and cloud security. Zhibin and Huang [9] have proposed the Privacy Protecting Ciphertext-Policy Attribute-Based Encryption [20] (PP-CP-ABE) method to protect the sensing data. It is pictorially depicted in Figure 8.12.

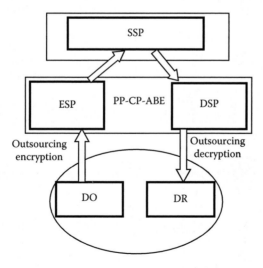

FIGURE 8.12
System architecture of security for light-weight devices. *Note:* SSP, storage service provider; ESP, encryption service provider; DSP, decryption service provider; DO, data owner; DR, data receiver.

Using PP-CP-ABE, heavy encryption and decryption operations can be securely outsourced from lightweight devices to cloud service providers. A data owner (DO) can request or store information from or in the cloud. This information is first encrypted by the encryption service provider before going to the storage service provider. When the data receiver (DR) requests for that data, the data are sent to the DR after the decryption is done by the decryption service provider.

8.7.1 Security in Social Network

A social network is like a virtual communication medium where users share multimedia data with others. In these sites, users provide their name, address, date of birth, gender, place of birth, school, interest, and other personal information. This information is shared with other users. Hence, attackers can gain personal information easily by using the social networking sites, which helps them in a wide range of network crimes such as identity theft.

8.7.1.1 Purpose of Attackers

Attackers have found social networking sites a better way to commit network crimes. Targets of attacks can be the following [6]:

- *Access control*: The attacker aims to get control of the computers of other users with malicious intention. The adverse effect is that the controlled computers are organized to perform some types of attacks such as denial of service attacks.
- *Personal information*: Some attackers are looking for important personal information such as bank account details, passwords, and social security number to commit further crimes.
- *Jokes*: Some users just want to play jokes with other users to improve their reputation. These types of attacks sometimes cause network congestion and affect the users' quality of experience.
- *Company information*: Users are business customers in most cases. In the past, attackers could not easily break the intranet because companies had strict protection measures. In contrast it is easier for attackers to obtain the trust of others with the advantage of social networks, which help them to gain professional information of users and customers.
- *Money*: Attacks on social networking sites have increasingly become financially driven. Most attackers target to gain bank account details, financial secrets, and private information.

By the end of 2008, the Kaspersky Lab collected a number of programs that attack social networking sites [8], as shown in Figure 8.13.

According to Figure 8.13, the number of malicious programs is increasing daily. From 2001 to 2005, the number of malicious programs was almost negligible. But since 2005, malicious programs have increased rapidly. MySpace has been attacked by higher numbers of malicious programs than other social networking sites. Though Facebook is comparatively secure compared with other social networking sites, it is still not free from malicious attacks.

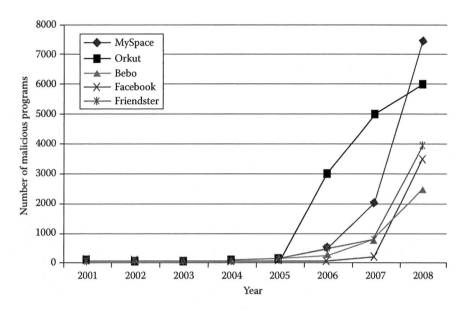

FIGURE 8.13
Attack on social network by malicious programs. (From Luo et al., An analysis of security in social networks, *Eighth IEEE International Conference on Dependable, Autonomic and Secure Computing*, pp. 648–651, 2009.)

8.7.1.2 Method of Attacks

Attacks on social networks are performed in different ways. Traditional spams are spread via e-mails, but now they spread fast via friend lists in social networks. The primary objective of spammers in online social networks is to reach a large number of social friends by spreading malicious code. Worms can self-replicate and spread automatically for stealing private information such as passwords and bank account numbers. Attackers can also use vulnerabilities that can be generated into the web page code to steal COOKIE, run FLASH, hijack accounts, force users to download malware, etc. [6,12]. Some plug-ins such as Flash are permitted to run on browsers, which brings a new threat to social networks. We must admit that both users and social networking sites have an impact on the security of social networks; therefore, social networking sites could provide enough security supports. Users should also increase their security awareness to combat the growing number of attacks.

8.7.2 Resource Allocation for Security Services

Numerous challenges exist in the mobile social cloud, including data replication, consistency, unreliability, availability of cloud resources, trust, security, and privacy. Research organizations and academia have undertaken a massive amount of work to secure a cloud computing environment. To attract potential consumers, the cloud service provider must target all security issues to provide a complete secure environment. Cloud security services are classified into two categories: normal security (NS) services and critical security (CS) services. CS service involves more complex security implementations such as stronger authentication and encryption algorithms, longer key size, strict security access policies, and so on. The NS service uses only basic security approaches such as authentication to validate users and access control tasks. NS service also involves low-complexity

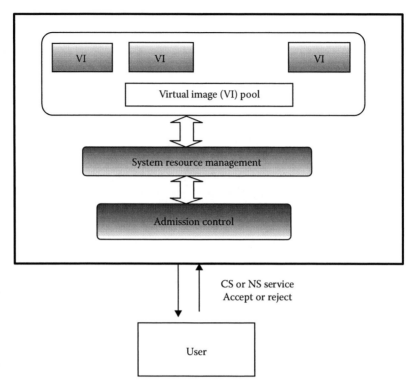

FIGURE 8.14
Resource virtualization in cloud for security services.

computing and resource isolation as the first requirement in order to provide these security services. It comprises a virtual image (VI), which manages a portion of the cloud system resources such as CPU and storage. Figure 8.14 shows how resource virtualization is done to provide various security services to mobile cloud users. When a security service request comes from the user, the system admission control consults with the system resource management model about the availability of resources, that is, VIs. If VIs are available, then the request is accepted and one or more VIs will be allocated to that security service. If resources are not available, the request is rejected and increases the probability of blocking [7]. But the cloud decides whether to accept or reject a security service request based on the currently available cloud resources and the arrival rate of future security service requests.

8.8 Trust in Mobile Social Cloud

With the explosive growth of cloud computing, the demand of social networking sites is increasing sharply. Users are mutually connected through social computing to form a social network. Trust is the foundation of all social interactions among society members. Users make decisions based on the trust between users and their friends and are more willing to accept information from trusted friends. Users also share their personal data with trusted friends, and the data are stored in the cloud. A social network connected by

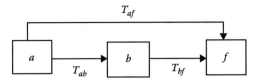

FIGURE 8.15
Trust transmission in friend path.

trust scores of people and a model propagating those trust scores are the basic building blocks in many of today's recommendation systems. Therefore, estimating how much one user will trust another is the most challenging issue in the social cloud environment. In social networks, users trust their friends and the friends of their friends; this is how trust can be transmitted [11]. With the help of trust transmission, users can obtain their indirect friend's trust value. Trust transmission is shown in Figure 8.15.

Cloud users are more concerned about whether data center owners will misuse their data by releasing them to third parties. Kai and Li [11] have proposed a trust management scheme augmented with data coloring, which can help address this issue. In this method, cloud drops, that is, data colors are added to the input data such as image, video, and document.

In a social cloud, trust plays a vital role as a collaboration enabler. However, trust is not trivial to define and observe; it represents an analysis to understand exactly what role trust plays in enabling the collaboration. This is done through the definition of the structure of a social cloud as a sequence of social and cognitive processes.

8.8.1 Trust Inference in Social Networks

In the context of social networks, trust is nothing but a commitment to an action based on belief that the future action of a person will lead to a good outcome. It gives indications for a participant's decision making in various activities on social networking sites. Trust inference actually aims to infer a trust value accurately, which may exist between two people without direct connections, based on trust transitivity [10,12]. A participant can give better recommendations to participants who have intimate social relationships with him/her. An intimate degree value r ($0 \le r \le 1$) of social relationships between participants and a role impact value ρ ($0 \le \rho \le 1$) are defined in trust-oriented social networks to calculate the trust value [12]. This value indicates the impact of the participant's recommendation roles if the participant is an expert or beginner in a specific domain.

From Figure 8.16, it is observed that if X trusts Y and Y trusts Z, then X can trust Z to some extent. The probability that X can trust Z is calculated as

$$P(T_{XZ} \mid r_{YZ}, \rho_Y) = \frac{\Pi(\theta) \cdot P(r_{YZ} \mid T_{XZ}) \cdot P(\rho_Y \mid T_{XZ})}{P(r_{YZ}) \cdot P(\rho_Y)} \tag{8.3}$$

where
$\Pi(\theta) = (T_{XY}) \cdot (T_{YZ})$ is the prior probability of trust inference
$P(r_{YZ} \mid T_{XZ})$ represents the probability of intimate degree value $r = r_{YZ}$ with the given condition that there is a trust relation between X and Z
$P(\rho_Y \mid T_{XZ})$ represents the probability of role impact value $\rho = \rho_Y$ with the given condition that there is a trust relation between X and Z

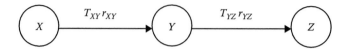

FIGURE 8.16
Trust inference in single trust path.

But in mobile cloud computing, trust may not be symmetric. Participants have different purposes in social networks, such as making friends, advertising, and carrying out business. In certain circumstances, these mechanisms may not deliver realistic trust values between those who trust and those who are trusted. So, before inferring trust, some constraints of recommendation roles and social relationship preferences need to be specified.

8.8.2 Trust Contextualizing in Social Clouds

The notion of trust actually needs an analytical perception as an idea of context and social action. So to address the provocation of understanding and defining trust, the social cloud can be griped like an exchange enabler, investigated, and specified experimentally [1]. To understand the concept of trust and social cloud context, there is a brief description about the social cloud structure.

8.8.3 Social Cloud Exchange Structure

A social cloud is a sequence of cognitive and social processes. It is represented in three stages [18]:

1. Prior expectation
2. Social interchange
3. Completion

The social cloud exchange structure [18] is shown in Figure 8.17.

8.8.3.1 Prior Expectation

Prior expectation deals with why a user would join or use a social cloud and how it would fulfill users' expectations for their contribution. Motivation, demand, and supply are expected by users. Motivation can be expressed through expected outcome, which includes a gain inutility, goal fulfillment, task completion, and feeling of usefulness or inclusion. These also include a sense of togetherness or belonging through participation in a cloud. The social context specifies special features of the social cloud and its users, for example, relationships such as close friend, family, colleague, and acquaintance, special features of the social graph such as connectivity and centrality, and the implicit trust between the users of social cloud and their friends. The interaction history describes the completion of all the three stages. Supply is the inferential availability of some useful resources in a cloud. Demand is the monitoring of individual requirements that motivate users to furnish to their squint, for example, the demand being persuaded in social capital. In this case, the requirements of the users of social cloud are compelled by one or more

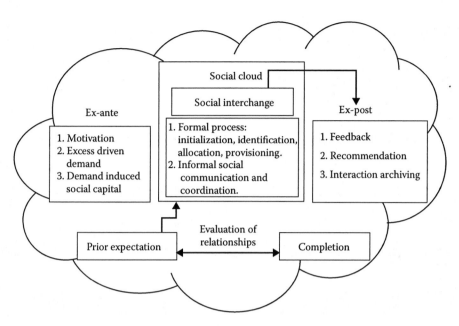

FIGURE 8.17
Social cloud exchange structure.

other users so that their capabilities and their resources can be exceeded. These resources are available in social capital form.

8.8.3.2 Social Interchange

Social interchange entails the abetment of exchange and collaboration among socially connected peers. It includes formal process, informal social communication, and coordination. In the formal aspects, we define collaboration like economic systems. In informal aspects, we define context specific and social structures, which help us to facilitate exchange like a socially driven procedure. The main aspects at this stage are to focus around allocation and identification of supply and demand. These can be implemented through socio-economic processes, messaging processes for coordination and communication, and the delivery and provisioning of exchange facts.

8.8.3.3 Completion

Completion is a process that addresses the actions that resolve an exchange and includes major components such as feedback, archiving, and recommendation. Feedback is a social distribution for exchange like an exercise in communication and reflection through two modes: (1) public feedback, which is the means of social channels such as notification, Facebook timeline, and newsfeed; and (2) local feedback, which is for potentially private users. Feedback includes possibly a reward disposed to providers or a thank you message.

Recommendation is a proposal for an activity after a negative or positive exchange result, that is, whether users are rewarded or have to pay penalty. This recommendation process is mainly dependent on a social context from collaboration. Archives act like a repository for governing social cohesion, that is, often an interaction history result and collaboration performance based on good expectation.

8.9 Applications of Mobile Social Cloud

Numerous applications have been developed with the integrated mobile social cloud environment.

8.9.1 Cloud-Assisted Adaptive Video Streaming

Disruptions and buffering delays are the major problems in video streaming, nowadays. To address these issues, a framework of cloud-assisted services has emerged [13]. Traditional adaptive streaming frameworks, such as Microsoft's smooth streaming technique as well as Adobe's and Apple's HTTP adaptive live streaming, had to maintain multiple replicas of the video with various bit rates, thus putting a huge storage burden on the server. Therefore, the recent H.264 Scalable Video Coding (SVC) technique has gained much attention. SVC defines a diverse profile of video streaming with one base layer (BL) and multiple enhancement layers (ELs). If only the BL is delivered, a video can be decoded and displayed at the lowest quality. When more ELs are delivered, a better quality of the video stream can be achieved. These sub-streams can be encoded by exploiting three scalabilities:

1. Spatial scalability by layering image resolution, that is, screen pixels
2. Temporal scalability by layering the frame rate
3. Quality scalability by layering the image compression, which can offer videos for a high variety of qualities with relatively less storage overhead

Cloud-assisted adaptive video streaming is shown in Figure 8.18.

The whole video storing and streaming system in the cloud is called Video Cloud (VC). There is a large-scale Video Base (VB) in VC. The VB stores most of the popular video clips from video service providers (VSPs). A temporary Video Base (tempVB) stores new candidates for popular videos to serve as a cache. The VC also keeps running a collector to seek popular videos from VSPs, re-encode the collected videos into SVC format,

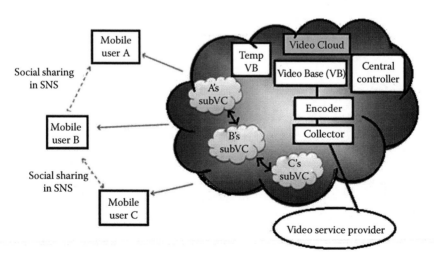

FIGURE 8.18
Cloud-assisted adaptive video streaming.

and then store in the tempVB [13]. When a mobile user requests video streaming, a sub-Video Cloud (subVC) is dynamically created for each active mobile user. Each subVC has a sub-Video Base (subVB) that stores the recently fetched video segments. If the mobile user requests for a new video that is not in the subVB or in the VB, the subVC fetches, encodes, and transfers the video. During the video streaming, mobile users will always periodically report link conditions to their corresponding subVCs. Then, the subVCs predict the available bandwidth of the next time window and adjust the combination of BL and ELs adaptively.

8.9.2 Personal Emergency Preparedness Plan

With the increasing growth of mobile computing, cloud computing and social services have become an integral part of the society during the event of an emergency or disaster. The Department of Health and Human Services (HHS) had sponsored a challenge for software application developers to design a Facebook application that will help people to prepare for emergencies and to obtain support from friends and families during its aftermath.

A dynamic application has been developed, which identifies and connects friends on Facebook who are willing to be "lifelines." HHS identifies a lifeline as someone who will act as the point of contact in the aftermath of an emergency or disaster. The department claims that a tremendous number of people use Facebook to share information about those who are potentially affected by the disaster. Social networking sites help people to connect with each other. In the event of an emergency, people can connect with their friends and family through these sites. There are web applications that provide a registry and message board for survivors, family, and loved ones who have been affected by a natural disaster. The application allows users to post and search for information about another person's well-being and location. The application operates by allowing agencies to collaborate with local officials to share information such as maps, reports, pictures, and videos. This information can be streamed in real time from a command and control center to the hands of first responders with mobile devices.

The motivation behind the Personal Emergency Preparedness Plan (PEPP) [10] application is to improve upon Google's Person Finder and Lockheed Martin's Open911 web application by incorporating a social networking platform such as Facebook. The PEPP app integrates features that would provide critical information for an emergency responder such as the geographical information system. In the aftermath of a disaster, traditional channels of communication such as cell phones and land line networks are frequently overwhelmed. Although SMS messaging has been an invaluable source of communication, it is reliable. Therefore, integrating with social networking sites such as Facebook and Twitter will decompress those channels of communication.

8.9.2.1 System Architecture

A high-level architectural overview of the PEPP app [10] is shown in Figure 8.19. It shows two interconnected hybrid clouds consisting of a social network platform on the left and cloud on the right. The composition of the clouds remains its unique entities and is joined by configuration settings that enable data sharing between the cloud entities.

Social networking platforms such as Facebook provide the core features that are available for integration, such as news feeds, notifications, platform dialogs, and the social graph. Facebook does not provide a hosting service for apps. Therefore, a developer who

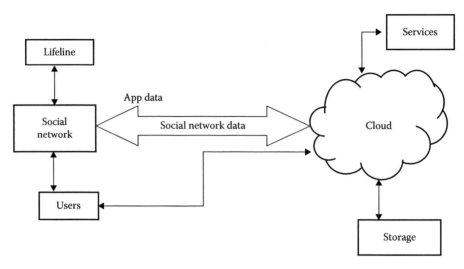

FIGURE 8.19
Architecture of personal preparedness emergency plan.

is creating a Facebook app must find his or her own hosting service. After subscribing to the hosting service and registering the app's URL, the next step is to sign up for a Facebook developer account. Specifically, the app will be loaded within an HTML <iframe> element. The web interface for the PEPP application appears as a native Facebook app. The app is integrated into the Facebook canvas while leveraging core features of the platform such as notifications, authorization dialogs, and the social graph.

8.9.3 Massively Multiplayer Online Games

The massively multiplayer online game (MMOG) [11] represents a beautiful application of social cloud computing. Millions of users are playing various kinds of games through the Internet daily. For example, Farmville and other similar games have more than 10,000,000 constant players. These players turn into a collaborative community to exchange information such as game review, news, advice, and expertise. Third parties such as volunteers and small businesses have built online communities that provide all this information to users of respective games or groups of similar types of games. Now these communities need to analyze various data such as news and advice to provide users' demands such as player reports and clan statistics. An analyzing architecture is depicted in Figure 8.20 using the cloud service called Continuous Analytics for Massive Multiplayer Online (CAMEO) games.

The CAMEO is the architecture for MMOG analytics in cloud. It mines information from the web and collects information by web 2.0 interfaces provided by various MMOG operators and their collaborators. Then, it integrates the information into comprehensive and time-spanning MMOG datasets. Finally, it analyzes the dataset and presents application-specific results. To analyze this huge amount of data, cloud is used, for example, Amazon EC2 service.

8.9.3.1 Challenges in CAMEO Architecture

Challenges in the CAMEO architecture are described in the following sections.

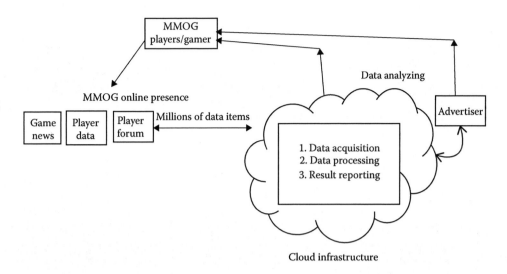

FIGURE 8.20
CAMEO architecture for analyzing MMOG.

8.9.3.1.1 Understanding User Community Need

Various kinds of data are generated during online gaming. CAMEO needs to analyze data of numerous types. It can do various kinds of analysis such as the following:

- Analyze skill, experience points, rank of a player and process information from single or multiple data snapshots, allowing for single time point and evolution analysis.
- Rank players according to one or more skills, extract the statistical properties for the whole community for one or more skills, extract the characteristics of the key players who improved most during a period such as a week, and compute the evolution of the top players during a period.

8.9.3.1.2 Data Management and Storage

Data management and storage have always been important issues in CAMEO. For simplification, CAMEO stores data centrally using a single administrator. For data management, it interacts automatically with the storage administrator. Three main solutions to store data are available to CAMEO:

1. Store the data on the same machine that acquires or generates it
2. Store the data outside the cloud
3. Store the data using the dedicated cloud storage services such as Amazon S3

By default, CAMEO uses the third solution, which is centralized, reliable, and scalable. It provides the highest transfer speed between storage and processing nodes for continuous analytics workload.

8.9.3.1.3 Performance, Scalability, and Robustness

All these characteristics are very important to maintain mainly the quality of service of CAMEO. As we use cloud service, it provides the scalability and robustness to CAMEO.

8.9.4 Geosmart: Social Media Education

Geosmart [12] is another application of social cloud computing. It is an interactive, informative, and communicative social media for education purposes. Geosmart is the combination of social media and could compute. It aims to increase intelligence in the Indonesian society and is a kind of online education system that uses the cloud and has become more interactive and interesting to users.

8.9.4.1 Entities of Geosmart

There are four main entities of Geosmart: user entity, technology entity, feature entity, and materials entity.

1. *User entity*: Any person involved directly or indirectly with the education system is a user of Geosmart. There can be six kinds of users involved in the system: students, teachers, lectures, campus students, parents, and the public. Students can collaborate and share their knowledge within the school or with other school students. Teachers can collaborate and share their knowledge of good teaching methodology with other teachers as well as students. Lectures found in video and power point presentation format in Geosmart are very interactive and useful for students. Through Geosmart students can collaborate and share with other students across campus. Geosmart can be a beneficial platform for discussion, communication, and information sharing among students. Parents can interact with teachers and various authorities to monitor their students. The media for alumni or education practitioners help to communicate and provide information about their school or campus through Geosmart. The entities of Geosmart are shown in Figure 8.21.

2. *Technology entity*: Technology is a vital fact of any social system to increase reliability and availability. The technology entities are web platform and mobile platform. Geosmart application represents the web 2.0 generation where the Internet is utilized to run the web applications. To increase the utility and portability of the

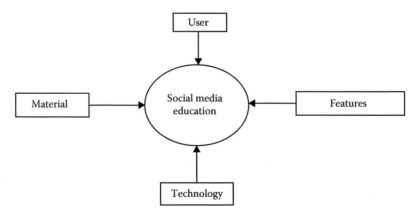

FIGURE 8.21
Conceptual diagram of Geosmart.

application, a mobile platform is needed. Massive uses of mobile devices led to the development of Geosmart messenger, which is an android app to control and use the application from mobile devices also.

3. *Feature entity*: Geosmart has various features for users such as discussion forum, chatting, educational materials, competition, album, ranking, try out, school page, badge, and Geosmart mobile.

4. *Materials entity*: Numerous educational services are available in Geosmart on demand, such as ebooks, videos, games, materials, try outs, and tutorials.

8.9.4.2 Architecture of Geosmart

Geosmart uses social cloud, which involves many third parties to contribute resources. Every service should be accessible to Geosmart registered users. The objective is to provide interaction between users and services. The architecture is given in Figure 8.22.

The users or contributors of Geosmart's service should perform the registration process first to get a Geosmart ID. The Geosmart API is used as an interface for third parties to deliver both content and application services to the site Geosmart. Users who upload the service to the Geosmart site are given a page with the URL interface (http://apps. goesmart.com/socialcloud/) access, which will be forwarded to the server and will result in a response page as interface/preview content. The Monitoring and Discovery System (MDS) is a component that provides information services that are contained in the resource server. Services are updated periodically through the MDS and stored in the server side. Telecommunication providers offer the content to system services. The system communicates with the existing services on the Geosmart's web transaction processing conducted for Geosmart's registered users only. For example, a teacher would like to contribute paid material/content/application education to the user community with a charging mechanism from Telco provider.

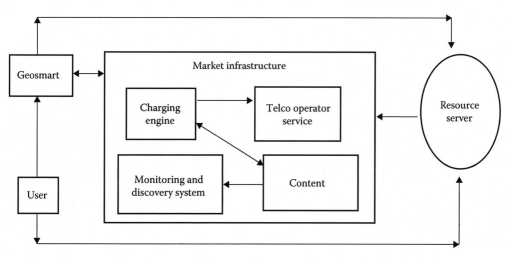

FIGURE 8.22
Geosmart architecture.

8.10 Conclusion

The tremendous popularity of mobile devices and social networks as well as the significant utility of cloud is behind the creation of mobile social cloud computing. This chapter has focused on various aspects of mobile social cloud computing. The social network platform can be used to gather a huge amount of user data for resource sharing and data analysis. Facebook, Twitter, online gaming, online video streaming, online TV streaming, and online education systems are using the concept of mobile social cloud computing and have become very popular nowadays. Although handling big data is challenging, the mobile social cloud has become feasible due to the growth of social network users. Security plays a major role in this field since a huge amount of social data is stored in the cloud. So, to provide security services to users, efficient resource allocation is required. Trust is another important aspect of mobile social cloud. The mobile social cloud leverages the existing relationships of people in social networks for service and resource sharing. Social networks are based on digital relationships mapped from real-world relationships. These relations are all about trust between people. In a social network, the question of whom to trust and whom to not is now of vital importance. Though there are many trust models, it is not always true that a real-world person is the same person on a social network. On the other hand, we send our social and personal data to the cloud. Therefore, data integrity and privacy are very important concerns. Consequently, the trust of the mobile cloud provider is also a vital issue, which has been discussed in this chapter.

Questions

1. Describe social mobile cloud computing with an example. Explain its architecture.
2. Explain resource sharing in social mobile cloud computing.
3. What is a social market? Explain all its protocols.
4. What is warehousing and analysis of social data? Explain their architecture.
5. What is tweet mapping and mood probability of social mobile cloud computing?
6. What is social compute cloud? Explain its architecture.
7. Discuss the 3D visualization of social network data.
8. Discuss the security issues in mobile social cloud computing.
9. Explain trust in mobile social cloud.
10. What is trust contextualizing in social cloud?
11. What are the applications of mobile social cloud?

References

1. S. R. Yerva, H. Jeung, and K. Aberer, Cloud based social and sensor data fusion, in *15th International Conference on Information Fusion*, Singapore, pp. 2494–2501, 2012.
2. K. Chard, S. Caton, O. Rana, and K. Bubendorfer, Social cloud: Cloud computing in social networks, in *IEEE Third International Conference on Cloud Computing*, Miami, FL, pp. 99–106, 2010.

3. K. Chard, K. Bubendorfer, S. Caton, and O. Rana, Social cloud computing: A vision for socially motivated resource sharing, *IEEE Transactions on Services Computing*, 5(4), 551–563, 2012.
4. I.-H. Ting, C.-H. Lin, and C.-S. Wang, Constructing a cloud computing based social networks data warehousing and analyzing system, in *International Conference on Advances in Social Networks Analysis and Mining*, Taiwan, China, pp. 735–740, 2011.
5. N. Ejaz, I. Mehmood, J. J. Lee, S. M. Ji, M. H. Lee, S. M. Anh, and S. W. Baik, Interactive 3D visualization of social network data using cloud computing, in *International Conference on Cloud Computing and Social Networking*, Bandung, Indonesia, pp. 1–4, 2012.
6. W. Luo, J. Liu, J. Liu, and C. Fan, An analysis of security in social networks, in *Eighth IEEE International Conference on Dependable, Autonomic and Secure Computing*, Chengdu, China, pp. 648–651, 2009.
7. H. Liang, D. Huang, L. X. Cai, X. Shen, and D. Peng, Resource allocation for security services in mobile cloud computing, in *IEEE Conference on Computer Communications Workshops*, Shanghai, China, pp. 191–195, 2011.
8. S. Ji, B. Seok Lee, K. Kang, S. G. Kim, C. Lee, O.-y. Song, J. Y. Choeh, R. Baik, and S. W. Baik, A study on the generation of OLAP data cube based on 3D visualization interaction, in *International Conference on Computational Science and Its Application*, Santander, London, United Kingdom, pp. 231–234, 2011.
9. Z. Zhibin and D. Huang, Efficient and secure data storage operations for mobile cloud computing, in *Eighth International Conference on Network and Service Management*, Las Vegas, NV, pp. 37–45, 2012.
10. S. Ghosh, G. Korlam, and N. Ganguly, Spammers networks within online social networks: A case-study on Twitter, in *Proceedings of 20th International Conference Companion on World Wide Web*, Hyderabad, India, pp. 41–42, 2011.
11. H. Kai and D. Li, Trusted cloud computing with secure resources and data coloring, *IEEE Internet Computing*, 14(5), 14–22, 2010.
12. L. Guanfeng, Y. Wang, and M. Orgun, Trust inference in complex trust-oriented social networks, in *International Conference on Computational Science and Engineering*, pp. 996–1001, 2009.
13. X. Wang, T. T. Kwon, Y. Choi, H. Wang, and J. Liu, Cloud-assisted adaptive video streaming and social-aware video prefetching for mobile users, *IEEE Wireless Communications*, 20(3), 72–79, 2013.
14. J. Greer, B. Melvin, and J. W. Ngo, Personal emergency preparedness plan (PEPP) Facebook app: Using cloud computing, mobile technology, and social networking services to decompress traditional channels of communication during emergencies and disasters, in *IEEE Ninth International Conference on Services Computing*, pp. 494–498, 2012.
15. I. Alexandre, A. Lăscăteu, and N. Ţăpuş, CAMEO: Enabling social networks for massively multiplayer online games through continuous analytics and cloud computing, in *Ninth Annual Workshop on Network and Systems Support for Games*, pp. 1–6, 2010.
16. A. Nugraha, S. H. Supangkat, and D. Nugroho, Goesmart: Social media education in cloud computing, in *International Conference on Cloud Computing and Social Networking*, pp. 1–6, 2012.
17. http://en.wikipedia.org/wiki/Online analytical processing.
18. S. Caton, C. Dukat, T. Grenz, C. Haas, M. Pfadenhauer, and C. Weinhardt, Foundations of trust: Contextualising trust in social clouds, in *Second International Conference on Cloud and Green Computing*, pp. 424–429, 2012.
19. S. Caton, C. Haas, K. Chard, K. Bubendorfer, and O. F. Rana, A social compute cloud: Allocating and sharing infrastructure resources via social networks, *IEEE Transactions on Services Computing*, 3, 359–372, 2014.
20. J. Bethencourt, A. Sahai, and B. Waters, Ciphertext-policy attribute-based encryption, in *IEEE Symposium on Security and Privacy*, Berkeley, CA, pp. 321–334, 2007.

9

Privacy and Security in Mobile Cloud Computing

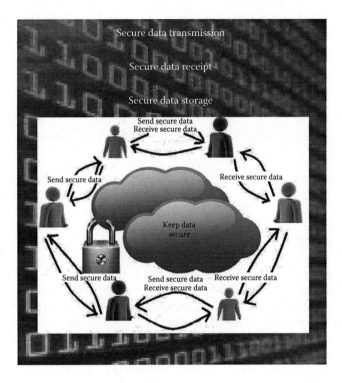

ABSTRACT   In mobile cloud computing, data are stored in the cloud. As both data access and storage are done outside the mobile device and into a remote cloud server, the user has no personal control over it. Therefore, security and privacy are the major challenges for the network security managers. With the rapid growth of social media and mobile web users (e.g., smartphone or tablet users), malicious threats have also increased. In this chapter, various aspects of security issues and solutions are analyzed in the field of mobile cloud computing.

KEY WORDS: *security, privacy, authentication, intrusion.*

9.1 Introduction

In this twenty-first century, mobile devices and wireless communication technologies are gaining popularity. Now, handheld devices such as smartphones, personal digital assistants (PDAs), and tablets are used to perform a number of tasks which were previously done by PCs. Such devices are resource-limited and do not provide the same efficiency and results as PCs. On the other hand, cloud has unlimited resources and works on "pay-as-you-consume" mode. A "cloud" is an elastic execution environment of resources involving multiple stakeholders and providing a metered service at multiple granularities for a specified level of quality of service. Cloud computing is a method of computing in which dynamically scalable and virtualized resources are provided as services over the Internet. Cloud computing consists of a number of servers and switches running in a physical layer to provide various cloud services, more commonly known as infrastructure as a service (IaaS), software as a service (SaaS), platform as a service (PaaS), data storage as a service (DaaS), communication as a service (CaaS), hardware as a service (HaaS), business as a service (BaaS), and security as a service (SecaaS) [2]. The cloud service providers (CSPs) are responsible for running, maintaining, managing, and upgrading cloud hardware to meet the increasing requirement of users. Mobile cloud computing (MCC) incorporates cloud computing into the mobile environment. MCC is a recent development in the field of mobile networks. Mobile computing is a term used to describe technologies that enable people to access network services anyplace, anytime, and anywhere. MCC refers to "an infrastructure where both data storage and the processing happen outside of the mobile device. Mobile cloud applications move the computing power and data storage away from mobile phones and into the cloud, bringing application and mobile computing to not just smartphone users but a much broader range of mobile subscribers" [1]. MCC provides simple and easy infrastructure for mobile applications and services. It enables users to flexibly utilize resources on demand, takes full advantage of cloud computing, and brings new types of services and facilities to users by providing ubiquitous service access. This integration of two different technologies, namely cloud computing and mobile networks, faces many technical challenges, such as low bandwidth, availability, heterogeneity, computing offloading, data accessing, security, privacy, and trust [1]. This chapter is mainly focused on privacy and security issues in MCC.

There has been a drastic increase in the use of smartphones in recent years. According to ABI Research predictions, the number of mobile cloud users will grow from 42.8 million (1.1% of total mobile users) in 2008 to 998 million (19% of total mobile users) in 2014 [2]. The security of smartphones is becoming increasingly important because they offer advanced services such as web browsing, instant massing, e-commerce, as well as personal and economical information storage. The proliferation of mobile malware increased by 46% in 2010 compared to 2009 [3]; 74% of chief information officers and IT executives are not willing to adopt cloud services because of the risks associated with security and privacy [2].

9.2 Security Needed in Different Levels for Securing Mobile Cloud Computing

In the mobile cloud environment, the user employs his or her mobile devices, such as smartphones, tablets, PDAs, and so on, to store data in the cloud with the use of communication channels. Gharehchopogh et al. [4] discuss security of mobile device data using cloud, with

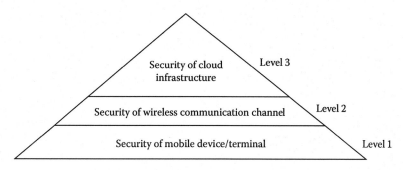

FIGURE 9.1
Security levels in mobile cloud computing.

the main focus on securing the cloud environment. In Ko et al. [5], security issues are divided into two categories, namely mobile network user's security and cloud security. In [6–8], the risk and issues related to mobile cloud computing are divided into three levels: mobile device/terminal, mobile network or wireless channel, and the cloud. Hence, we can say that the security and privacy risk should be deeply analyzed in three aspects:

1. Security of mobile device/terminal
2. Security of wireless communication channel
3. Security of cloud infrastructure

Three different levels of security are needed, as shown in Figure 9.1.

9.2.1 Level 1: Security Issues in Mobile Devices

In Level 1, the focus is on security and risk issues related to mobile devices. In [4–8], the main concern is the security of handheld devices. These devices have open operating systems, third-party applications, and wireless access to the Internet anywhere, anytime. Today, mobile phones are getting smarter with the advancement of hardware performance, technology, and communication bandwidth (3G, 4G, WiMax, etc), not only providing voice calls or Internet access but also applications and services that are possible by PCs and laptops. Thus, smartphones are now used as an enterprise tool in businesses, which helps increase the productivity of employees by allowing them to interact with their customers, partners, and colleagues. As smartphones are capable of supporting services and applications equal to PCs and desktops, it is vulnerable to the threats and risks as in PCs and desktops. Security issues in mobile devices with examples are given in Table 9.1. They mainly include malware, worms, Trojan horse, vulnerable applications, and OS. Even data can be compromised if the device is stolen, lost, or tampered with.

TABLE 9.1

Security Issues in Mobile Devices with Examples

Security Levels	Security Issues	Examples
Level 1: Mobile devices/ terminal	Information-stealing malwares, spam, phishing, data loss from lost or stolen devices, data leakage from poorly-written applications, vulnerabilities in hardware or OS, unsecured Bluetooth or Wi-Fi	Zimto and NickspyTrojans are information-stealing malwares, fake websites, digital wallet hacking, unwanted message from unknown vendors.

9.2.1.1 Approaches to Mitigate Security Issues Related to Mobile Devices

There are different approaches to lessen the security issues related to mobile devices. Anti-malware programs are run on the devices to identify and delete Trojan horse, viruses, and worms. Periodically updating the OS and downloading applications from known vendors such as Google, Apple, and Microsoft can also help. Also, unexplained links should not be tried, receiving data transmission from strange phones should be avoided, new, unauthorized software should not be installed, and the interface of Bluetooth, Wi-Fi, and so on, should be shut down. If a device is stolen or lost, there must be some remote data wiping technique so that the data cannot be misused. The mobile device can also use hardware-based encryption techniques for internal and external memory support.

9.2.2 Level 2: Security Issues in Communication Channels

Level 2 deals with issues related to securing the wireless communication channel [6–8] between the mobile and cloud servers. Mobile devices access their resources and services through communication channels from cloud servers. This increases the number of WAP (wireless application protocol) gateways and IMS (IP multimedia subsystem) equipment in the IP network, giving rise to many new security threats in the mobile Internet. These mobile terminals access phone service, short message services (SMS), and other Internet services using 3G, Wi-Fi, WiMax, and Bluetooth. Such broad access methods result in an increase in security risks associated with networks, causing information leakage and malicious attacks. A number of attacks have been identified in the communication between wireless mobile devices and the cloud environment. When mobile devices communicate with the cloud, they are more vulnerable to communication threats. With rapid increase in cloud usage, there is an increase in the number of security issues related to the communication channel. Even the free Wi-Fi connections in public places (e.g., airport, cafés) are prone to attacks because of the weak encryption techniques used in Wi-Fi. The attacker can break wireless interface and steal sensitive data. Illegal terminals can access the network with fake ID and carry out malicious activities. Security issues in the communication channel with examples are given in Table 9.2. There are a number of attacks identified in communication channels such as access control attacks, confidential attacks, integrity attacks, authentication attacks, and availability attacks.

9.2.2.1 Approaches to Mitigate Security Issues Related to Communication Channel

For protecting data from leakage while being transmitted to servers, a number of approaches are available. The mobile users mainly encrypt data while transmitting into cloud so that an adversary cannot understand or even be able to get the data. Secure transmission protocols such as https and SSL can be used to transfer data; even VPN (virtual private network) can

TABLE 9.2

Security Issues in Communication Channel with Examples

Security Levels	Security Issues	Examples
Level 2: Communication channel/mobile network	Access control attacks, confidentiality attacks, integrity attacks, authentication attacks, availability attacks.	War driving, rough APs, MAC spoofing, WEP cracking, Man-In-The-Middle attack, Evil Twin, AP phishing, frame injection, reply attacks, guessing, VPN login cracking, LEAP cracking, DoS, Beacon flood, etc.

be used. Socket programming is also used for secure transmission of sensitive data in a cloud environment. Also, public key encryption is used for protecting Man-In-The-Middle (MITM) attacks. Strong password and biometric authentication should be used to enhance data security during transmission. Even the rough access points at public places should be avoided for security reasons. Switching off the wireless interfaces, such as Wi-Fi and Bluetooth, after using the mobile device will also help.

9.2.3 Level 3: Security Issues in Cloud Computing

Level 3 contains the most important issues in mobile cloud computing and prevents a large number of mobile users from using cloud services [4–8]. The users offload their data to the cloud and lose control over those data. An increasing combination of smartphones with cloud infrastructure raises the possibility of threat attacks in the cloud. Cloud computing is based on virtualization technology, and if there is some vulnerability in the virtualization software, the data of one user on the same physical server can be leaked to that of other users. There is also a need for proper access control and data management according to the needs of the consumer. Security issues of cloud infrastructure with examples are given in Table 9.3. In [4,8], isolation of data from other users where the user data are stored and cloud server flexibility are discussed. In Ko et al. [5], the integrity of offloaded data, authentication, and digital rights management (DRM) are dealt with as the main issues related to the cloud environment. In Hui et al. [6], user data and privacy protection, platform reliability from insider and outsider attacks, and access control are the main issues discussed. A number of cloud security issues, such as attacks on virtual machines, availability and single-point failure, phishing, authorization and authentication, and security management in hybrid cloud, are evaluated by Morshed et al. [7].

9.2.3.1 Approaches to Mitigate Security Issues in Cloud Infrastructure

In [4–8], the authors provide different techniques and mechanisms for the protection of data in the cloud. To increase the trust of customers for storing data in the cloud server, it should provide privacy, authentication, confidentiality, and availability of services. These security mechanisms must be strong enough to handle attacks by adversaries and hackers. If a user subscribes to a cloud service, then the CSPs should provide the location of the stored data. The location should be free from geopolitical issues. Thus, the privacy and security of data should be maintained. In the cloud, security is promised using current security technologies such as VPN, access control, encryption, and other such means. There must be a mechanism to recover the user's data if the data is lost or erased by an attacker. There should also be a secure and efficient key management mechanism for the cloud environment. Cloud should use an implicit authentication technique to reduce the risk of fraud in a mobile cloud.

TABLE 9.3

Security Issues in Cloud Infrastructure with Examples

Security Levels	Security Issues	Examples
Level 3: Cloud environment	Integrity, digital rights management, virtual machine attacks, phishing, authentication and authorization attacks, platform level attacks	Data and application integrity, pirating and illegal distribution of digital contents, side channel attacks, SQL injection, etc.

9.3 Security Issues in Mobile Cloud Environment

The integration of mobile devices with cloud infrastructure has given rise to a number of security issues. This includes authentication, authorization, data security, application security and integrity, privacy, digital right management, and so on. We will discuss all these issues with their existing schemes.

9.3.1 Application Security

A number of applications are run on mobile devices. These applications are mainly used for managing personal information and for business needs. These applications include Internet chatting, electronic mails, games, schedulers, and so on. Mobile applications now use cloud services hosted on cloud servers. Providing security of such mobile cloud application is of great concern. Most of these applications are downloaded from Google store, Apple store, Nokia store, or a third-party application store. Such stores do not have any scheme to remove malware from applications. According to Dinh et al. [1], 10 billion applications were downloaded from the Android market in 2010, and 250,000 applications contained malware. Good applications are modified using malicious codes and are spread through unofficial repositories. Such applications leak private data, dial premium numbers, and are backdoor-triggered via SMS, for example. So, there must be some schemes that should take care of the security of such applications and the threat related to mobile cloud applications. In the next section, we discuss some schemes used for applications security and the related threats.

9.3.1.1 Existing Schemes for Application Security

An application-specific firewall was proposed, called WallDroid, in Kilinc et al. [9]. WallDroid is mainly a firewall for Android applications with extra functions. The key component used for providing security in this architecture is VPN technology and the cloud-to-device messaging (C2DM) framework for Android. In this framework, cloud is used to keep track of millions of applications with their reputation and to compare the traffic with the list of known malicious IP servers. Every application has its unique ID, which is a combination of a certificate and a hash value. Android applications are classified into three categories according to their reputation: The Good, The Bad, and The Unknown. Well-known applications are The Good applications, while known to be malicious applications are those that are not known to be The Good and The Bad.

If the application is good, it is directly connected, while if it is a known bad application, the Internet connection is blocked. For unknown applications, VPN service is used for Internet connection via a VPN server, as shown in Figure 9.2. The VPN then monitors the data traffic to see whether it is malicious or not and whether it is sending personal data in clear text or not. If the VPN server determines it as malicious, then it blocks its traffic. The earlier framework is proposed to determine malicious and unknown Android applications [9]. A secure web referral service is proposed by Xu et al. [10], which uses a secure search engine (SSE) for mobile devices to protect the mobile website against phishing and SSL strip-based MITM attack. In this, a cloud-based virtual computing is used for providing each user a VM as personal sec-proxy to analyze the web traffic. In VM, the SSE uses web crawling to check a valid IP address and certification chain. A phishing filter is also used for checking URLs with optimized execution time.

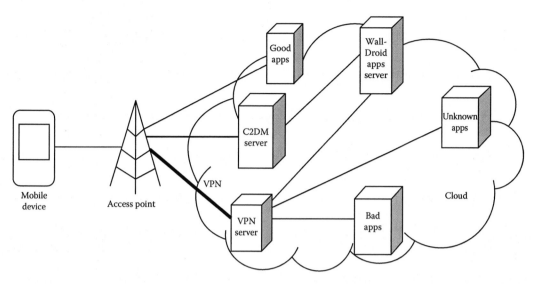

FIGURE 9.2
Architecture of WallDroid.

The components of the SSE service model include URL services, SSL verifier, phishing filter, SSE crawler, SSE services, DNS services, and storage services. Mainly SSL verifiers and phishing filters are used to provide secure web browsing.

The processing of SSE is shown in Figure 9.3. SSL verification is used to counter MITM attack, while a phishing attack is countered by the phishing filter. The performance of SSE

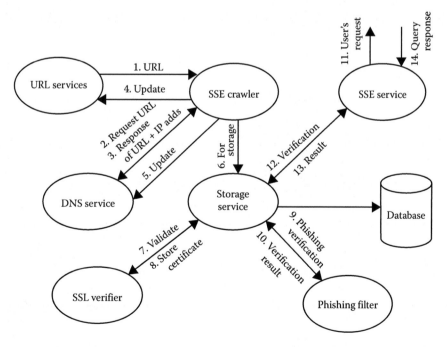

FIGURE 9.3
Working principle of secure search engine.

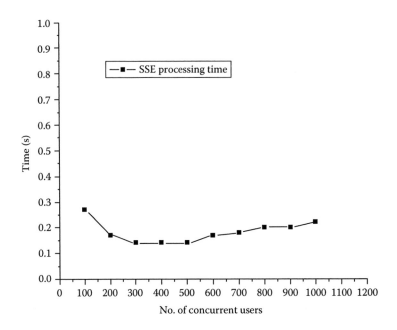

FIGURE 9.4
Processing time taken by SSE services.

is calculated for responding to a valid request. In this experiment, the x-axis represents number of concurrent users, and y-axis represents time taken to process (in seconds). The experiment is carried out with an empty cache and 100 users. The cache will be filled with inspected websites. Figure 9.4 shows that, at the beginning, the processing time is high but it drops as the time increases due to caching.

A framework is proposed that ensures the security of component-based applications. This model secures the data transmission between the component of the same application at the installation on the mobile device and when being updated [11]. Different security schemes are used for different types of data employed by the applications according to mobile device's energy consumption. It also ensures confidentiality and integrity of the application's components. This framework is also called secure mobile cloud (SMC) and includes five managers for securing mobile cloud applications. The managers are the mobile manager, the mobile and cloud security manager, the optimization manager, the application manager, and the policy manager. The mobile manager collects data and events on mobile devices and sends them to the appropriate manager. Security managers take care of composite security of mobile devices and the cloud. The optimizer manager collects and sends the sensor's data. The application manager checks the integrity of application at setup, while the policy manager determines which security component is required for different security levels. In this framework, application integrity is verified at installation and at updating. Integrity is ensured by checking the existence of applications in stores such as Amazon, Apple, Google store, and so on. Low-energy consuming and component based security architecture for mobiles or LECCSAM [8] is a flexible security scheme that allows terminal users to specify the extension of security properties that they will prefer to integrate with the data using HTTPS.

9.3.2 Authentication Issues

The migration of private and enterprise data to the cloud raises security and privacy issues. To access these sensitive data only by the legitimate users, an authentication protocol is used. Traditionally, a user provides his or her password to the requested server for authentication, which may be attacked. In mobile cloud, legal user authentication becomes an important issue. In Figure 9.5, a simple example of the authentication process is shown.

In the following section, different authentication schemes are proposed to authenticate users in cloud using their mobile devices.

9.3.2.1 Existing Authentication Schemes

An authentication scheme that does not need to enter password, user name, biometric data, and so on, for authentication is proposed in Chow et al. [12]. This framework simply utilizes TrustCube for authentication and generates a score according to user's behavior. The generated probabilistic authentication score is then compared with a threshold value to determine whether the client is authentic or not. The authentication score is not fixed, and can be varied for different applications. This scheme consists of four modules: client devices, authentication consumer, authentication engine, and data aggregator, as shown in Figure 9.6. The client device generates the noticeable context and action such as Internet browsing history, call records, location history, MMS, SMS, phone information, and so on.

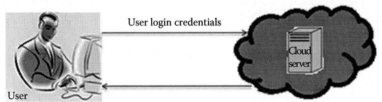

FIGURE 9.5
Simple example of authentication process.

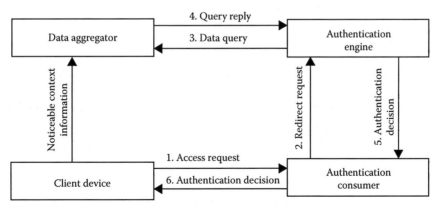

FIGURE 9.6
Authentication architecture.

The data generated by the client device are stored in the local cache until collected by the data aggregator. The authentication engine extracts noticeable context information from the data aggregator and the authentication policies from the authentication consumer to authenticate the mobile device. The authentication policies depend on the client's request. Finally, the authentication engine responds to the client according to the data provided to it through the authentication consumer.

A next-generation authentication scheme for mobile and CE devices, which uses a zero knowledge proof (ZKP) technique for authentication ID, is proposed in Grzonkowski et al. [13]. This scheme is anti-phishing and does not reveal the user's password to the visiting website. The user is not redirected to other web pages after login. This scheme is called SeDiCi 2.0, which consists of three entities: client (C), services (S), and authentication services (AS). Figure 9.7 shows the data flow between the three entities. Client creates an account in AS using his or her password in client application to generate the public key.

The client registers to the service. Then the service verifies the client and records its login detail. The client login and public key are send to AS. To authenticate to service, the client again visits to service and gains Auth_ID from service as response. Auth_ID and URI are then sent to AS, and AS verifies the URI corresponding to Auth_ID. Now the URI is exposed with the given Auth_ID, then the client sends login to service, and service verifies client using Auth_ID. If the verification is successful, the client is authenticated.

An advance protocol for authentication based on biometric encryption is proposed in Zhao et al. [14]. This can be used for future authentication when the mobile devices are equipped with biometric sensors, as shown in Figure 9.8.

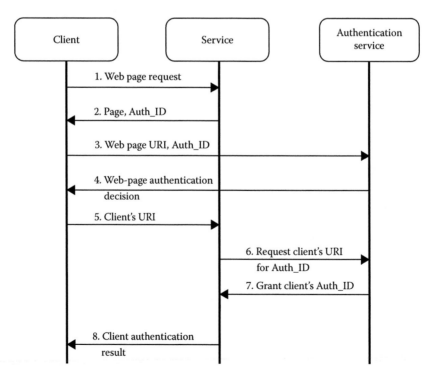

FIGURE 9.7
Dataflow of SeDiCi 2.0 for authentication.

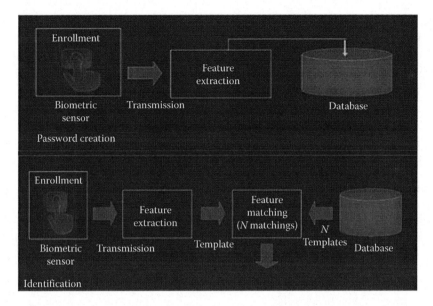

FIGURE 9.8
Biometric encryption-based authentication scheme.

This scheme is more reliable than the traditional password-based schemes because biometric data is difficult to forget, forge, share, or lose. User has his record of biometric features stored in the cloud database.

9.3.3 Data Security

In this section, we will mostly focus on the security of mobile device data that is offloaded to the cloud storage. Mobile devices contain private, commercial, financial, and enterprise data. Leakage of such sensitive data to others may lead to personal and economic loss. More threats are involved when such important data are offloaded to the cloud. The user loses his control over off-premises data. So, such offloaded data must be kept safe and confidential. There must be some mechanism to know data integrity and data to be available when needed. The user also should be aware of where his data are located. Some schemes are discussed in the next section, which focuses on data security in mobile cloud computing.

9.3.3.1 *Existing Schemes for Data Security*

Mobile device data security is discussed in [19–27], with focus on integrity and confidentiality. Different cryptographic algorithms are used to protect the data from adversaries. These algorithms include incremental cryptography, attribute-based encryption, digital signature, identity-based encryption, message authentication code, and hashing functions. Some authors have also considered the mobile's resource limitation while implementing their schemes.

An incremental cryptography-based trusted computing is used to provide integrity to mobile device files (F). This system consists of three elements: mobile client, cloud service provider (CSP), and trusted third party (TTP) [15]. CSP provides the management, operation, and allotment of cloud resources and services. TTP manages the configuration and installation of a secure coprocessor on the remote cloud. It distributes the secret key K_S to the associated mobile client and generates a message authentication code (MAC) on behalf of the mobile clients. It has three phases: the initialization phase, the data updating phase, and the integrity verification phase. In the initialization phase, the mobile device data is prepared with incremental authentication. For every file block, an incremental MAC_{FX} is created using K_S, as given in Equation 9.1. The file block is then transferred to the cloud.

$$MAC_{FX} = \sum_{i=1}^{k} HMAC(F_i, K_S) \tag{9.1}$$

The data updating phase consists of three main operations: creation of file blocks, insertion of file blocks, and deletion of file blocks. If the file is available in the cloud, then it sends a copy to the mobile device as well as to the trusted crypto coprocessor (TCC). TCC generates the MAC'_{FX} using K_S and sends it to the mobile client. The integrity is checked by comparing MAC_{FX} with MAC'_{FX}. If both are equal, the file is inserted at B block at the ith position of the file, and again incremental MAC is generated using old MAC and K_S. The block deletion operation is same as block insertion at the ith position of the file. The difference in deletion is that MAC updating depends on the deleted block and old MAC. The main integrity mechanism is offloaded to TCP, which saves the processing power of mobile devices. The coprocessor generates the MAC according to the request and sends it back to the mobile device. The MAC of the coprocessor and MAC of mobile device are compared and, if they are equal, the integrity is successfully verified, as given in Equation 9.2:

$$MAC(Mob) = MAC(TCP) \tag{9.2}$$

A scheme for authentication of mobile user and the integrity of mobile device data is discussed in Hsueh et al. [16]. In this scheme, standard encryption algorithm, hash function, digital signature, random number, and secret value are used to provide overall mobile device data security when offloaded to the cloud. SSL is used for secure access, and access lists (ACLs) are also available for individuals and the group. This framework consists of mainly four modules: mobile device (MD), cloud service provider (CSP), certification authority (CA), and telecommunication module (TM), as shown in Figure 9.9.

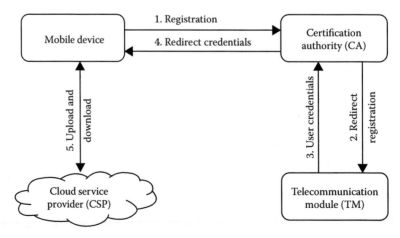

FIGURE 9.9
Architecture of secure data storage in cloud.

In this model, the certifying authority is responsible for the authentication of mobile devices. The telecommunication module generates and keeps records of the mobile device's passwords and other information used to access cloud services.

In this framework, it is assumed that the secure key (*SK*), public key (*PK*), and session key (*SEK*) are distributed securely among the mobile devices, telecommunication model, and CA. To access the services of the cloud, the mobile user has to register onto the cloud through CA. After successful registration, TM generates a password (*PWD*) for mobile devices to use the cloud resources. This is given in Equation 9.3.

$$MD \rightarrow CA: E_{PKTM}E(MU, Num, TK), U_N, S_{SKMU}(MU, Num), H(MU, Num) \tag{9.3}$$

where

MU represents mobile user's name

Num represents mobile user's number

TK is combination of *Num* and cloud *PWD* random

U_N is the random number generated for identity proof

H is the standard hash function, EPKTM represents encryption with the *PK* of TM

S_{SKMU} generates a signature for mobile user using a cryptographic function on the passed value and *SK* of the mobile device

When the message is received at CA, it authenticates the user with the received signature. If the user is a valid one, the following message is sent to TM, as given in Equation 9.4.

$$CA \rightarrow TM: E_{PKTM}(MU, Num, TK), U_N, S_{SKCA}(H(MU, Num)) \tag{9.4}$$

The TM authenticates the CA using the S_{SKCA} key. If the CA is authenticated, the TM registers the mobile user and saves the mobile user's information in the local database. The data is used for future verification. The TM generates *PWD* for the mobile device and encrypts it with the mobile device's *PK* for secure transmission. *PWD* is again encrypted with TK to ensure that only an authorized user can decrypt it on receiving. TM forwards the secure information to the mobile device through CA, as given in Equations 9.5 and 9.6.

$$TM \rightarrow CA: E_{PKMU}(MU, Num, U_N, E_{TK}(PWD)) \tag{9.5}$$

$$CA \rightarrow MD: E_{PKMU}(MU, Num, U_N, E_{TK}(PWD)) \tag{9.6}$$

Now, the mobile device encrypts the file with SEK and uploads the file along with *PWD*, *MU*, and S_{SKMU} on the cloud as given in Equation 9.7.

$$MD \rightarrow C: PWD, MU, E_{SEK}(Data), S_{SKMU}(H(MU \| SV \| E_{SEK}(Data))) \tag{9.7}$$

where *SV* represents the secret value generated by the mobile device and is known to *MD*, CA, and TM. To upload a file to the cloud, *MD* has to send *PWD*, *MU*, and *H(MU∥SV)*. Cloud regenerates the hash value using *MU* and *SV*, and then compares the result with the received signature for authentication. Then, the cloud sends the encrypted file to *MD* along with a signature as given in Equation 9.8.

$$C \rightarrow MD: E_{SEK}(Data), H(E_{SEK}(Data) \| SV) \tag{9.8}$$

The mobile device receives the signature and decrypts the file using *SEK*.

Secure data processing is achieved through trust management and private data isolation. In this model, identity-based cryptography and attribute-based data access control are used for trust management [17]. It ensures security and privacy for mobile devices with the help of multi-tenant secure data management and trust management, and extended semi-shadow image (ESSI)-based data processing model. This architecture consists of three domains: cloud public service and storage domain, cloud trusted domain, cloud mobile and sensing domain. Cloud service and storage domain provides SaaS, while the cloud trusted domain manages the certificate distribution, key distribution, and identity management. In attribute-based identity management, publicly known attributes are used for private key generation for secure communication. This identity can be used as a signature to authenticate the user. ESSI is also called clone of mobile devices running in the cloud trusted domain. ESSI increases the storage and processing power of the mobile devices. It also provides security and privacy to the data and information of the device. Secure policies and rules are used in the cloud trusted domain with the help of a distributed firewall, which is used to check the incoming and outgoing packet for the malware.

The data management system is divided into two groups: critical and normal data. Critical data are encrypted with the user-generated key, while the normal data are encrypted using the cloud-generated key. The incoming data received by ESSI are classified as normal or critical. If the data are identified as critical data, then they are passed through the encryption, decryption, and verification (EDV) module and stored in the secure storage of ESSI. The masking procedure preserves the privacy of the data depending on user preference. This gives scalability, protection to critical data, computation distribution, and resistance to single-point failure.

In Table 9.4, three data security schemes are compared according to their encryption methods and security features, and also their drawbacks.

In [18–24], secure storage of mobile device data to cloud is discussed with consideration of mobile device resource constraints. The mobile device creates a file and processes it and finally uploads it to single cloud or multiple clouds. Three schemes are proposed to achieve security of mobile device data: encryption-based scheme (EnS), code-based scheme (CoS), and sharing-based scheme (ShS) [18]. The link between the mobile device and cloud is secured using media access protocols such as SSL, IPSec, and so on. This

TABLE 9.4

Comparison of Different Data Security Schemes

Encryption Scheme/Method/ Principle	Security Features			Other Features	Drawbacks
	Integrity	Confidentiality	Authentication		
Incremental cryptography (MAC) [15]	Yes	No	No	Energy efficient	Single-point failure, less scalable
Standard hash function, digital signature [16]	Yes	Yes	Yes	Simple and easy to implement	Less energy efficient and scalable
Attribute-based encryption and identity based cryptography [17]	No	Yes	Yes	Resistance to single-point failure, scalable, energy efficient	Mobile device is compromised if somehow ESSI is attacked and manipulated

framework provides data confidentiality and data integrity. Each of the three schemes deals with secure data uploading to the cloud and secure data downloading from the cloud to mobile devices.

A user must have the password, the encryption key (*EK*), and the integration key (*IK*) to upload and download files, as given by Equations 9.9 and 9.10:

$$EK = H(PWD \| FN \| FS) \tag{9.9}$$

$$IK = H(FN \| PWD \| FS) \tag{9.10}$$

In EnS, file *F* is encrypted using *EK* and produces *F′*, and message authentication code (MAC) is generated by using the hash function to *IK* and file for integrity check, as given in Equations 9.11 and 9.12.

$$F' = E(F, EK) \tag{9.11}$$

$$MAC = H(F, IK) \tag{9.12}$$

The mobile device (*MD*) sends concatenation of the encrypted file *F′* and hashing of File name (FN), and sends it to cloud storage (CS), as given in Equation 9.13:

$$F' \| H(FN) \| MAC \tag{9.13}$$

MD stores only FN and deletes *EK* and *IK*.

To download the file, *MD* sends hash FN, and CS searches the file using FN and sends the corresponding file to *MD*. PWD is needed to get *EK*, *IK*, and decrypt *F′* to get *F*, as shown in Equation 9.14:

$$F = D(F', EK) \tag{9.14}$$

MD generates MAC of decrypted file F using *IK* to check the integrity.

A scheme incremental cryptography refers to the improvement of existing schemes [18], i.e. encryption-based scheme, coding-based scheme, and sharing-based scheme. This scheme improves block insertion, deletion, and modification operation in terms of resource utilization of a mobile device [19]. This scheme also requires considerable energy at the initial stage but improves file modification in terms of turnaround time.

In this scheme, the file is divided into d blocks of n bits; hence, FS satisfies Equation 9.15:

$$FS\%d = 0 \tag{9.15}$$

Each file is encoded separately, and finally the encrypted file (EF) is generated by performing the concatenation operation, as given in Equations 9.16 and 9.17:

$$C_j = E_{EK}(F_j) \tag{9.16}$$

where $1 \leq j \leq d$

$$EF = C_1 \| C_2 \| C_3 \| ..... \| C_d \tag{9.17}$$

MAC is generated for each block, and the final MAC is generated by concatenation of all MACs, given in Equations 9.18 and 9.19:

$$MAC_{Fj} = HMAC_{IK}(F_j) \tag{9.18}$$

$$MAC = HMAC_{IK}(MAC_{F1} \| MAC_{F2} \| ..... \| MAC_{Fd}) \tag{9.19}$$

MD only keeps FN and d, while uploads H (FN), blocks MAC, and final MAC to the cloud. *MD* deletes original file, *F*, *IK*, *EK*.

For downloading the file *F*, *MD* sends H (FN) to the cloud, and the cloud searches for the FN and sends the file *F* to *MD*. *MD* divides *F* into d blocks, and each block is decrypted using *EK*. Finally, each decrypted block is concatenated to produce *F*, as given in Equations 9.20 and 9.21:

$$F_j = DEK_j(C_j), \quad 1 \leq j \leq d \tag{9.20}$$

$$F = F_1 \| F_2 \| ... \| F_d \tag{9.21}$$

Integrity verification is done by comparing the new MAC calculated using decrypted *F* and the old MAC sent by cloud.

The incremental version of the other two (i.e., coding-based and sharing-based) schemes works in the same way to produce the block MAC and final MAC. The incremental version of EnS, CoS, and ShS consumes more resources because of the extra MAC and the final MAC generation. But the overall turnaround time balances this extra computation while block insertion, deletion, and modification are done.

In Ren et al. [18], some modifications are suggested when data is transmitted between *MD* and cloud storage. Attackers may also monitor the link and recover the file. It is also impossible for the *MD* to guarantee secure storage of data on the cloud. Hence, schemes suffer from MITM attacks.

All three schemes, namely EnS, CoS, and ShS, are improved with respect to security and are called the secure encryption-based protocol (SEnP), secure coding-based protocol (SCoP), and secure sharing-based protocol (SShP) [20]. In SShP, file *A* equal to the size of file *F* is generated by *MD* to participate in XOR for privacy of data. Digital signature and public key encryption are used for data integrity. Acknowledgment is sent back to *MD* from cloud to confirm secure storage. A random number *N* is used for acknowledgement.

MD uploads file *F*, MAC, digital signature, and *N* to the cloud, as given in Equation 9.22:

$$F'[j] \| \{\{H(H(FN+j) \| MAC \| HF'[j] \| N)_{PR_{MD}}\}_{PU_{CSj}}) \tag{9.22}$$

where
  PR is a private key
  PU is the public key of *MD* and *CS*

CS computes *N* from the digital signature and sends it back to *MD* using its private key and *MD*'s public key, as given in Equation 9.23:

$$\{\{N\}_{PR_{CSj}}\}_{PU_{MD}} \tag{9.23}$$

MD decrypts using its private key and compares the value of the received *N* with the stored *N*. If the values of *N* are equal, then secure storage of data is verified.

The other two schemes follow the same procedure for upload of data using SCoP scheme; from *MD* to cloud is given by Equation 9.24:

$$F'[j] \parallel \{\{H(FN + j) \parallel MAC \parallel H[F'[j]) \parallel N\}_{PR_{MD}}\}_{PU_{CSj}} \tag{9.24}$$

Uploading of data through SEnP scheme from *MD* to cloud is given by Equation 9.25:

$$F' \parallel \{\{H(FN) \parallel MAC \parallel H(F') \parallel N)_{PR_{MD}}\}_{PU_{CSj}}\} \tag{9.25}$$

Downloading is done in same way as discussed in Zhou et al. and Shin et al. [21,22], respectively.

A scheme with the core idea of outsourcing data in the cloud provides trust-based security management in mobile cloud and also provides data confidentiality and fine-grained access control. This framework uses identity-based proxy-re-encryption (PRE) technique to provide security to the mobile cloud [23]. This mechanism protects the data from the untrusted cloud as the data is encrypted and also provides the benefits of cloud storage. The integrity is achieved based on MAC and public signature-based scheme. It consists of three entities: data owner, data server, and data sharer, as shown in Figure 9.10.

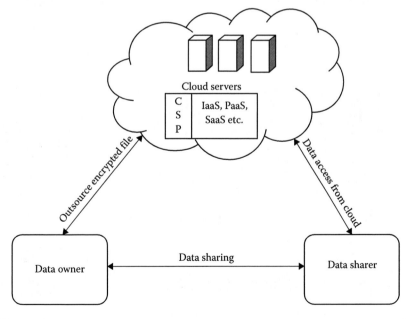

FIGURE 9.10
Data storing and sharing on cloud.

Data owner and data sharer use mobile devices to access cloud services through the Internet. Data owner forwards encrypted data on the cloud server and allows authorized data sharers to access data and decrypt the file. In this protocol, each user has its unique identity and secret key. The cloud stores data, which are delegated to the ciphertext encrypted with data owner's ID to requester ID.

A trusted platform module (TPM) used for all encryption key and key sharing among legal users for data protection is proposed in Shin et al. [22]. TPM function is developed in the secure domain of ARM TrustZone because most of ARM-based mobile devices are not equipped with the TPM chip. This DFCloud framework defines TPM-based secure channel setup, TPM-based key management, remote client attestation, and a secure key sharing protocol across multiple users/devices. In this scheme, client- and server-side encryptions are used. SSL is used for securing data transmission between the client and the server. DFCloud architecture consists of three components: client, DFCloud server, and commodity cloud storage services, as shown in Figure 9.11. TPM emulator is used instead of hardware TPM and provides the same security as hardware components do.

The cloud storage services in the client secure side consist of three subprocesses to maintain security. They are attestation components, key management components, and file handlers. The TrustZone monitor provides context switching between secure and normal worlds. The server acts as a proxy between the client and cloud storage. It consists of a file metadata store, which contains owner information and data block location, and provides services of file uploading and downloading with the use of the metadata store.

There are different protocols used for secure cloud storage services, such as user login, remote attestation, key creation, key sharing, and data uploading and downloading. The user employs his ID and password to read store files by a retrieving a key that is generated only after passing attestation services. Remote attestation protocol (RAP) is used for login and attestation. Platform configuration registers (PCRs) with the PCR number are sent to client TPM as a result of login and attestation. After login, key creation and management are done by use of PCR value, which contains the attestation value H. After passing the attestation test, the sealed EKA is unsealed and used. Key sharing is done between

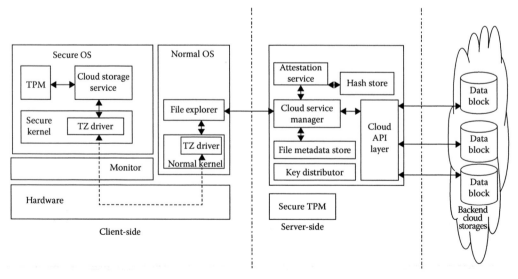

FIGURE 9.11
DFCloud architecture.

device A and device B. In key-sharing, the server distributes the nonce value to A and B. In this, asymmetric encryption is used for sharing the keys between TPM of device A with TPM of device B. A encrypt key EKA uses the public key of B and sends it to B. Then, B decrypts it using its private key. The sealed key is then stored in persistent storage.

Data uploading and downloading are done through file explorer of the client side and cloud service manager of the server side. File handler performs encryption or decryption using keys. Then file handler copies the files to local storage and a download session is established between file explorer and cloud service manager. File explorer sends a download request to server, and cloud service manager searches the metadata of the file stored. After searching in the store, the server sends the location of cloud storage to file explorer. Encrypted data is then decrypted using keys and stored locally. Uploading is the reverse of downloading.

Performance evaluation of files is done on the basis of three categories: without encryption, with encryption in normal world, and with encryption in secure world. Figure 9.12a shows the file uploading time, while Figure 9.12b shows the file downloading time of different sizes. Secure world can only perform cryptographic operation to 512 bytes at a time.

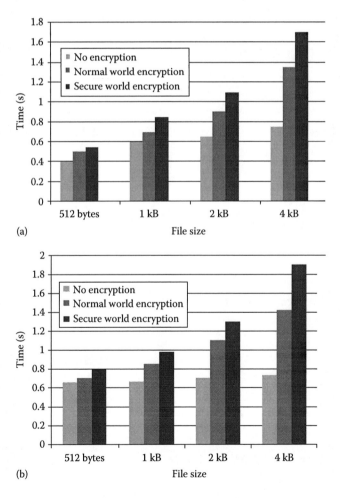

FIGURE 9.12
File upload/download performance. (a) File uploading. (b) File downloading.

Time increases with the increase of file size due to the context switching between the secure world and normal world. Time also increases with the cryptographic operation with increase of file size.

9.3.4 Digital Rights Management

Digital contents such as e-books, images, audios, videos, and so on, are now stored in cloud. Mobile users can access these contents using the Internet from the cloud servers. These digital contents could be pirated and distributed illegally. Digital rights management checks this abuse by regulating content usage. The DRM system allows only authorized users who have the license to access such contents.

9.3.4.1 Existing DRM Scheme

A cloud-based SIM DRM (CS-DRM) scheme for the mobile cloud environment has four main entities: a SIM card, a DRM agent, a custom player, and a CS-DRM-compliant browser [24], as shown in Figure 9.13.

The SIM card is used to provide subscriber identity, to authenticate between cloud clients, and also to verify the integrity of the license. The DRM agent is used to communicate between cloud clients and to implement logical rules. Custom player is used to

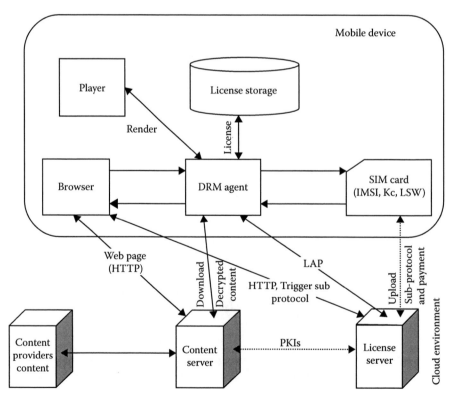

FIGURE 9.13
Architecture of digital right management.

play digital contents that cannot be illegally distributed. The CS-DRM browser is used to browse the website of the backend and also to notify the DRM agent of the next action according to a response or event. This scheme consists of five phases: preparation, rights management, license acquisition, play, and download/upload. The preparation phase initializes the backend, generates the keys for symmetric encryption, transfers content ID to licensed server, and so on. The rights management phase customizes digital contents, while the license acquisition phase acquires the license of that digital content from the license server. After getting the license of the digital content, the user decrypts the content to play. In the download phase, the user can change the device and download the license to his or her new device to enjoy the digital contents. Upload phase is used to guarantee the integrity of the digital content. Implementation of the CS-DRM scheme called phosphor, which shows that this scheme is efficient, secure, and practicable, is also discussed.

9.3.5 Intrusion Detection

Mobile devices are now called smartphones because they are used not only for phone calls but also for browsing, reading news, watching videos, and many more things, as done by PCs. With the increase in software complexities of smartphones, the number of bugs and exploitation of vulnerabilities also have increased. Smartphones use the same software architecture as PCs, and they are attacked by same class of viruses, worms, malware, and Trojan horses. These malware programs get into the smartphones through unofficial repositories and carry out their malicious activities. The compromised device can send the user's private data and call logs, premium messages, financial transactions, location, and so on. To prevent these attacks, the best way is to install an antivirus such as AVG-Mobilation, Kaspersky, or Avastetc on the smartphones for intrusion detection.

9.3.5.1 Drawbacks of Intrusion Detection

Intrusion detection has a number of drawbacks such as storage requirements, the need for larger CPUs, and battery consumption. Since the antivirus software is based on loading signatures that require storage, and as they run on mobile instruments, they draw a large amount of CPU and battery power. Smartphones have limited resources, so such solutions are not efficient for these devices. Hence, some lightweight intrusion detection schemes are proposed by researchers, which are based on the integration of the mobile device and cloud computing.

9.3.5.2 Existing Schemes for Intrusion Detection

There are two approaches used for detecting malware in mobile devices. The first is the static approach, which uses an antivirus installed on the smartphones with the signature of malwares database looking for suspicious patterns. Second is the dynamic approach, in which behavior-based detection is involved and runs in a controlled and isolated environment for detecting any malicious codes. In this chapter, we discuss mostly the dynamic approach for intrusion detection of mobile devices using cloud servers.

9.3.5.2.1 *Crowdroid*

Behavior-based malware detection for Android phones, called Crowdroid, is a lightweight client application that is downloaded from Google store and installed in the device [25]. Crowdroid is used to send all the preprocessed system calls to a central server. Data acquisition, data manipulation, and malware analysis and detection are the three components of this architecture. The data manipulator parses the received data and creates a system call vector for each interaction; hence, a dataset is prepared according to its behavior. These datasets are then clustered for determining benign application or malware application with the help of a behavior-based malware detection server. Experiments show that clustering the results enables successful detection of self-written and real malware. The full architecture is given in Figure 9.14.

A malware detection scheme using a lightweight host agent that runs on mobile devices and an off-device network service that runs on cloud and receives files from the agent for detection have been proposed [26]. Host agents trap the file, divert it to the handler routine, and generate a unique ID by hashing the file. This ID is then compared with the cached ID,

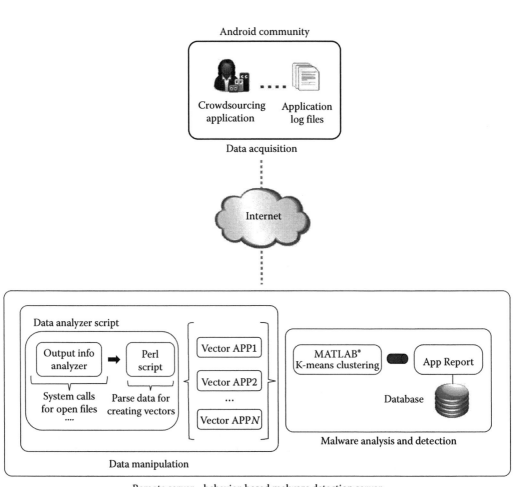

FIGURE 9.14
Crowdroid architecture.

and if the ID is not present, it is sent to the cloud for analysis. Network service receives the file sent by the host agent and determines whether the file is malicious or not. The network service consists of a number of antivirus engines running in parallel on a virtual machine on the cloud server. This scheme provides SMS spam filtering, phishing detection, and a centralized blacklist to provide security to mobile devices.

9.3.5.2.2 Paranoid Android

Paranoid Android (PA) is a versatile security, which offers a wide range of security measures for smartphones [27]. A prototype is also implemented. Called PA for intrusion detection, it is scalable and flexible. In this scheme, a synchronized replica of the phone runs on the cloud security server. The tracer on the phone mainly records the data between the kernel and user space employing system calls and replays on the replica. The full architecture is given in Figure 9.15.

9.3.5.2.2.1 Architecture

1. The tracer records all information needed to accurately replay its execution.
2. The replayer receives the trace and faithfully replays the execution within the emulator.
3. The proxy intercepts and temporarily stores inbound traffic, and the replayer can access the proxy to retrieve the data needed for replaying.

The recorded information is transmitted to the replica using the proxy server. The cloud replays the tracer information on the concurrent replica, and if attack is detected, PA needs to warn the user about the threat. When the user receives the notification of an attack, it starts a recovery process using the data stored in the replica and brings the device back to a safe or clean state.

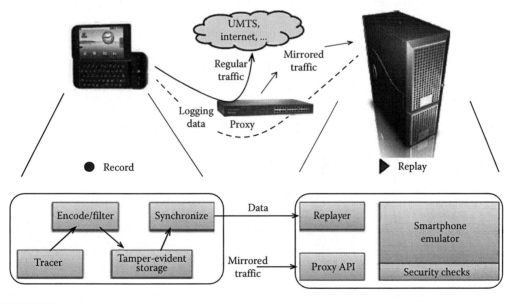

FIGURE 9.15
Paranoid Android architecture.

A cloud-based intrusion detection system for smartphones has been proposed in which a response engine continuously performs in-depth and forensic analysis of smartphones to detect any misbehavior [28]. The response engine will take appropriate action if any misbehavior is detected. In this scheme, the smartphone must be registered with the intrusion detection system. After registration, the smartphone has to provide device information such as the company name, model number, operating system, software, applications, and so on. The emulator, after receiving all the information about the device, creates a replica of the same smartphone on the cloud. The cloud provides a lightweight client agent to be installed on the device for proxy setting. The proxy server duplicates the incoming and outgoing traffic. The data are then provided to the emulated platform for detection. The client agent gathers all input and sensed data of the device, sends them to emulated replica, and waits for the reply. When any misbehavior is detected, the client is notified and appropriate action is taken to bring back the system to the normal condition. In this, a protocol is prepared for the intrusion analysis engine for Linux kernel, which uses two sources of information, namely a set of intrusion detection systems and system call tables.

9.3.5.2.3 *Secloud*

Zonouz et al. [29] extended the work done in Houmansadr et al. [28], and a prototype called Secloud was proposed and implemented as a comprehensive security solution for Android smartphones. Secloud is a powerful intrusion detection scheme that uses the cloud to protect smartphones. Secloud uses fewer resources of the mobile device and provides real-time security as a service to the smartphone, as shown in Figure 9.16. Secloud is kept synchronized to the actual device, and if any intrusion is detected, the emulated replica sends a notification to take appropriate action to the threat. The practical deployment of Secloud takes care of the file consistency by hashing the folders, and only those hash values, that are not present in the device to save resources, are sent. User privacy, encryption, and alternative ways for notification are used to make the system robust.

The encryption method is used by Secloud to encrypt and send credential data to replica, where these data are decrypted and analyzed. Experiments show that Secloud accurately detects known and unknown threats. It is also efficient in CPU and memory utilization.

In Table 9.5, different existing intrusion detection schemes are compared according to the type of intrusion detection, platforms, prototypes, and features.

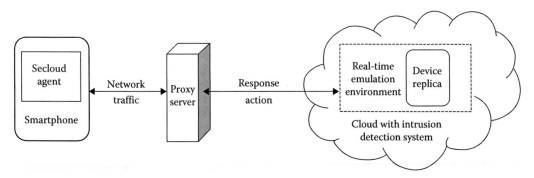

FIGURE 9.16
Secloud architecture for intrusion detection.

TABLE 9.5

Comparison of Different Intrusion Detection Systems

Authors	Intrusion Detection Approach	Platform	Prototype	Other Features
Burguera et al. [25]	Dynamic	Android OS	Crowdroid	Lightweight, scalable
Oberheide et al. [26]	Dynamic	Android OS	CloudAV (Mobile version)	Less complex, energy efficient, scalable
Portokalidis et al. [27]	Dynamic	Android OS	Paranoid Android	Highly scalable and flexible, energy efficient
Houmansadr et al. [28]	Dynamic	Android OS	Secloud with minimum function	Energy efficient, scalable
Zonouz et al. [29]	Dynamic	Android OS	Secloud with full function	Powerful, energy efficient, secure

9.4 Conclusion

Cloud computing is an emerging field of wireless network, and integrating it with handheld devices solves numerous issues such as storage, battery power, and computational processes, but raises a number of security threats both in the cloud and in the devices. These threats are mostly a combination of mobile device threats, communication threats, and cloud environment threats. Different types of framework related to security issues in mobile cloud computing have been explained in this chapter. The issues were related to ensuring privacy, authentication, security, trust, and so on, to data and applications that are offloaded to the cloud from mobile devices. Also discussed were how the mobile user authenticates them to cloud, how location-based service privacy is achieved, and how the real-time intrusion detection is achieved. Different comparison tables were used to analyze the features and drawbacks of the proposed schemes. It was observed that most of the frameworks offload processor-intensive computation jobs to the cloud because of the resource limitation of mobile devices. New security threats, which are due to lack of isolation among various virtual machine instances running on the same physical server, need to be handled. To provide security in the MCC environment, the cloud provider should ensure data security, network security, data locality, data integrity, application security, data segregation, data access, data breach issues, and various other factors.

Questions

1. Comment on security and privacy of mobile cloud computing.
2. Draw and explain the security models of mobile cloud computing.
3. Compare the different intrusion detection systems used in MCC.
4. What are the security issues in the mobile cloud computing environment?
5. What is digital right management in MCC?
6. What are the approaches to mitigate security issues related to mobile devices?

References

1. H. T. Dinh, C. Lee, D. Niyato, and P. Wang, A survey of mobile cloud computing: Architecture, applications, and approaches, *Wireless Communications and Mobile Computing*, 13(18), 1587–1611, 2013.
2. A. N. Khan, M. L. M. Kiah, S. U. Khan, and S. A. Madani, Towards secure mobile cloud computing: A survey, *Future Generation Computer Systems*, 29(5), 1278–1299, 2013.
3. X. Lin, Survey on cloud based mobile security and a new framework for improvement, in *IEEE International Conference on Information and Automation*, Shenzhen, China, pp. 710–715, 2011.
4. F. S. Gharehchopogh, R. Rezaei, and I. Maleki, Mobile cloud computing: Security challenges for threats reduction, *International Journal of Scientific and Engineering Research*, 4(3), 8–14, 2013.
5. S. K. V. Ko, J. H. Lee, and S. W. Kim, Mobile cloud computing security considerations, *Journal of Security Engineering*, 9(2), 143–150, 2012.
6. S. Hui, Z. Liu, J. Wan, and K. Zhou, Security and privacy in mobile cloud computing, in *Ninth International Wireless Communications and Mobile Computing Conference*, Sardinia, Italy, pp. 655–659, 2013.
7. M. S. Morshed, M. M. Islam, M. K. Huq, M. S. Hossain, and M. A. Basher, Integration of wireless hand-held devices with the cloud architecture: Security and privacy issues, in *International Conference on Parallel, Grid, Cloud and Internet Computing*, Barcelona, Spain, pp. 83–88, 2011.
8. S. Resondry, K. Boudaoud, M. Kamel, Y. Bertrand, and M. Riveill. An alternative version of HTTPS to provide nonrepudiation security property: A flexible component-based approach for secured transactions in a mobile environment, in *Wireless Communications and Mobile Computing Conference (IWCMC), 2014 International*, August, Nicosia, Cyprus, pp. 536–541, 2014.
9. C. Kilinc, T. Booth, and K. Andersson, WallDroid: Cloud assisted virtualized application specific firewalls for the Android OS, in *IEEE 11th International Conference on Trust, Security and Privacy in Computing and Communications*, Liverpool, England, pp. 877–883, 2012.
10. L. Xu, L. Li, V. Nagarajan, D. Huang, and W. T. Tsai, Secure web referral services for mobile cloud computing, in *IEEE Seventh International Symposium on Service Oriented System Engineering*, Redwood City, CA, pp. 584–593, 2013.
11. D. Popa, M. Cremene, M. Borda, and K. Boudaoud, A security framework for mobile cloud applications, in *11th RoEduNet International Conference*, Sinaia, Romania, pp. 1–4, 2013.
12. R. Chow, M. Jakobsson, R. Masuoka, J. Molina, Y. Niu, E. Shi, and Z. Song, Authentication in the clouds: A framework and its application to mobile users, in *Proceedings of the ACM Workshop on Cloud Computing Security Workshop*, New York, pp. 1–6, 2010.
13. S. Grzonkowski, P. M. Corcoran, and T. Coughlin, Security analysis of authentication protocols for next-generation mobile and CE cloud services, in *IEEE International Conference on Consumer Electronics-Berlin*, Berlin, Germany, pp. 83–87, 2011.
14. K. Zhao, H. Jin, D. Zou, G. Chen, and W. Dai, Feasibility of deploying biometric encryption in mobile cloud computing, in *Eighth China Grid Annual Conference*, Changchun, China, pp. 28–33, 2013.
15. W. Itani, A. Kayssi, and A. Chehab, Energy-efficient incremental integrity for securing storage in mobile cloud computing, in *International Conference on Energy Aware Computing*, Cairo, Egypt, pp. 1–2, 2010.
16. S. C. Hsueh, J. Y. Lin, and M. Y. Lin, Secure cloud storage for convenient data archive of smart phones, in *IEEE 15th International Symposium on Consumer Electronics*, Singapore, pp. 156–161, 2011.
17. D. Huang, Z. Zhou, L. Xu, T. Xing, and Y. Zhong, Secure data processing framework for mobile cloud computing, in *IEEE Conference on Computer Communications Workshops*, Shanghai, China, pp. 614–618, 2011.

18. W. Ren, L. Yu, R. Gao, and F. Xiong, Lightweight and compromise resilient storage outsourcing with distributed secure accessibility in mobile cloud computing, *Tsinghua Science and Technology*, 16(5), 520–528, 2011.
19. A. N. Khan, M. M. Kiah, S. U. Khan, S. A. Madani, and A. R. Khan, A study of incremental cryptography for security schemes in mobile cloud computing environments, in *IEEE Symposium on Wireless Technology and Applications*, Kuching, Malaysia, pp. 62–67, 2013.
20. X. Liu, R. Jiang, and H. Kong, SSOP: Secure storage outsourcing protocols in mobile cloud computing, in *IEEE 14th International Conference on Communication Technology*, Chengdu, China, pp. 678–683, 2012.
21. Z. Zhou and D. Huang, Efficient and secure data storage operations for mobile cloud computing, in *Proceedings of the Eighth International Conference on Network and Service Management*, Laxenburg, Austria, pp. 37–45, 2012.
22. J. Shin, Y. Kim, W. Park, and C. Park, DFCloud: A TPM-based secure data access control method of cloud storage in mobile devices, in *IEEE Fourth International Conference on Cloud Computing Technology and Science*, Taipei, China, pp. 551–556, 2012.
23. W. Jia, H. Zhu, Z. Cao, L. Wei, and X. Lin, SDSM: A secure data service mechanism in mobile cloud computing, in *IEEE Conference on Computer Communications Workshops*, Shanghai, China, pp. 1060–1065, 2011.
24. C. Wang, P. ZouPeng, Z. Liu, and J. Wang, CS-DRM: A cloud-based SIM DRM scheme for mobile internet, *EURASIP Journal on Wireless Communications and Networking*, 14, 1–19, 2011.
25. I. Burguera, U. Zurutuza, and S. Nadjm-Tehrani, Crowdroid: Behavior-based malware detection system for Android, in *Proceedings of the First ACM Workshop on Security and Privacy in Smartphones and Mobile Devices*, New York, pp. 15–26, 2011.
26. J. Oberheide, K. Veeraraghavan, E. Cooke, J. Flinn, and F. Jahanian, Virtualized in-cloud security services for mobile devices, in *Proceedings of the First Workshop on Virtualization in Mobile Computing*, New York, pp. 31–35, 2008.
27. G. Portokalidis, P. Homburg, K. Anagnostakis, and H. Bos, Paranoid Android: Versatile protection for smartphones, in *Proceedings of the 26th Annual Computer Security Applications Conference*, New York, pp. 347–356, 2010.
28. A. Houmansadr, S. A. Zonouz, and R. Berthier, A cloud-based intrusion detection and response system for mobile phones, in *IEEE/IFIP 41st International Conference on Dependable Systems and Networks Workshops*, Hong Kong, China, pp. 31–32, 2011.
29. S. Zonouz, A. Houmansadr, R. Berthier, N. Borisov, and W. Sanders, Secloud: A cloud-based comprehensive and lightweight security solution for smartphones, *Computers and Security*, 37, 215–227, 2013.

10

Trust in Mobile Cloud Computing

ABSTRACT Trust of an object refers to the goodness, honesty, reliability, and faithfulness of it. Trust a system that works according to our expectation. In mobile cloud computing, trust is a vital parameter because in this case personal data storage and data processing occur remotely within the long-distance cloud. In this chapter, we have discussed different types of trust in mobile cloud computing. We have also discussed how the trust of the entire system can be increased with the removal of malicious users.

KEY WORDS: *cloud, mobile cloud computing, trust, QoS.*

10.1 Introduction

Cloud computing is an emerging computing paradigm. It offers substantial cost saving in the IT budget. According to the National Institute of Standards and Technology [1], cloud computing is a model for enabling convenient, on-demand network access to a shared pool of configurable computing resources (e.g., networks, servers, storage, applications, and services) that can be rapidly provisioned and released with minimal management effort or service provider interaction.

Cloud computing has some good features such as low investment, easy maintenance, flexibility, fast deployment, reliable service, availability, scalability, pay-per-use model, elasticity, broad network platforms, and multi-tenancy. There are three kinds of cloud service models: software as a service (SaaS), platform as a service (PaaS), and infrastructure as a service (IaaS).

Mobile cloud computing (MCC) is the integration of cloud computing with mobile devices and mobile networks. So, MCC offers mobility over cloud computing. It uses the cloud computing technology through mobile devices. Basically, MCC refers to an infrastructure where both data storage and data processing operations happen outside the mobile device and in the cloud servers. Although cloud computing as well as MCC seems very attractive, it is still not welcome in most places. One of the main reasons behind this is the issue of trust. Trust is a challenge in cloud computing, MCC, sensor network, MANET, and e-commerce-related areas.

According to dictionaries, trust means firm belief in the reliability, goodness, honesty, effectiveness, faithfulness, or ability of someone or something. Trust provides confidence and reliance in something that is expected to behave or deliver as promised. We trust a system that we know thoroughly and that behaves according to our expectation. We trust a system less if it gives us insufficient information about its services and with which it is not secure to work.

The issue of trust is one of the biggest obstacles for the development of cloud computing. In 2010, Fujitsu Research Institute found in a survey that 88% of potential cloud consumers were worried about the security of their data [2]. Cloud consumers are always worried about who has access to their data in the cloud servers. They demand more awareness of what goes on in the backend physical server. Such surveys demonstrate the urgency for practitioners and researchers to quickly remove the obstacles to trust. Trustworthiness between cloud service providers (CSPs) and cloud consumers is very important for the wide acceptability of MCC. CSPs must evaluate and encourage trustworthy customers as well as remove malicious customers from the system. This will lead them to provide reliable and effective services. Cloud consumers should also follow trustworthy CSPs and avoid unethical CSPs. To gain consumer trust, CSPs must offer better transparency, more consumer control of data and processes, and clearly stated security provisions. A trustworthy relationship is a must between the CSPs and the cloud consumers for the effective worldwide deployment of MCC. If mobile users do not trust the CSPs and keep themselves away from availing cloud computing services, the entire MCC technology will become useless.

10.2 Trust Properties

Some important properties of trust are as follows [3,4]:

- Trust is field specific. It has different attributes in different application areas. So, the meaning of trust should be confined within a few specific fields.
- In mathematical sense, trust is not symmetric. If entity X trusts entity Y, it does not mean that entity Y will also trust entity X.
- Trust is not transitive. This means, if entity X trusts entity Y and entity Y trusts entity Z, it does not guarantee that entity X trusts entity Z.
- Trust may change dynamically. For example, a cloud consumer's degree of trust on a specific CSP is dynamically adjusted with the change in the CSP's performance and quality of service.
- Trust is a probabilistic value about an entity, and it is generally within the range of real numbers between 0 and 1. In initial condition, trust value is 0.5.
- Trust is multidimensional. An example may be when we calculate the trust value of a transaction entity in e-commerce. Here, the trust value may be evaluated using the trust attributes such as product quality, product price, product delivery speed, etc.
- Trust is personal belief in some entity. Different people may come from different backgrounds, and they may have different criteria to judge a single entity. So, the confidence level may vary from person to person on the same system or entity.

10.3 Components of Trust

We may classify the components of trust in the following categories, as shown in Figure 10.1 [2].

10.3.1 Security

Security makes it difficult or uneconomical for an unauthorized person or hacker to access confidential information or resources. An example is encryption technique.

10.3.2 Privacy

Privacy is every individual's fundamental right. Privacy techniques protect against leakage or exposure of personal data.

FIGURE 10.1
Components of trust.

10.3.3 Auditability

Auditability is the ability to evaluate an organization, system, process, project, or product. Auditing means a systematic examination of operations, statements, data, records, and performances of a system or an enterprise for a specific purpose [5]. In the auditing process, the auditor collects evidence, evaluates it, and makes a judgment, which is communicated through the audit report. Auditing is done by an independent body.

10.3.4 Accountability

Accountability guarantees that all operations carried out by individuals, systems, or processes can be uniquely identified and that the trace to the author and the operation is kept [6]. It is also called traceability.

10.4 Types of Trust

Broadly, there are three types of trust [7]: direct trust, indirect trust, and hybrid trust.

10.4.1 Direct Trust

If an entity trusts another through direct association or direct communication, this type of trust is called direct trust. In Figure 10.2, entities X and Y are in direct interaction.

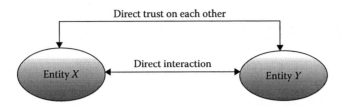

FIGURE 10.2
Direct trust.

10.4.2 Indirect Trust

An entity can indirectly trust another based on the recommendations of others. If two entities are not directly associated and know each other through a third entity, this type of trust may occur. Indirect trust is required when two entities have no prior interaction. Hence, the trust value will be calculated based on observation and recommendation. In Figure 10.3, node Z has indirect trust on Y based on the recommendation of the trusted entity X.

10.4.3 Hybrid Trust

Hybrid trust is computed based on both direct experience and recommendation information, as shown in Figure 10.4. Here, entity Z directly trusts Y and X. Moreover, X may have recommended Z to trust Y. So, Z may also indirectly trust Y. At the end, we can say that Z has hybrid trust on Y.

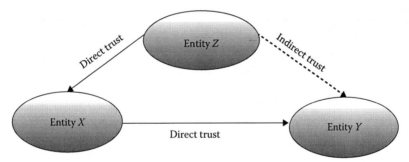

FIGURE 10.3
Indirect trust based on recommendation.

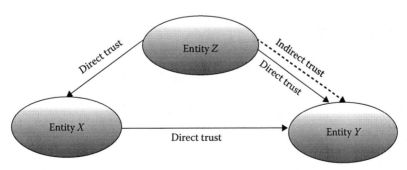

FIGURE 10.4
Hybrid trust from Z to Y.

10.5 Trust Issues

Trust is hard to develop but easy to lose. A single mistake can destroy years of trustworthiness. Reputation and brand image are all important parameters of trust. In general, the level of security and cryptography increases the trustworthiness of a communication system. The following are the issues related to trust [8].

10.5.1 Weak Chain of Trust

Subcontractors may be present between the CSP and the customer. These subcontractors are needed for the quick delivery of services to the customer. They may even contract others, whose identity, reputation, and trustworthiness cannot always be checked. A reason may be the deadline or time boundary. However, throughout this chain of subcontractors, proper and legal agreements may not be followed. Customers may not know all the subcontractors or even they may be totally unaware of the service chain. This produces a weak chain of trust.

10.5.2 Lack of Control and Visibility

Customers do not know who will process their personal information and whether their data will be adequately protected in the cloud. This lack of control and visibility produces distrust, and ultimately customers will keep themselves outside the cloud domain, especially when sensitive information is involved.

10.6 Ways of Trust Establishment

Some approaches help consumers in identifying trustworthy or dependable CSPs. We may classify these approaches as follows [9].

10.6.1 Service-Level Agreement

Service-level agreements (SLAs) are a way to establish trust on CSPs. SLAs state what exactly the CSP can provide, what the features are, what security mechanism is applied, which assurances are actually met, etc. In an MCC environment, users monitor violations of SLA and can apply for compensation to the CSP. Unfortunately, SLAs have been standardized recently, and they are far from implementation. CSPs take advantage of this situation. They make SLAs in such a way that they can deprive the customer of getting compensation.

10.6.2 Audit

Many audit standards are available, and different CSPs use different audit standards, for example, FISMA, SAS70 II, ISO 27001, etc. This assures cloud consumers about their offered services. For example, the audit SAS 70 II covers the operational performance of the system and relies on a specific set of goals. Audit reports are not alone sufficient to alleviate users' concern. Moreover, most of the CSPs do not want to share their audit reports. This leads to lack of transparency.

10.6.3 Measuring and Rating

Recently, a new cloud marketplace named Spot Cloud has been launched. It provides a platform where cloud consumers can choose among potential providers in terms of location, quality, and cost. This supports cloud consumers in identifying dependable CSPs. The CSPs are rated based on a questionnaire filled by the current cloud consumers. There is a provision for combining technical measurements with consumer feedback for comparing and assessing the trustworthiness of CSPs in the future.

10.6.4 Self-Assessment Questionnaire

The Cloud Security Alliance has provided a questionnaire for ensuring security features of CSPs, called the Consensus Assessments Initiative Questionnaire (CAIQ) [10,11]. The CAIQ provides ways to assess the capabilities and competencies of CSPs in terms of different features such as information security, governance, and compliance. The CAIQ is filled up by CSPs. It is basically a set of questions a cloud consumer may wish to ask a CSP. The questions are about the security implementation by CSPs in their IaaS, PaaS, and SaaS delivery models. However, the CAIQ evaluation strategy is not yet standardized. The evaluation is necessary for comparing potential CSPs, through which we can be assure that the services offered by the CSPs comply with industry-accepted security standards, regulations, and audits.

10.6.5 Trust and Reputation Model

The approaches mentioned in Sections 10.6.1 through 10.6.4 are time consuming and cumbersome. Moreover, these trends lack a unified approach where all these criteria can be combined and evaluated to support customers in selecting the most effective and dependable CSP.

To assist customers in understanding the differences and selecting the most trustworthy CSP, trust and reputation models represent a promising and essential basis. These models have some parameters known as QoS+ parameters to support customers in selecting the most appropriate cloud providers before actually interacting with them. These parameters should be measured and analyzed properly according to their importance.

10.6.5.1 QoS+ Parameters for Trust and Reputation Models

The following are the standard parameters used in trust and reputation models [9,12]:

- *Service-level agreement*: SLAs are done between CSPs and cloud consumers.
- *Compliance or accreditation or certification*: CSPs use different audit standards to prove themselves.
- *Portability*: Portability means the ability to run in different platforms and operating systems. CSPs should provide services to all platforms.
- *Geographical location*: CSPs provide information about the geographical location of their data centers.
- *Customer support*: Generally, CSPs provide information regarding customer support in the SLA.

- *Performance*: Performance-related data regarding cloud providers can be found through service monitoring technologies. Performance consists of availability, elasticity, latency, bandwidth, and reliability.
- *Federated identity management*: This type of information is found through SLAs.
- *Security measures*: Cloud consumers are always worried about the security of their data, and CSPs should provide the relevant information. Security measures consist of cryptographic algorithms, key management, physical security support, data security support, and network security.
- *User feedback*: User feedback, recommendation, publicly available reviews, etc., are very important in cloud marketplace. Feedback regarding CSPs can be given as a whole or as the basis of individual criteria.
- *Service deployment and delivery models*: The deployment models (e.g., private, public, and hybrid clouds) and service-delivery models (e.g., IaaS, PaaS, SaaS) used by CSPs are also very important.

10.6.5.2 Promising Trust and Reputation Models

There are many existing trust and reputation models. Some promising models are eBay, RFSN, Beta Reputation, Tidal Trust, Buchegger's model, Epinions, Certain Trust, Hang's model, BNTM, Unitec, Abawajy's model, TESM, FIRE, Grid Eigen Trust, Eigen Trust, social REGRET, and Billhardt's model [9].

10.7 Trust Evaluation

The trust model is mainly concerned with trust representation, trust measurement, and trust evaluation. Trust evaluation is the core of the trust model. Trust can be evaluated in the ways described in the following sections [9].

10.7.1 Black Box Approach

In the black box approach, the trustworthiness of an entity is evaluated by considering only the observed output; no knowledge about the internal architecture of the system or service is required. For example, the evaluation can be made by considering only the user feedback. A trust and reputation model that follows this approach to evaluate trustworthiness considers CSP as a black box.

10.7.2 Inside-Out Approach

In the inside-out approach, the trustworthiness of an entity is evaluated by considering the internal architecture of the system and trustworthiness of its subsystems.

10.7.3 Outside-In Approach

The outside-in approach combines the black box and inside-out approaches. In this approach, the trustworthiness of an entity is evaluated based on the knowledge of the internal architecture of its components as well as the observed behavior of the overall service.

10.8 Detailed Study of Various Aspects of Trust in MCC

Various aspects of trust in MCC are essential for CSPs, cloud consumers, and other related entities. The trustworthy frameworks in MCC are very important as they form the essential basis for the secure and trustworthy operation of MCC. Both CSPs and cloud consumers are benefitted through these frameworks. A brief study has been conducted on these and is discussed in this section.

10.8.1 User Behavior Trust

Cloud providers offer their services to cloud users. Users may be authentic or malicious. Authentic users are interested in doing their job with the service provided by the CSPs. Malicious users, who might be competitors, always try to harm the CSPs. They might run an application that may take maximum resources. Malicious codes may interfere with other genuine users. So, CSPs must evaluate the trustworthiness of users by analyzing user behavior. Traditional user identification and authentication are not enough. User behavior trust evaluation allows CSPs to provide services more reliably to genuine users. CSPs should eliminate malicious users from their domain by evaluating user behavior trust for the sake of their own performance. The details of user behavior trust have been provided by Li-qin et al. [13].

10.8.1.1 Evaluation Principles of User Behavior Trust

Some evaluation principles of user behavior trust are as follows:

- Trust value attenuates over time. Old or out-of-date user behavior record or trust value has the natural decreasing property in evaluation.
- Recent behavior plays an important role in trust evaluation. While abnormal behavior has a great influence in trust evaluation, conventional behaviors are less significant.
- The user behavior trust evaluation is based on a large number of historical behaviors of the user.
- The trust value should increase slowly. It is very risky if we assign a high trust value to a user who has accessed the cloud only a few times.
- At the time of finding cheating behavior, reduced trust value has greater importance than that gradually increased. It can prompt the user to reduce fraud.

10.8.1.2 Decomposition of User Behavior Trust

User behavior trust can be subdivided into four categories:

1. *Contract behavior sub-trust*: Contract behavior sub-trust tells us whether the user behavior complies with the legal agreement between CSPs and cloud consumers. Setting up a proxy server or excessive downloads may not be according to SLAs.
2. *Security behavior sub-trust*: Security behavior sub-trust implies whether the user has the tendency to attack the CSP infrastructure. An attack on cloud resources or hacking other user's information belongs to this category.

3. *Identity reauthentication sub-trust*: Identity reauthentication sub-trust refers to authenticating the user again. The user may employ a mobile device for using cloud resources, and these resources are easy to lose. If the device is lost and the CSP identifies some abnormal activity such as excessive downloads, authenticating the user again is the central theme of the identity reauthentication sub-trust.

4. *Expense behavior sub-trust*: Expense behavior sub-trust means whether the user is using the cloud resources optimally. Suppose the user has finished working and is still occupying the resources. This will lead to denial of service to other users. So, the CSP must monitor the expense behavior sub-trust and reward good users.

All the user behavior sub-trusts are further divided into behavior evidences. Figure 10.5 shows the hierarchy of user behavior trust evaluation.

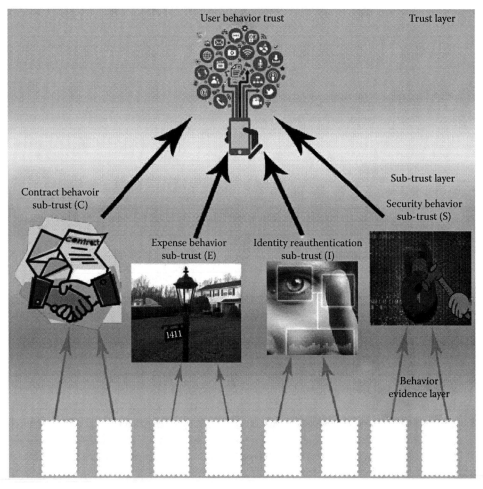

FIGURE 10.5
Hierarchy of user behavior trust evaluation.

10.8.1.3 Acquirement of Behavior Trust Evidence

True and reliable trust evidence can be obtained by the following methods:

- Network flow detection tool is a good choice.
- Invasion detection system can get trust evidences such as transfer delay, time of access, rate of buffer overflow, and time of operation failure.
- Log and audit trails such as system log, application log, auditing record, and network management log are another option.
- Data collecting tools provide real-time traffic monitoring. An example is Cisco's Net Flow Monitor.
- Cisco Works Software is among the providers of network management software.
- Using specialized hardware, such as NetScout, can provide evidence directly.
- Security products, such as firewalls, access control system, and so forth, produce valuable information.

10.8.1.4 User Behavior Trust Evaluation Methods

When users access the cloud, we can get behavior evidence after analyzing the access pattern. Trust evaluation is a hierarchical process, in which trust is subdivided into behavior evidences. The main difficulty in the evaluation process is assigning weights to each subpart of the evaluation hierarchy.

10.8.1.4.1 AHP-Based Evaluation

The evaluation strategy based on the analytic hierarchy process (AHP) is for each access. AHP simplifies the problem and depends on the evaluator's knowledge and experience.

10.8.1.4.2 FAHP-Based Evaluation

The AHP result is an exact real number, but in practical situations, the evaluation may not be like that. So, another evaluation method named fuzzy analytic hierarchy process (FAHP), based on fuzzy numbers, has been developed, and it makes the evaluation result more real. FAHP-based evaluation strategy is also for each access.

10.8.1.4.3 FANP-Based Evaluation

In AHP, the lower layers affect the upper layers in the user behavior evaluation hierarchy, but the reverse does not happen. In actual scenario, it is possible that the upper layers are also affecting lower layers. So, we use another method named analytic network process (ANP), which greatly depends on experts' knowledge. Hence, another evaluation method named fuzzy analytic network process (FANP) has been developed, which combines the advantages of ANP and FAHP. The FANP strategy is for each access.

10.8.1.4.4 Double-Sliding, Window-Based Evaluation

The double-sliding, window-based evaluation strategy is for long access. A sliding window is used here to evaluate behavior trust. The trust value is mainly related to time and the window's size. Original evidences are retained in this process. When a node does not exchange information with others for a long time, its overall trust value is decreased. In this method, recent and abnormal behavior has greater importance in evaluation.

10.8.2 Trustworthy Mobile Sensing Framework

A trustworthy mobile sensing framework is very essential. Nowadays, smartphones are embedded with mobile sensors. With the mobile sensing technology, it is possible to get information about smartphone users in each and every moment. The sensor information includes call log, running apps, browsing history, SMS log, battery status, GPS, screen on/off state, camera, cell tower, ID location, WLAN, contacts, microphone, Bluetooth, to name a few. Data consumers need all this information for the statistical analysis of human behavior. The analysis data are necessary to develop new products or create human behavior models by different organizations, for example, health organizations, insurance companies, and the like. Some behavior patterns may be mobile phone usage patterns, calling patterns, user mobility patterns, and so on. These types of statistical analyses are required in application development, such as health behavior monitoring, future activity prediction, trend forecasting, psychological status estimation, anomaly detection, and authentication through behavior modeling. But getting user data causes privacy issues. Mobile users may not be comfortable with sharing their personal all-day data with a trusted or untrusted third party, the data consumer, who will do the actual data analysis. This framework has been discussed in detail by Zhangy et al. [14].

To overcome the privacy problem in this framework, the data collected by the mobile sensor are directly sent to the private cloud. The private cloud is implemented in the user's place, and the user has full physical control over the cloud server. So, there are no issues of privacy or stolen data. Widgets can be found in the widget market. The widget market is a place where all data widgets are kept for viewing and installing by the user. When releasing a widget on the widget market, the developer must clarify, for example, "what kind of data is going to be processed by the widget," and "what part of the data the widget will transmit to the data consumer." The smartphone user installs widgets in his private cloud. These widgets are data extraction applications. They collect relevant user data, perform data analysis, and send the statistical data to the data consumer. The data consumer collects all statistical data and creates a report or does the data modeling task.

In this framework, a trusted third party is used to certify the widgets so that the users can be sure of their activity. The trusted third party is responsible for maintaining the market and keeping it trustworthy. Figure 10.6 shows the architecture of a trusted mobile sensing framework.

10.8.3 Mobile Agent-Based Trustworthy Infrastructure for MCC

In a cloud computing environment, both the CSP and the cloud consumer should be trustworthy. To achieve this, a mobile agent-based trustworthy infrastructure is required. A detailed description of such a framework is given by Ramaswamy et al. [15].

In this model, the cloud broker (CB) maintains the trustworthiness of the cloud environment. CB is a trusted third party. Every CSP has to first register with the CB. The CB gives the CSP a dummy job and invokes a mobile agent. It evaluates the trustworthiness of the CSP according to the report of the agent. If the CSP is performing well, its trust index is increased, else decreased. There is a threshold value after which the CSP is purged out of the cloud environment by the CB.

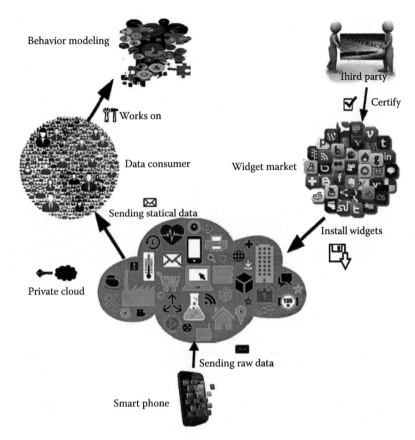

FIGURE 10.6
Architecture of trustworthy mobile sensing framework.

The key provider, a subpart of the CB, provides each customer with a public-private key combination. The cloud consumer request is first sent to the CB, which then forwards the request. After each service, the cloud consumer returns a gossip message to the CB. This message, containing user experience, is forwarded to all cloud consumers. If the user provides actual experience, the CB awards the cloud consumer, and the trust index of the cloud consumer is increased accordingly. If malicious users provide wrong experience, they are penalized with decreasing trust index. If the trust index goes below a certain value, the cloud consumer is removed from the environment. The CB distinguishes between wrong and right messages by previous experience. The gossip message monitoring is done by another mobile agent of the CB. User experience can be +1 or −1 based on good experience or bad experience.

The trust value (TV) of the CSP, according to Ramaswamy et al. [15], may be evaluated in the following way:

$$\text{TV}_{\text{CSP}} = e^{\pm t} * \left(\frac{\text{VMl}}{\text{VMr}} + \sum_{i=1}^{n} U_i + \sum_{i=1}^{m} \text{PN} + \text{PS} \right) \tag{10.1}$$

The TV of the *i*th cloud consumer may be evaluated as

$$TV_{CC} = e^{\pm t} * (U_i * (PNS) + PC_i) \tag{10.2}$$

where
   n is the number of cloud consumers who have been serviced
   m is the tolerance threshold
   U_i is the user experience of the *i*th customer
   VMl is the number of virtual machines launched for a job
   VMr is the number of virtual machines required for a job

Penalty (PN) is the punishment given due to malicious behavior, and prize (PS) is the total number of awards given to the CSP by the CB for trustworthy behavior. Prize points (PPi) are the awards for trustworthiness given to the *i*th customer by the CB, and PNS is the PN or PS awarded to the CSP by the CB. PC_i is the award given to *i*th customer by the CB. If positive trust value shows up, it is multiplied with e^{-t} to reduce the trust value with time. If negative trust value shows up, it is multiplied with e^{+t} to further reduce the trust value with time.

Figure 10.7 shows the message packets of the cloud consumers.

Here, Message ID is the unique message identifier. The source cloud consumer ID is the identity of the message sender. The destination cloud consumer ID is the identity of the customer who receives the message. *B* stands for broadcasting. *S/G* field specifies whether the message is a service request or a gossip message. The user experience field can have the values +1 or –1 according to user satisfaction. The time stamp field is given to avoid message reply attacks. Figure 10.8 shows the architecture of a mobile agent-based trustworthy infrastructure.

10.8.4 Building Trustworthy Social Network Based on Call Behavior

Since interest in smartphones has increased tremendously, people are aware of the newly developed applications for them. Social networking applications are in the frontline of these applications. We may define the concept of friend in a social networking context

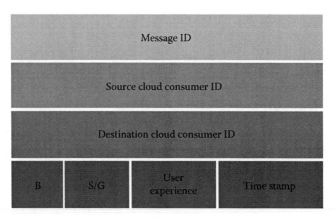

FIGURE 10.7
Message packets of cloud consumers.

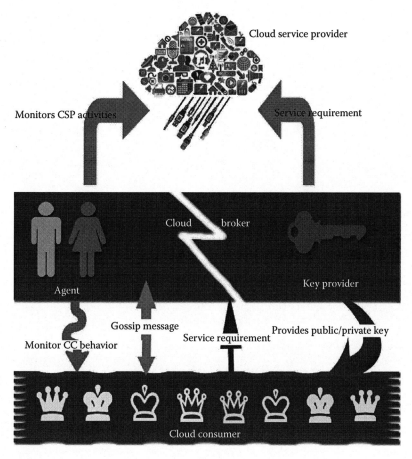

FIGURE 10.8
Architecture of mobile agent-based trustworthy infrastructure.

in a new way. By analyzing the calling pattern of the smartphone user, we may calculate the trust value of different persons. The trust value is dependent on frequency, intimacy, and recency. Based on that trust value, a local social network can be made where only the people directly connected through call are involved. Next, based on the different local social networks of different users, a global social network can be made. In the global social network, all the local social networks are connected. This aspect of MCC is given in detail by Kim and Park [16] and Chen et al. [17].

For implementation, we may extract the call details of a smartphone user. Then, we may create a local social network of each user by calculating the trust value based on the call behavior pattern. Also, the rank of each associated user is made based on the trust value. Next, we need to send each user's data to the cloud server for the creation of a global social network. A global rank is made there to facilitate the creation of the global social network.

This particular kind of application may be interesting and effective for social network providers as well as mobile users. Figure 10.9 shows the dependencies of call behavior trust.

Table 10.1 describes the different parameters used to determine the trust values of different users in the existing model.

Call frequency is related to the total number of calls between two smartphone users.

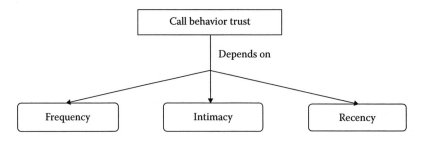

FIGURE 10.9
Call behavior trust dependencies.

TABLE 10.1

Parameters Used in Trust Calculation

Parameters	Description
$\text{Talk}_{X,Y}$	It calculates the total number of calls between X and Y.
$D_{X,Y}$	It means the total time of each call made by X to Y.
$\text{Recency}_{X,Y}$	It means how current the call between X and Y was. Trust decreases with time. So, a recent call is more important than a past call.

Formally, call frequency can be defined as follows:

$$\text{Frequency}_{X,Y} = \frac{\text{Talk}_{X,Y}}{\sum_{P=1}^{n} \text{Talk}_{X,P}} \tag{10.3}$$

Intimacy is related to the duration of calls among users. Call duration is again divided into two parts: call send duration and call receive duration.

Call send duration is defined as follows [16]:

$$\text{Duration } S_{X,Y} = \frac{\sum D_{X,Y}}{\sum_{P=1}^{n} D_{X,P} + \sum_{P=1}^{n} D_{P,x}} \tag{10.4}$$

Call receive duration is defined as follows [16]:

$$\text{Duration } R_{X,Y} = \frac{\sum D_{Y,X}}{\sum_{P=1}^{n} D_{X,P} + \sum_{P=1}^{n} D_{P,x}} \tag{10.5}$$

Finally, intimacy, which is based on duration, is defined as follows:

$$\text{Intimacy}_{X,Y} = \theta * \text{Duration } S_{X,Y} + (1-\theta) * \text{Duration } R_{X,Y} \tag{10.6}$$

where θ is the weight factor and $0 \leq \theta \leq 1$.

The trust value ($T_{X,Y}$) of user X to Y is calculated as follows:

$$T_{X,Y} = \alpha * \text{Frequency}_{X,Y} + \beta * \text{Intimacy}_{X,Y} + \delta * \text{Recency}_{X,Y}$$
$$(\alpha + \beta + \delta = 1, 0 \le \alpha, \beta, \delta \le 1)$$

(10.7)

where α, β, and δ are three weight factors for the evaluation of the trust value.

Another term that may be relevant in the context of building social networks is "tendency." Tendency may be formally defined as follows:

$$\text{Tendency}_A = \frac{\sum_{P=1}^{n} D_{A,P}}{\sum_{P=1}^{n} D_{A,P} + \sum_{P=1}^{n} D_{P,A}}$$

(10.8)

where
$0 \le \text{Tendency} \le 1$
$\sum_{P=1}^{n} D_{A,P}$ is the total time that user A called to all other n number of users

Analyzing all these phone call data, we may get interesting results, which may help us to develop interesting social networking applications.

10.8.5 Trust-Based Mobile Commerce

Mobile commerce (m-commerce) can be regarded as a subset of e-commerce. With the rapid development of wireless technology, we can now conduct business or make purchases from our home. Mobile technology adds mobility to e-commerce. It is now possible to do online business everywhere and anytime with the help of smartphones or notebooks. This aspect of trustworthy mobile commerce is discussed in detail by Liu [18].

But there is a big problem with mobile commerce: the issue of trust. How two strangers can trust each other and proceed for online transactions to conduct business is a big issue to resolve.

10.8.5.1 Automated Trust Negotiation

Automated trust negotiation (ATN) is a way to solve the trust problem between two strangers. In this process, digital certificates and credentials are exchanged between two strangers for each other's identification. The certificates contain attributes for negotiation. A series of digitally signed policies and certificates are disclosed iteratively to establish trust between the two parties [19].

If we integrate ATN into mobile commerce, the problem of trust is solved, but a new problem arises. Mobile devices have power constraints, and they are also not computationally efficient. The ATN process will result in high computation and communication overhead on mobile device.

One solution can be to use cloud computing. The digital certificates and credentials may be kept in cloud server. Whenever the mobile user instructs, the cloud server will do the ATN process on behalf of the mobile user.

Here, again comes the trust problem. All digital certificates, asymmetric keys are kept in the cloud server. What is the guarantee that someone sitting in the server will not misuse them?

10.8.5.2 Proxy Certificate

Proxy certificate is a good solution to the aforementioned problem. Instead of keeping long-term credentials, users can generate short-lived proxy certificates and store them in the cloud servers. The temporary certificates have a very short lifetime, typically 6 h. They follow RFC3820 and can be used efficiently in the ATN process. Moreover, if the user wants, any restriction can also be placed on how these certificates can be used. The X.509 extension assures that the digital credentials can be used only if the user provides any request to the cloud server.

10.8.5.3 CBTN Protocol for Trust-Based Mobile Commerce

The Cloud-Based Trust Negotiation (CBTN) protocol for mobile commerce works with the aforementioned approach. Through cloud-based, short-lived proxy certificates and using ATN, CBTN protocol establishes a trustworthy relationship between two strangers willing to do online business with mobile devices. At the end of the service, a sequence of operations is sent to users through which they can check if the off-loaded trust negotiation was performed correctly or not.

10.9 Conclusion

The evaluation of cloud service provider's trustworthiness is important with respect to cloud users. It helps customers to choose between a number of CSPs and also to access cloud services without any question of data security hazard in their mind. With respect to CSPs also, evaluation of the trustworthiness of the user behavior is essential. It helps the CSPs to encourage actual users and remove malicious users from the system. This is important for the integrity and reliable service delivery of the system.

Questions

1. What do you mean by "trust"? What are the properties of trust?
2. Explain the components of trust with respect to MCC.
3. How is trust established between two entities? What are the types of trust?
4. Explain the relationship of QoS with trust and reputation.
5. Describe the evaluation approaches of trust in MCC.
6. What is proxy certificate?

7. Explain the protocol for trust in the mobile commerce environment.
8. What do you mean by "trustworthy mobile sensing"?
9. Explain AHP-, FAHP-, and FANP-based trust evaluation models.

References

1. P. Mell and T. Grance, The NIST definition of cloud computing, 2011 [Online]. http://csrc.nist.gov/publications/nistpubs/ 800-145/SP800-145.pdf.
2. R. K. L. Ko, P. Jagadpramana, M. Mowbray, S. Pearson, M. Kirchberg, Q. Liang, and B. S. Lee, TrustCloud: A framework for accountability and trust in cloud computing, in *IEEE World Congress on Services*, Washington, DC, pp. 584–588, 2011.
3. J. Golbeck, Computing with trust: Definition, properties, and algorithms, in *IEEE Securecomm and Workshops*, Baltimore, MD, pp. 1–7, 2006.
4. A. K. Singh, Trust and trust management models for ecommerce & sensor network, *International Journal of Engineering Research and Applications*, 2(6), 585–619, 2012.
5. http://en.wikipedia.org/wiki/Audit.
6. http://www.privacycommission.be/en/glossary/accountability, 2015.
7. K. Govindan and P. Mohapatra, Trust computations and trust dynamics in mobile ad hoc networks: A survey, *IEEE Communications Surveys & Tutorials*, 14(2), 279–298, 2012.
8. S. Pearson and A. Benameur, Privacy, security and trust issues arising from cloud computing, in *Second IEEE International Conference on Cloud Computing Technology and Science*, Indianapolis, IN, pp. 693–702, 2010.
9. S. M. Habib, S. Hauke, S. Ries, and M. Muhlhauser, Trust as a facilitator in cloud computing: A survey, *Springer Journal of Cloud Computing*, 1(1), 1–18, 2012.
10. https://cloudsecurityalliance.org/research/cai/, 2015.
11. Y. Delmar, Cloud assessment try the cloud security alliance consensus questionnaire, 2011. http://yogrc.typepad.com/yo_delmars_grc_and_beyond/2011/01.
12. S. M. Habib, S. Ries, and M. Muhlhauser, Cloud computing landscape and research challenges regarding trust and reputation, in *IEEE, Ubiquitous Intelligence & Computing and Seventh International Conference on Autonomic & Trusted Computing* (UIC/ATC), Xian, China, pp. 410–415, 2010.
13. T. Li-qin, L. Chuang, and N. Yang, Evaluation of user behavior trust in cloud computing, in *IEEE International Conference on Computer Application and System Modeling*, Taiyuan, China, vol. 7, pp. V7-567–V7-572, 2010.
14. J. Y. Zhangy, P. Wuy, J. Zhu, H. Hu, and F. Bonomi, Privacy-preserved mobile sensing through hybrid cloud trust framework, in *IEEE Sixth International Conference on Cloud Computing* (CLOUD), Santa Clara, CA, pp. 952–953, 2013.
15. A. Ramaswamy, A. Balasubramanian, P. Vijaykumar, and P. Varalakshmi, A mobile agent based approach of ensuring trustworthiness in the cloud, in *IEEE International Conference, Recent Trends in Information Technology* (ICRTIT), Chennai, India, pp. 678–682, 2011.
16. M. Kim and S. O. Park, Trust management on user behavioral patterns for a mobile cloud computing, *Springer Cluster Computing*, 16(4), 725–731, 2013.
17. S. Chen, G. Wang, and W. Jia, A trust model using implicit call behavioral graph for mobile cloud computing, in *Cyberspace Safety and Security*, in *5th International Symposium, CSS*, November, Springer International Publishing, Zhangjiajie, China, vol. 8300, pp. 387–402, 2013.
18. B. Liu, Cloud-based trust establishment protocol towards mobile commerce, in *International Conference on Computer, Networks and Communication Engineering*, Atlantis Press, Beijing, China, pp. 700–702, 2013.
19. W. H. Winsborough, K. E. Seamons, and V. E. Jones, Automated trust negotiation, in *IEEE DARPA Information Survivability Conference and Exposition*, Hilton Head, SC, vol. 1, pp. 88–102, 2000.

11

Vehicular Mobile Cloud Computing

VCC is a new hybrid technology that has a remarkable impact on traffic management and road safety by instantly using vehicular resources, such as computing, storage and Internet for decision making.

Md Whaiduzzaman et al.

ABSTRACT   Mobile cloud computing involves the study of several mobile agents such as vehicles and robots. All these mobile agents collaborate and interact to feel the environment, process data, propagate outputs, and mostly share resources. The vision of vehicular mobile cloud computing (VMCC) is a nontrivial argumentation with different dimensions along with conventional mobile cloud computing. In VMCC, the underutilized resources of vehicles such as storage, Internet connectivity, and computing power are shared among drivers as well as rented on the Internet for other customers/users. VMCC is a new hybrid technology, which has a marvelous impact on road safety and traffic management as it uses instant vehicular resources such as storage, computing power, and the Internet for decision making. In this chapter, VMCC is discussed with its simulation performed using the network simulator Qualnet 7.1. According to the analyzed metrics, it is shown that vehicle-to-vehicle (V2V) communication is preferable than vehicle-to-infrastructure (V2I) communication as data rate is higher in V2V communication.

KEY WORDS: *VANET, cloud computing, mobile computing, vehicular computing.*

11.1 Introduction

Recently, vehicular ad hoc networks (VANETs) have come into the view. A VANET is used to yield suitable wireless network services [1]. The evaluation and deployment of wireless communication systems have completely altered the life of ordinary people by offering Internet services and applications, which have made human life easier and more flexible. Today, vehicles are no longer just a means of transportation. Now vehicles are equipped with smartphones that have more sensing, processing, and communicating capabilities [2]. In mobile phones, a vehicular track (Vtrack) system is installed, which improves sensor reliability and energy efficiency. The mapping scheme for a Vtrack system is used for estimating travel time [3].

In a VANET, the environment changes constantly, which makes the network susceptible to various kinds of threats. Wormhole attack is the most common type of attack, leading to denial of service, masquerading, data tampering, etc. In order to improve the performance of the network, the performance of various routing protocols has been tested in a simulation environment to determine which protocol is better suited as a defense in the case of a wormhole attack [4].

Cloud computing is based on the Internet, whereby IT resources, information, and software applications are provided to smart devices and these services are computed on demand. In recent times, accessing and storing of computer data and other applications are not done on the Internet browser by running the software installed in the personal computers. Mobile cloud computing (MCC) is defined as an architecture in which data processing and storing are performed outside the mobile devices. The mobile cloud applications move data storage and computing power from the mobile device to the cloud. Simply put, MCC allows users to function without investing in business or infrastructure by renting infrastructure or necessary software. It also provides several other features such as provisioning software and hardware with high-speed Internet at very low cost. MCC provides pay-per-use service for computing resources as per need.

The layered architecture of MCC is shown in Figure 11.1. The infrastructure as a service (IaaS) layer has several types of virtualization in which resources, computing, networking, hardware, and storage are included. At the bottom layer, infrastructure devices, virtualized

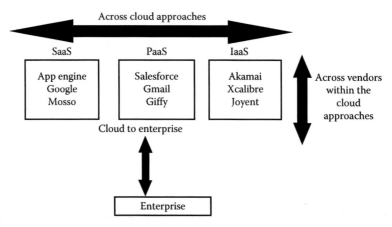

FIGURE 11.1
Layered architecture of MCC.

hardware, and operating system (OS) are installed. An example of IaaS is Amazon Web Services, which provides two types of services: (1) a Simple Storage Service (S3) or Elastic Book Store (EBS), which provides storage service, and (2) Elastic Compute Cloud (EC2) services for computing resources. The platform as a service (PaaS) layer includes several mobile OS such as Symbian, iPhone, Android, and many others. This layer provides the environment for parallel processing, distributed storage, and management. Microsoft Azure and Google AppEngine are examples for PaaS. In the software as a service (SaaS) layer, analytical, interactive, transaction, and browsing facilities are included. In SaaS, simple software programs with some applications as well as customer interface are delivered to users. IBM is an example of SaaS.

11.2 Vehicular Ad Hoc Network

VANETs are special types of vehicles equipped with digital maps, positioning systems, and wireless communication systems, which act as smart machines for safety and comfort, as shown in Figure 11.2.

Today, a simple truck or a car is expected to have some of the devices such as a GPS system, an onboard computer, a short-range radar device, a radio transceiver, a camera along with some special features, and also various types of sensing devices. In VANETs, mostly two types of communication take place: vehicle to vehicle (V2V) and vehicle to infrastructure (V2I). V2I is a connection between vehicles and roadside units (RSUs) that contain strong computing devices placed at several locations. RSUs are connected with base stations.

In VANETs, vehicles exchange and share traffic information. Primarily, VANETs are used for traffic efficiency improvement and traffic safety by communicating through messages. Hence, it results in less number of accidents and traffic jams. In VANETs, communication between vehicles takes place through a dedicated short-range communication (DSRC) wireless device. VANETs communicate by either V2I communication or intervehicular communication (i.e., V2V). V2V communication has some unique features such as multi-hop mobile ad hoc network (MANET). In VANETs, each node-presenting vehicle participates in forwarding packets. In reality, all vehicles do not willingly forward packets. Some vehicle drivers shut down their wireless devices according to their will.

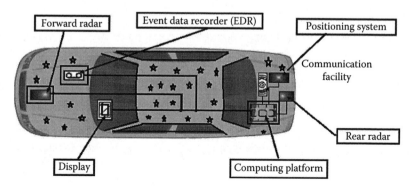

FIGURE 11.2
Components of smart vehicle.

In urban VANETs, vehicles move like clusters conditioned by traffic lights. An RSU acts as a router in VANETs. RSUs are the backbones of this network for providing several services. For designing RSUs, several approaches have been considered such as standalone-type RSUs or hybrid RSUs. RSUs are situated at the minimum distance of 2–5 km.

11.2.1 Working Principles of Vehicular Ad Hoc Networks

In the present world, the number of vehicles exceeds more than 750 million. Each vehicle needs an authority for governing it. V2V communication takes place by using DSRC, whose range is 5.9 GHz. Vehicular network is an ad hoc communication where each node is free to move and no wired connection is required. RSU is used as a router that connects the vehicles and network devices [5]. Connection between vehicles and RSUs is done by an onboard unit (OBU) present in each vehicle via DSRC. OBU provides different types of services such as entertainment, GPS navigation, and the like. VANETs can provide man-to-machine interaction with the help of short/wired range sensors and wireless technologies. A tamper-proof device holds the secrets of the vehicle, including information such as trip details, keys, routes, speed, and driver's identity.

Several routing algorithms are convenient for VANET stimulation, such as ad hoc on demand distance vector (AODV) and dedicated short range (DSR) [6]. The available protocols work in two phases. During phase I, it identifies the neighboring vehicles for the creation and broadcast of infrastructure, and in phase II, data transferring is done. In AODV protocol, two phases run in parallel. In the case of DSR, both phases are executed one by one, that is, phase I and then phase II. For this reason, AODV is preferable to DSR.

11.3 Architecture and Working Model of Vehicular Mobile Cloud Computing

In the present decade, a considerable amount of access to web, either from mobile phones or fixed devices, is available for mobile resources and services. Most of the queries are related to the environment or surroundings where users live, and the best probes for these environments are mobile agents such as vehicles and people. However, the data collected are stored in various mobile agents and aggregated by using specialized local context. The mobile agent forms a mobile cloud, which offloads the task to the Internet cloud. In vehicular mobile cloud computing (VMCC), drivers can send their queries to the mobile vehicular cloud [7] for finding the cause of road accidents, sudden traffic jams, etc. This type of query or information is maintained, created, or propagated inside the mobile cloud. It is essential to upload or send these inquiries since data of all traffic jams or road safety issues cannot be uploaded in the Internet. If somehow all information related to roads could be uploaded, it becomes expensive and time consuming. An efficient increase in processing capacity and storage and insufficiency of urban spectrum result in storing the locally pertinent content on mobile cloud rather on Internet cloud. The advantage of mobile cloud over Internet cloud is that it reduces communication delay. It also decreases the spectrum cost and enlarges the range of applications. At present, only those tasks that are energy consuming and too complex to run in the mobile cloud are uploaded in the Internet cloud.

Vehicles moving across roadways or standing in parking lots are treated as underutilized, but plentiful, computational resources that can be used for furnishing public services. Every day, a large number of vehicles spends lots of time in parking lots or traffic jams

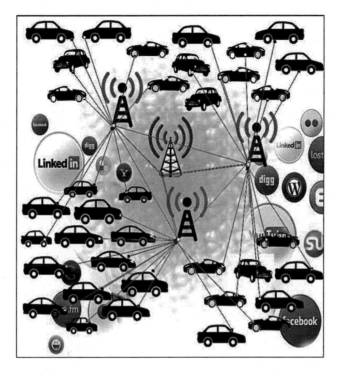

FIGURE 11.3
Vision of vehicular mobile cloud computing integrating vehicular ad hoc network, mobile computing, and cloud computing along with social networking sites.

where their untapped resources are simply wasted. To use these wasted resources, vehicles are treated as nodes in the cloud computing environment. Some users or vehicle holders rent their huge storage and computing facilities to get more capacity and advantage economically. Nowadays, a vehicle is based on autonomous self-organized resources where it serves on demand. It also helps to solve serious problems that occur unexpectedly. The key features that distinguish vehicular mobile cloud and conventional cloud are autonomy, agility, and mobility.

Vehicular cloud computing (VCC) integrates the advantages of cloud computing with VANET to provide the drivers with computational resources. The primary aim of VCC is to provide energy-efficient and real-time services to reduce traffic congestion and road accidents, ensure road safety, etc. It merges the services of MCC to realize the goal of an intelligent transportation system (ITS) [8]. VCC encompasses a large number of vehicles whose resources include communication, sensing, computation, and so on, which can be harnessed and allocated to other users on demand [9], as shown in Figure 11.3.

11.3.1 VMCC Architecture

VMCC consists of the following three layers, as shown in Figure 11.4:

1. Vehicular computing layer
2. Mobile computing layer
3. Cloud computing layer

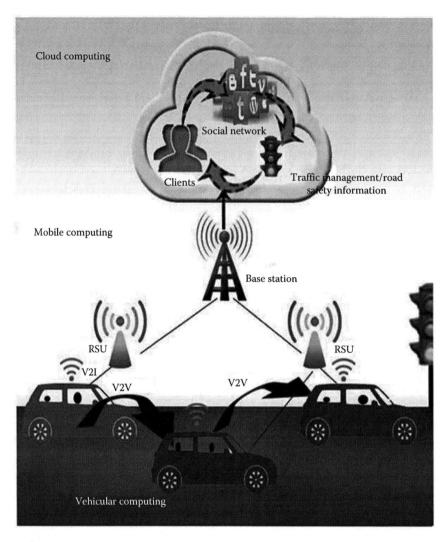

FIGURE 11.4
Three-tier VMCC architecture: vehicular computing, mobile computing, and cloud computing.

The vehicular computing layer consists of vehicles, RSUs, and base station. When the vehicles are within the range of each other, V2V communication takes place. When they are outside the range, they can communicate via RSUs and base station. In the first layers, the vehicle acts as a node, where it does not have random mobility but moves in random directions following a certain path. While moving on roads or standing in parking lots or in traffic, vehicles communicate with other vehicles in their range for either entertainment or conveying important messages. When they are not in the range of other vehicles, they communicate to RSUs, then to base stations through RSUs, and finally from base stations to cloud.

The second layer of the architecture involves mobile computing. This layer consists of drivers and copassengers, who can communicate using their mobile devices such as mobile phones and laptops. The drivers and passengers can participate in a social network and share their photos and views. This layer is the most important layer as it acts as a sandwich between the two, that is, VANET and cloud. The connection between cloud and vehicles takes place from this layer only; it is either V2V communication or V2I communication.

The third layer is the cloud computing layer. The resources of the vehicles are pooled to form a cloud in which information regarding traffic congestion, road accidents, and so forth, are processed, stored, and forwarded or sent to drivers or copassengers when they are in need. The cloud computing layer provides all the services at any time on demand.

In VMCC, passengers traveling through a common route can form a network to share their interests and essential information regarding road traffic. A cloud is formed by utilizing onboard vehicular resources such as processing power and storage to aid in VMCC [10].

Insufficient parking problems are getting worse day by day with the increase in vehicles. So for employing several multilevel parking garages, cloud computing or wireless sensor networks (WSNs) are required. It also provides dynamic parking facilities. Basically, cloud assists two types of parking services: traditional parking garages and dynamic parking services [11]. By using smart terminals such as smart phones, parking reservation systems can be supported. Here, VMCC can be used as VANETs' parking service system along with MCC capability. The emphasis is based on parking reservation service, the decision-making technique for traffic authorities, and cloud-assisted architecture. The following are the two types of parking spaces:

1. *Traditional parking garages*: Any information related to parking is detected by WSNs, and then it is forwarded to the cloud through the Internet. This cloud is called traffic cloud, which collects the data and transmits to users who get the information on their smart terminals. Sometimes, it is published on nearby billboards, especially for those users who do not have smart terminals.

2. *Dynamic parking services*: Vehicles face rush hour especially in the morning and the evening. So during that time, vehicles are temporarily kept on roadsides. Here, a dynamic arrangement of parking services on roadside is considered, with the help of MCC and WSNs. This technique is used for analyzing the reservation service processes, planning service of traffic authorities, and context-aware optimization.

Decisions made by traffic authorities depend on many factors such as weather conditions, road conditions, traffic congestion, and traffic flow forecasting. Decisions are made according to the decision tree. In some cases, fatal factors result in decision making, such as typhoon approach. Typhoons drive in a specified manner according to their eyes' mobility, where resource requester "moves" depending on the resource requester's mobility [12]. But presently, this typhoon approach is used in VANETs as a resource-sharing protocol where once the resource requester gets the resources from the resource holder, it becomes a new resource holder [12]. The approach enhances the spatial locality over time.

Parking space status is monitored and determined with the help of a special corresponding system and is immediately offloaded to the cloud from where drivers or copassengers get the updated information on their smart terminals such as smartphones. Anyone can also log on to the traffic cloud and get the information about parking space.

Context-aware optimization not only includes parking garage status and road conditions but also counts the duration of the vehicles standing in the parking lot. With the help of this information, the best parking location for drivers or passengers can be optimized. For each of the parked vehicles, the duration of their parking is uploaded in the cloud and is shared with drivers after analysis. In this way, drivers can acquire information on parking garages and acquire best services.

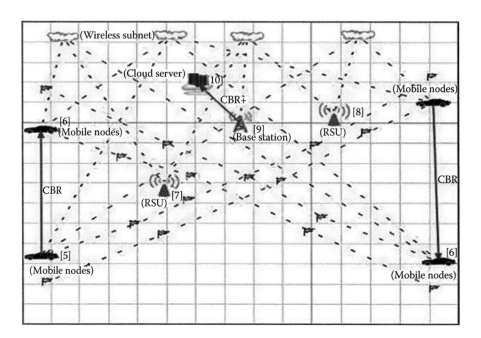

FIGURE 11.5
Architect module of QualNet showing the simulation scenario including four vehicle nodes, two CBRs, one base station, two RSUs, and one cloud server.

11.3.2 Performance Analysis of VMCC

To study the performance of VMCC, depending on parameters such as throughput, end-to-end delay, and jitter, a network simulation is performed using Qualnet 7.1, as shown in Figure 11.5. After the simulation is performed, the statistical analyses of these parameters are performed based on the movement of the vehicles. Quality of service parameters such as throughput are determined when the vehicles are communicating with each other, that is, when the vehicles are within their range or when the vehicles are outside the range and communicating through RSUs. The delay and jitter while the communication is taking place can be analyzed. This is an open research area in using network simulation tools. Various network parameters can be analyzed and optimized using different practical scenario.

11.4 Privacy and Security in Vehicular Mobile Cloud Computing

Privacy and security are two important concerns when some resources between two users are shared. The two constraints that should be maintained while sharing resources are as follows:

1. Privacy and security issues of the producer, that is, the vehicle owner should be preserved.
2. Privacy and security of who rented the resources, that is, the consumer [13] should also be preserved.

The types of attacks possible in VANETs include Sybil attack, Replay attack, Fabrication attack, and the like. There are also different types of attackers such as malicious attacker, selfish attacker, for example. For these types of attacks and attackers, some privacy and security requirements should be considered, such as availability, authentication, confidentiality, integrity, and real-time constraints [14]. The major problems of VMCC are to protect the information or data. In VMCC, messages are transferred between vehicles with the help of mobile devices. Users should be more concerned about which information should be sent and which should be kept private. Moreover, the data, functions, and trust affirmation of mobile phone applications can be deputed to MCC if mobile users or mobile devices become temporarily unavailable. MCC provides complete protection of devices compromised by adversaries and disruptive behavior.

11.4.1 Security and Privacy Attacks in VMCC

Presently, security systems are designed in such a way that they can stop attackers from entering systems. In the case of vehicular cloud, security handling is quite tough as multiple users with high mobility share the same physical machine/infrastructure as their targets. In VMCC, attackers are free to move anywhere as they are in moving nodes such as vehicles. Therefore, it becomes very difficult to track the attackers [15].

Attackers target the following:

1. *Confidentiality*: To find the identity of others, important documents, data, and locations of services that are executing.
2. *Integrity*: Attackers are always in search of important documents, executable coding, and results kept in cloud.
3. *Availability*: Attackers try to hamper the resources and machines, services, applications, and privileges.

11.4.1.1 Possible Forms of Attacks

The possible forms of attacks are as follows:

1. Finding the location of the target vehicle and then physically moving closer to that target machine.
2. By mapping the VC technology, finding the possible areas of user services.
3. Launching several experimental accesses on the cloud.
4. Pinpointing system leakage for collecting high privileged assets.

11.4.1.2 Threats in VMCC

Different threats in VMCC are as follows:

1. *Spoofing*: Here, the attacker acts like another person or holds the identity of other users to obtain important data and private information.
2. *Tampering*: Here, the malicious user changes or modifies the data and forges the data or information.
3. *Repudiation*: The malicious user manipulates the identification of the data, operations, and actions.

4. *Information disclosure*: The unauthorized person discloses all personal information such as identities, legality, medical, finance, residence, political, and biological traits, geographic records, and ethnicity.

5. *Elevation of privilege*: The attacker utilizes a bug, design flaw, or system leakage in an OS [16].

11.4.1.3 Authentication of Mobility Nodes

In VMCC, authentication includes identifying user's identity and integrity of messages. For conducting authentication, some metrics can be adopted [15,17] as follows:

1. *Ownership*: Each user has his/her personal identity such as security tokens, identity cards, and software tokens.

2. *Knowledge*: A user always has the knowledge of something unique such as passwords or security questions.

3. *Biometrics*: This includes face, voice recognition, fingerprint, and signature.

However, authentication of vehicles is not easy because of high mobility, which makes them very difficult with respect to location. In the case of an accident alert, signal, or message, finding the exact location where the accident took place is very difficult because the location of the vehicle is constantly changing. Another problem that may arise for high mobility and shorter range of transmission in the case of vehicular movement is that sometimes vehicles become out of reach. In that case, it is harder to update security tokens. Even identity authentication creates the possibility of Sybil attacks.

11.4.2 Solution to Secure VMCC

The main purpose of VMCC is to convey safety messages. Basically, there are three important types of safety messages:

1. Information related to public traffic conditions such as huge traffic jams, which indirectly prevent major accidents.

2. Accident avoidance application where vehicles exchange information or messages for cooperative safety.

3. Liability messages send after accidents [18,19].

Some major safety messages include speed, time-stamp, direction, acceleration, percentage of acceleration, and speed change since last message.

11.5 Limitations of Vehicular Mobile Cloud Computing

11.5.1 Mobility in VMCC

The VMCC differs from the ad hoc network with respect to mobility because each vehicle is treated as mobile nodes in the VMCC. The speed of the mobile nodes is very high. Even the connections between the vehicles are for a short duration and terminated soon.

11.5.2 Volatility in VMCC

VMCC lacks longer life context. For this reason alone, there is the need for a lifelong password for making personal contact. But this lifelong password is quite impractical.

11.5.3 Privacy

For avoiding Sybil attacks, each vehicle should have unique identity, but sometimes this solution cannot be implemented because some drivers do not want to expose their information. They are more concerned about privacy.

11.5.4 Liability

Liability provides a better opportunity for an investigation done legally, but these data or information cannot be ignored if they are related to accidents. But again the issue is privacy.

11.5.5 Scalability

The number of vehicles is increasing each day. There is no global authority for governing the growing vehicular nodes. Even the DSRC standard is different for different places.

Presently, a few cars are provided with the equipment needed for DSRC communication. In the future, thought must be given to dealing with numerous vehicles for getting financial benefits and encouraging firms for investing in the proposed technology.

11.6 Challenges in Vehicular Mobile Cloud Computing

The challenges in VMCC are as follows:

1. *Mobile flexible architecture*: The mobility of vehicles directly affects storage resources and computational power [1]; for example, a vehicle moving along a certain path is not constant. Even vehicles in a parking lot are always changing. Therefore, the change in mobility of vehicles affects the VMCC architecture.

2. *Robust architecture*: The fundamental structures and the building blocks of VMCC architecture are designed and engineered for facing the structural stress due to unstable working environment [1].

3. *Service-oriented architecture*: The present layered architecture of TCP/IP is not sufficient for supporting today's upcoming applications and technology. The use of service- and component-based architecture with adequate learning facilities and monitoring opportunity is best to cope with extensible reusable resources and applications, which are deployed as a common service.

11.7 Applications

Two major categories of applications of VMCC are as described in the following sections.

11.7.1 Safety Applications

According to Olariu et al. [20], about 80% accidents can be avoided if warning messages are provided to drivers. There are some scenarios where safety applications can be implemented:

- *Accidents*: Vehicles usually move at a very high speed on major roads. If any accident occurs, drivers get much less time to depress their brakes to prevent collision with the vehicle in front of them. As a result, the approaching vehicle crashes. A safety application can send warning information either to the reckless driver or to the approaching vehicle about the speed limit.
- *Intersection*: Due to the traffic flows, drivers face various challenges. Driving through or near intersections causes high collision possibilities. According to Aijaz et al. [19], 45% of road crashes occur at intersections. The number of accidents can be decreased if a warning is given to drivers before the impending collision.
- *Road congestion*: Safety applications can be provided to vehicles to find the best possible way to reach their respective destinations. It can lead to less traffic jams, smooth flow of traffic, and a smaller number of accidents.

11.7.2 User Applications

User applications provide passengers with advertisement, entertainment, and important information during a journey, such as the following:

1. *Internet connectivity*: Today, the Internet has become an important part of everyone's life. Even while traveling or standing in a parking lot, users search for an Internet connection. In VMCC, the mobile devices are always equipped with Internet, which is used for not only business but also entertainment, especially when a vehicle gets stuck in a traffic jam or when it is placed in a parking lot.
2. *Peer-to-peer application*: Communicating between vehicles or sharing information and ideas among each other is only possible when vehicles are in the range of each other.

11.8 Conclusion

Shifting of VANETs to the cloud is a novel paradigm. The three-layered architecture of VMCC is described here: cloud computing, MCC, and VANETs. Vehicles considered in this chapter are the moving nodes. V2V communication occurs when vehicles are in the range of each other. Otherwise, V2I communication occurs where the RSUs communicate with the base station and finally to the cloud for getting Internet connection. This enables the vehicles to access the Internet anytime anywhere. Vehicles on road often get stuck in traffic, and drivers get bored. VMCC contributes to the entertainment of passengers. Additionally, urgent information, such as traffic congestion, accidents, road blocks, natural calamities, as well as information related to facilities such as hospitals, schools, cafeterias, and the like, can be shared via this network. Thus, VMCC can help in mitigating several types of emergencies by providing prior information to unaware commuters. In the case of natural disasters, network infrastructure may collapse. In such situations, VMCC can significantly contribute in disseminating information. In the future, it is expected that VANETs will be supporting more services and applications for increasing road safety and efficient road transportation. Some protocols can also be implemented in the future for multilevel privacy preserving communication in VMCC.

Questions

1. What are the components of VMCC?
2. Explain the significance of VMCC.
3. Draw the block diagram of a vehicular ad hoc network and explain its significance.
4. Explain the working principles of vehicular ad hoc networks.
5. Analyze the performance of VMCC.
6. Comment on average end-to-end delay, average jitter, and throughput of VMCC.
7. Compare the performance of VMCC-based on vehicle movement.
8. What are security and privacy attacks in VMCC?
9. What are the threats in VMCC?
10. How is authentication of mobile nodes performed?
11. Explain the possible solution to secure VMCC.
12. What are the limitations of VMCC?
13. Explain the mobility management in VMCC.
14. Explain the volatility in VMCC.
15. Explain the authentication and privacy system in VMCC.
16. Comment on the liability of VMCC.
17. Comment on the scalability of VMCC.
18. What are the major challenges in VMCC?
19. What are the applications of VMCC?
20. Explain the safety application of VMCC.

References

1. A. Abhale, S. Khandelwal, and U. Nagaraj, Shifting VANET to cloud—Survey, *International Journal of Advanced Research in Computer Science and Software Engineering*, 3, 1056–1066, 2013.
2. S. Smaldone, L. Han, P. Shankar, and L. Iftode, RoadSpeak: Enabling voice chat on roadways using vehicular social networks, in *Proceedings of the First Workshop on Social Network Systems*, Glasgow, Scotland, ACM, pp. 43–48, 2008.
3. R. Fei, K. Yang, and X. Cheng, A cooperative social and vehicular network and its dynamic bandwidth allocation algorithms, in *IEEE Computer Communications Workshops (INFOCOM WKSHPS) Conference*, Shanghai, China, pp. 63–67, 2011.
4. T. Singh and S. K. Dhanda, Performance evaluation of routing protocols in VANETs, *International Journal of Advanced Research in Computer and Communication Engineering , Performance Evaluation*, 2(9), 3590–3594, 2013.
5. J. Wang, J. Cho, S. Lee, and T. Ma, Real time services for future cloud computing enabled vehicle networks, in *IEEE International Conference on Wireless Communications and Signal Processing (WCSP)*, Nanjing, China, pp. 1–5, 2011.
6. N. Abbani, M. Jomaa, T. Tarhini, H. Artail, and W. El-Hajj, Managing social networks in vehicular networks using trust rules, in *IEEE Wireless Technology and Applications (ISWTA) Symposium*, Langkawi, Malaysia, pp. 168–173, 2011.
7. M. Gerla, Vehicular cloud computing, in *IEEE Ad Hoc Networking Workshop (Med-Hoc-Net)*, Ayia Napa, Cyprus, pp. 152–155, 2012.
8. S. Al-Sultan, M. M. Al-Doori, A. H. Al-Bayatti, and H. Zedan, A comprehensive survey on vehicular Ad Hoc network, *Elsevier Journal of Network and Computer Applications*, 37, 380–392, 2014.
9. M. Eltoweissy, S. Olariu, and M. Younis, Towards autonomous vehicular clouds, in *Ad hoc Networks*, Springer, Berlin, Germany, pp. 1–16, 2010.
10. M. Whaiduzzaman, M. Sookhak, A. Gani, and R. Buyya, A survey on vehicular cloud computing, *Elsevier Journal of Network and Computer Applications*, 40, 325–344, 2014.
11. J. Wan, D. Zhang, S. Zhao, L. T. Yang, and J. Lloret, Context-aware vehicular cyber-physical systems with cloud support: Architecture, challenges, and solutions, *IEEE Communications Magazine*, 52(8), 106–113, 2014.
12. G. Y. Chang, J. P. Sheu, and J. H. Wu, Typhoon: Resource sharing protocol for metropolitan vehicular ad hoc networks, in *IEEE Conference on Wireless Communications and Networking (WCNC)*, Sydney, Australia, pp. 1–5, 2010.
13. S. Olariu, I. Khalil, and M. Abuelela, Taking VANET to the clouds, *International Journal of Pervasive Computing and Communications*, 1, 7–21, 2011.
14. H. Mousannif, I. Khalil, and H. Al. Moatassime, Cooperation as a service in VANETs, *Journal of Universal Computer Science*, 8, 1202–1218, 2011.
15. G. Yan, D. Wen, S. Olariu, and M. C. Weigle, Security challenges in vehicular cloud computing, *IEEE Transactions on Intelligent Transportation Systems*, 14, 284–294, 2013.
16. Microsoft, The stride threat model, Microsoft, 2005 [Online]. http://msdn. microsoft.com.
17. D. Wen, G. Yan, N. N. Zheng, L. C. Shen, and L. Li, Toward cognitive vehicles, *IEEE Intelligent System*, 3, 76–80, 2011.
18. A. B. Abhale, S. A. Khandelwal, and U. Nagaraj, Motivation towards cost effective roadside communication techniques vehicular cloud environment, *IEEE International Conference on Convergence of Technology*, Pune, India, 2014.
19. A. Aijaz, B. Bochow, F. Dötzer, A. Festag, M. Gerlach, R. Kroh, and T. Leinmüller, Attacks on inter vehicle communication systems—An analysis, in *Proceedings of WIT*, Hamburg, Germany, pp. 189–194, 2006.
20. S. Olariu, T. Hristov, and G. Yan, The next paradigm shift: From vehicular networks to vehicular clouds, in *Mobile Ad Hoc Networking: Cutting Edge Directions*, Wiley, pp. 123–167, 2009.

12

Business Aspects of Mobile Cloud Computing

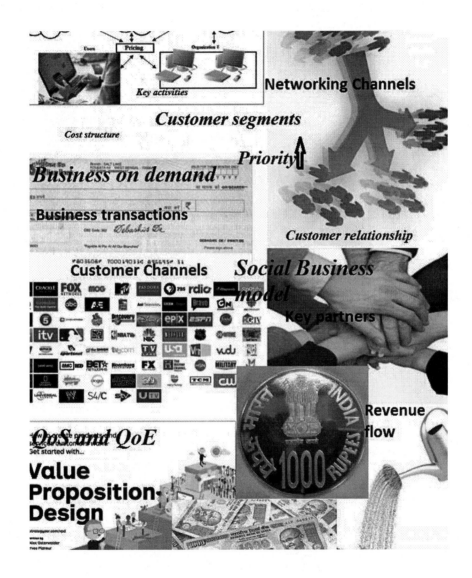

There is an equally urgent need for understanding the business-related issues surrounding cloud computing

Subhajyoti Bandyopadhyay

Professor of Information Systems, University of Florida

Cloud computing—The business perspective, *Decision Support Systems*, **Elsevier, 2011**

ABSTRACT   Mobile cloud computing (MCC) business management depends on the quality of service (QoS) of the cloud service providers (CSPs) to mobile users. Smiles on the customer's face and retaining premium customers are the primary targets of service providers for business development. This chapter discusses various economic and efficient business models having competitive advantages over each other on various aspects of quality-based parameters. Business in MCC is based on sharing revenue among mobile network service providers, CSPs, and Internet service providers. Mobile cloud pricing also depends on the level of security, which is called security mobile cloud pricing. A mobile cloud–based insurance policy ensures quality at a satisfactory level based on various categories of service-level agreements (SLAs), such as premium service level or ordinary service level. Today, new business models for the social mobile cloud are used to transform the business using social media over mobile Internet and MCC. The cooperative impact of social networking service, mobile Internet service, and cloud computing service is creating incredible online business opportunities. With the advent of the bring your own device (BYOD) program, companies are building various apps regardless of operating systems and devices to enable employees to respond to business requirements anywhere and anytime. Most organizations are building apps to support many mobile operating systems as a result of BYOD being in full swing. This model provides efficient, flexible, and cost-effective ways to access data and business applications on our own preferred devices. It also protects business data and distributes and manages enterprise content on personal digital devices.

KEY WORDS: *dynamic pricing, pricing strategies, cloud business models, pricing, revenue sharing, customer relationship, revenue streams, cost structure, key activities, brokering, price allocation, billing, planned downtime, unplanned downtime, availability.*

12.1 Introduction

Mobile communication technologies are increasingly gaining popularity. Due to the exponential growth in mobile businesses, recent researches have been focused on cloud computing for mobile devices for which MCC environments have been developed [1]. MCC enables running applications, which are beneficial, between resource-constrained devices and cloud [2]. But when resources are limited, it is critical to build a business model for mobile cloud services. If one knows how resources are shared between CSPs, it would be easy to understand the business models of MCC, which include economic models and resource-sharing models. MCC mainly deals with improvement in the performance of mobile applications and enhances resource utilization by service providers [2]. Multiple service providers offer mobile services such as online gaming in the mobile environment. They can form a coalition to create a resource pool to improve the efficiency and utilization of long-term reserved wireless access bandwidth and servers in data centers. In the mobile cloud business model, an economic model is presented. In the new model, MCC providers share resources with each other to overcome the resource limitation problem [2]. Companies are looking for a business world where every aspect is mobile enabled. Cloud computing provides a new horizon for the creation and consumption of next-generation business applications. Advanced levels of cloud service provide software as a service (SaaS) applications, which live in the cloud and can be used at any place from any device. SaaS companies such as Salesforce.com and Concur offer mobile versions of their services.

Earlier it was difficult to start a business without capital funding as infrastructure cost was very high. Today, anyone with talent can build a mobile application and test it using cloud services. Cloud computing provides scalability, helps companies in building applications, and supports millions of users.

Instead of building mobile and cloud strategies separately, companies should think of building a combined strategy. In the mobile cloud world, a company must build the following:

- *Network-aware service*: The company must understand how its applications and services will react while connecting with erratic environment.
- *Sensor-aware service*: The company should be aware of what type of data are available. The Internet of Things devices offer a wide range of data sources such as orientation, motion, and environmental conditions.
- *Device-aware service*: The company should know what features the device offers.
- *Identity-aware service*: The same identity and access management service should work seamlessly for both cloud services and mobile applications.

In the mobile cloud world, companies have to design a big data and analytics strategy to deliver the right information at the right time to a person's device of choice, such as tablets, smartphones, and PCs. This concept is known as the "next generation of IT."

12.2 Cloud Business Models

A business model is a blueprint for a strategy to be implemented by organizational structures, processes, and systems. It describes how an organization creates, delivers, and captures value. It is described by nine basic building blocks as follows [1]:

1. *Customer*: One or several customers should be served by an organization.
2. *Value propositions*: Helps to solve problems of customers and tries to satisfy customer needs.
3. *Channels*: Delivers solutions to problems and satisfactory factors.
4. *Customer relationship*: Relationship is established and maintained with each customer segment.
5. *Revenue streams*: Results from value propositions successfully offered to customers.
6. *Resources*: Assets required to offer and deliver the aforementioned elements.
7. *Activities*: The most important initiatives an organization must perform to make its business model work.
8. *Partnership*: The network of service providers.
9. *Cost structure*: Describes all costs incurred to operate a business model.

An MCC business model satisfies all these blocks in operation during real-world business implementation.

12.2.1 Cloud Computing Architecture

Many cloud companies are increasingly motivated to create optimized business models for various aspects of cloud. Here, a hierarchical classification of business model is presented [3]. The business model consists of the following three technical layers:

1. *Infrastructure as a service*: This layer helps to enable technologies. There are two types of infrastructure business models: The first model provides storage capacity and the second model provides computation power. Amazon offers storage capacity (S3) and computing power (EC2). A majority of pricing models of cloud computing are pay per use or based on subscription of services to facilitate or map virtualization technologies. CSPs organize infrastructures in a cluster-like structure. The cloud providers who offer pure resource services can enrich their offers via value-added services, which can manage the hardware underlying this layer.

2. *Platform as a service*: This layer is located above the infrastructure layer in the cloud, providing value-added services from technical and business perspectives. In the development platform, application codes are written by developers and the applications run in a web-based manner, without the need to worry about scalability when application usage grows. Morph lab and Google App Engine provide platform for managing grails. Different business platforms such as Microsoft and Salesforce have gained attention for managing tailored business applications.

3. *Software as a service*: This is the top layer and works as the interface between users and cloud. It delivers the application via the infrastructure layers. An example of SaaS or B2B sector is SAP, which is used to deliver service-oriented business solution over the web. Xignite offers services on a pay-per-use basis.

12.2.2 Current Offerings of Cloud Computing

To fulfill user demands, cloud providers such as IBM, Google, and Oracle have to extend their resources to provide services for computation, storage, database, and application. The offered cloud services can be categorized into storage, billing, accounting, business process management, storage database, marketplace, data sharing, data processing, and web service. Pay per use is the simplest pricing model in which customers pay a fixed price for the usage of a fixed unit. If mass production and widespread delivery make price negotiation impractical, then the pay-per-use model is employed [3]. Another pricing model is the subscription model in which users subscribe to a combination of services for a longer time (monthly or yearly) for a fixed price. A simple pricing model is always preferred by users and providers. Durkee [3] has shown that dynamic pricing policies are more efficient than static pricing policies. A market in which cloud providers have scarce resources and high demand, capacity allocation depends on customer choice, classification, and appropriate pricing. Cloud providers can gain higher revenue by offering customized products with additional services based on the same commodity.

12.2.3 Pricing Strategies of Cloud

Every CSP offers services to customers at the lowest possible price because many cloud users select price as their primary decision criteria. In this way, many CSPs entering the cloud market provide services at a lower price than other CSPs. In this way, a price competition grows up between CSPs such that [3]

- Products are advertised artificially by differentiating from other products.
- Hidden prices are not mentioned to cloud users.
- Products are presented to customers by obfuscation of specifications.
- Product quality and pricing are lowered to promote increased sales.
- Customers are kept waiting for a long time without benefits.

Hence, price competition threatens cloud users from taking long-term subscriptions. Now we will focus on how competition between CSPs affects the cloud market.

Many cloud providers use virtual infrastructure, resulting in variable and unpredictable outcomes. Furthermore, because many cloud providers have no idea of the underlying hardware and software, overcommitment is apt to result. Overcommitment techniques include the following [3]:

- Resource allocation, reallocation, and allowing customers to access the memory.
- Always allowing shared resources instead of private resources.
- Performance of service should be in range.
- Performance is reduced due to overallocation of resources on the physical server.

On the other hand, vendors choose low-priced, low-performance infrastructure to provide cloud computing at lower prices. Some issues that prevent cloud providers from guaranteeing enterprise-grade performance include the following [3]:

- *Traffic shaping*: Sometimes, data rates are dropped due to the transfer of virtual private data center, and that is called traffic shaping.
- *Using fast Ethernet*: As the cloud computing component and servers are connected via fast Ethernet switches, customers have to pay more price for faster Ethernet.
- Maximum disk drives in the server are about 3 years old.

12.2.4 Method of Dynamically Pricing Resources

Cloud computing provides resources as a service over the Internet, and users have no knowledge of the underlying infrastructure. Cloud providers offer infrastructure as a service, communication as a service, and software as a service based on resource types. Here, cloud means both the hardware and the software that enable services. There are mainly two types of clouds: public cloud, which is available to all users, and private cloud, which is available to limited users because it provides services to users within an organization. Now most companies [3–13] use the pay-per-use fixed pricing model. However, the usage of cloud computing has been increasing in both services offered and resource providers. Many cloud providers offer similar services due to the increasing number of cloud users. To improve scalability and reliability, users utilize the same service across the cloud with interoperability features. The main objective of federated cloud is to integrate the resources of many cloud providers and transparently offer services to users.

Cloud providers and users are rational, self-interested parties who tend to maximize their benefits. Accordingly, both of them will not reveal their true resource valuation when allocating resources. Figure 12.1 depicts a strategy-proof dynamic resource pricing scheme suitable for allocation of resources on federated clouds where pricing is used to manage rational users [4]. A rational user may be an individual or organization depending on application. In federated clouds, users may ask for more than one type of resources from various providers. On the other hand, in fixed pricing, users manually aggregate resources from different providers, and a dynamic pricing scheme allocates a request for multiple resource types. In federated clouds, supply and demand of the resource vary with users joining and leaving the system. Mihailescu and Teo [4] have shown that using a dynamic pricing scheme, user welfare, successful request percentage, and allocated resources percentages increase when compared to fixed pricing. Standalone and federated CSPs are shown in Figures 12.1 and 12.2, respectively.

12.2.4.1 Dynamic Pricing Scheme

As the demand and supply change continuously, the fixed pricing strategy does not reflect the current market price, which leads to lower user welfare and imbalanced market. Figure 12.3 shows the loss of welfare by sellers due to the use of fixed pricing. There are two types of demands:

1. *Under demand*: In this case, the fixed price becomes higher than the market price and users may look for alternative resources.

2. *Over demand*: The seller welfare becomes limited by a fixed pricing scheme, which could be increased when the resource price is higher, as shown in Figure 12.3.

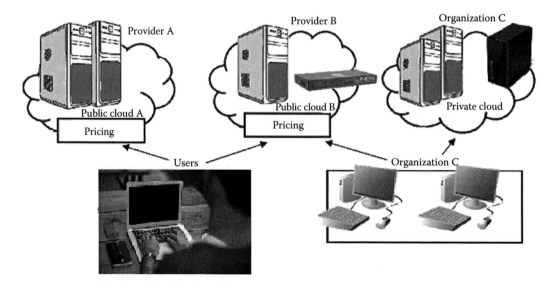

FIGURE 12.1
Standalone cloud service providers use fixed pricing. (From Mihailescu, M. and Teo, Y.M., Dynamic resource pricing on federated clouds, *Tenth IEEE/ACM International Conference on Cluster, Cloud and Grid Computing*, 2010, pp. 513–517.)

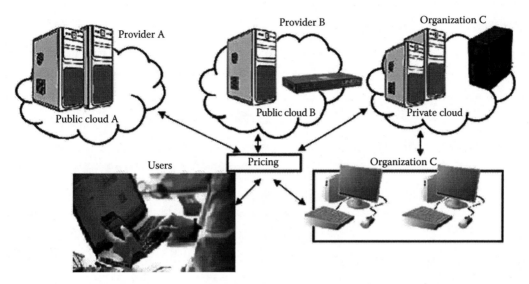

FIGURE 12.2
Federated cloud service providers use dynamic pricing. (From Mihailescu, M. and Teo, Y.M., Dynamic resource pricing on federated clouds, *Tenth IEEE/ACM International Conference on Cluster, Cloud and Grid Computing*, 2010, pp. 513–517.)

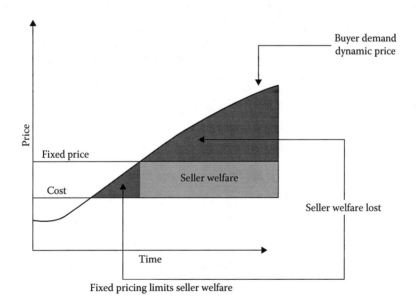

FIGURE 12.3
Fixed pricing scheme limits seller welfare. (From Mihailescu, M. and Teo, Y.M., Dynamic resource pricing on federated clouds, *Tenth IEEE/ACM International Conference on Cluster, Cloud and Grid Computing*, 2010, pp. 513–517.)

In a market of federated clouds, the pricing strategy sets resource payment based on the demand and supply. This dynamic pricing strategy has invoked sellers to provide multiple resource types. Unlike earlier cloud services, recent cloud services have introduced more resource types such as bandwidth and storage. Currently, Amazon has enlarged its offer to 10 different virtual machine configurations with different pricing strategies [4].

For simplicity, a centralized market maker has been used to compare the efficiencies of dynamic and fixed pricing. The centralized implementation has the advantage of measuring economic and computational efficiencies using a simple setup for a large simulated network.

12.2.4.2 User Welfare in Cloud Computing

The difference between user utility and payment is called user welfare [4]. Here user utility and advertisement price are same. Thus, both buyers and sellers are truthful. In the case of fixed pricing, a truthful buyer is considered, that is, advertising request price to represent the buyer's utility. In the case of sellers, it has the same resource price but it may differ from the seller's utility. Thus, only average buyer welfares are compared with respect to fixed and dynamic pricing. A balanced market is considered where supply and demand are equal [4]. Buyer price varies according to the same percentage as depicted in Figure 12.4, which shows that average buyer welfare increases in dynamic pricing scheme.

In the case of fixed pricing, the successful buyer request percentage is almost 50% for all market conditions (over demand, balanced market, under demand) since the buyer's cost is uniformly distributed with mean equal to the seller's cost. This successful buyer request percentage decreases with the increment of the number of resource types as the number of sellers assigned to fulfill buyer requests also increases.

On the other hand, in dynamic pricing this successful buyer request percentage varies under different market conditions. For instance, when supply is greater than demand, that is, in the underdemand condition, the buyer request percentage is higher than that in the balanced market condition when supply and demand are equal. However, in the

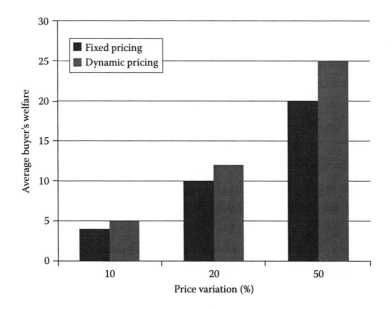

FIGURE 12.4
Comparison of price variation for fixed pricing and dynamic pricing. (From Mihailescu, M. and Teo, Y.M., Dynamic resource pricing on federated clouds, *10th IEEE/ACM International Conference on Cluster, Cloud and Grid Computing*, 2010, pp. 513–517.)

overdemand condition when demand is greater than supply, the successful buyer request percentage decreases. The dynamic pricing scheme shows a higher percentage of successful buyer requests [4].

12.2.5 Cooperation among Cloud Service Providers

In the federated cloud concept, one cloud provider shares the resources of another cloud provider and also dynamically shares the revenue. Cloud computing can reduce the total cost of ownership. Hence, this service can be used efficiently and flexibly [5]. With the available resources, cloud providers can generate more revenue to share resources. They can generate a virtual organization where they can create a coalition and offer services to public cloud users. But how resources and revenues are shared in the coalition and which coalition structure is desirable for all cloud providers are critical issues. The resource pool is shown in Figure 12.5.

A hierarchical cooperative game model is composed of two interrelated cooperative games to analyze the decisions of data center owners (i.e., cloud providers) to support internal users and offer services to public cloud users. First, in the lower level, a stochastic linear programming game model is developed to study the resource and revenue sharing for a given group of cloud providers (i.e., coalition). This game model takes into account the random internal demand of cloud providers to determine the optimal services, that is, available virtual machines to be offered to public cloud users. This stochastic linear programming game is equivalent to the linear programming game whose solution can be obtained as the core of cooperation. Second, in the upper level, the formulation of a coalitional game, for which cloud providers can form groups to share resources and revenues, is shown. The analytical model is based on the Markov chain issued to obtain a stable coalitional structure [5].

12.2.5.1 Brokering Mechanism in Mobile Cloud Business Model

The layered cloud architecture (SaaS, PaaS, IaaS) naturally leads to a design in which intercloud federation takes place at each layer mediated by a "broker" concerning the parties corresponding to that layer [6]. The federation among CSPs leads to an increment in consumer value. The cloud business model is consistent with broker-mediated supply and service delivery in finance and manufacturing.

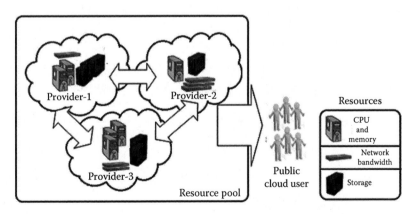

FIGURE 12.5
Resource pool in cloud computing service. (From Niyato, D. et al., Resource and revenue sharing with coalition formation of cloud providers: Game theoretic approach, *Eleventh IEEE/ACM International Symposium on Cluster, Cloud and Grid Computing*, 2011, pp. 215–224.)

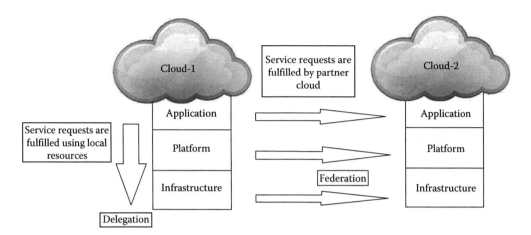

FIGURE 12.6
Federation and delegation in cloud application support. (From Villegas, D., *J. Comput. Syst. Sci.*, 78, 1330, 2012.)

Figure 12.6 shows the actual federated cloud structure mediated by brokers [6]. Here, two independent clouds are shown supporting service layer offerings from application or SaaS at the top of platform or PaaS to operating system or IaaS. Service requests are fulfilled using local resources through delegation or done by resources provided by partner cloud through federation. Users submit requests to the application layer, which assesses if sufficient local resources are available to service the requests within a specified time. If the application layer cannot meet its service goals, it can optionally fulfill the requests through an independent SaaS layer provider of the same service as indicated by the horizontal arrow (i.e., the federation connecting Cloud 1 to 2). Results are returned to the user as if locally produced by the application executing in Cloud 1. An application layer under stress also has a second option to increase capacity through delegation. In this service abstraction, the application layer works together with its underlying layers to provide the required computing needs. In delegation, the application layer asks the PaaS layer in the local cloud for additional resources. The request for more resources may be fulfilled in multiple ways depending on the availability in the current cloud. The PaaS layer can delegate a request for more raw virtual machines to the local IaaS layer and can provision the necessary platform software. If sufficient resources are not available locally, the PaaS layer can attempt to acquire them from another cloud in the federation through brokering at the PaaS layer. While attractive from a business perspective, this federated cloud model requires new technologies to work efficiently. As it is a layered model, an important part of the design is to maintain isolation of concerns between layers. For example, the SaaS application delivers a result to the customer in a certain response time. It is aware of the aggregate processing and network transmissions necessary to meet the delivery time. But the application does not need to know the details of the underlying infrastructure.

The federation model is shown in Figure 12.7. For this purpose, the Weather Research and Forecasting (WRF) SaaS is used. WRF is a batch mode service in which customers request weather forecasts over a region with a specified level of detail/resolution. It provides a good case study for cloud hosting as it is a high-performance computing application for which private and government agencies would like to leverage their joint resources through cloud services. Here, a single site hosts the WRF interface to the customer at the SaaS layer, and additional PaaS or IaaS resources are brought under the control of that site when needed to meet performance requirements.

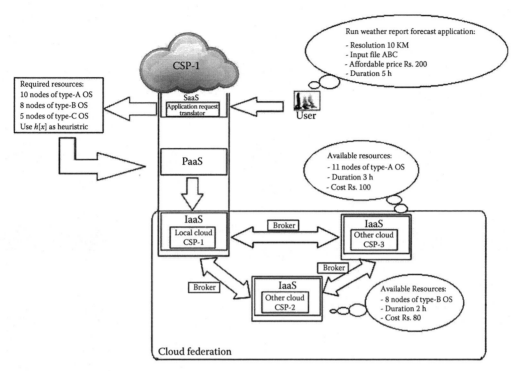

FIGURE 12.7
Conceptual model for brokering mechanism. (From Villegas, D., *J. Comput. Syst. Sci.*, 78, 1330, 2012.)

The user requests the CSP (i.e., CSP-1) to run an application named Weather Report Forecast providing the required specifications such as resolution, duration time, and affordable cost. The request is received by the SaaS layer of CSP-1. An "application request translator" in the SaaS layer translates the user request into the requirement of resources. If the required resources are not available in the SaaS layer, it passes the request to the local PaaS layer in search of additional resources, which again delegates the request for more resources to the IaaS layer in the local cloud to provide the required computing needs. After all these, if sufficient resources are still unavailable to fulfill the requirement, the IaaS layer looks for other clouds (say CSP-2, CSP-3) in the cloud federation through a "broker."

12.2.5.2 Broker as a Service

Several clouds are federated to enlarge the resource pool and extend the portfolio of resources via a cloud broker. Cloud brokers perform arbitration or interoperability-related tasks to federate resources. The main purpose of a cloud broker is to help a customer choose the most suitable CSP and service with respect to customer needs [7,12]. The broker has a cost management system for handling commercial services [7], as shown in Figure 12.8. The cloud broker has application programming interfaces (APIs) and standard abstract APIs to manage resources from different clouds. Another abstract API is used to negotiate with customers about cloud service facilities.

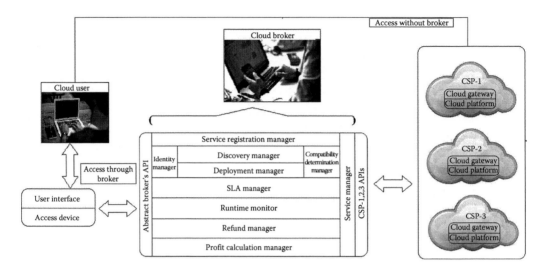

FIGURE 12.8
Broker's architecture and communication scenarios. (From Aazam, M. and Huh, E.N., Advanced resource reservation and QoS based refunding in cloud federation, *Proceedings of 33rd IEEE GLOBECOM*, 2014, pp. 8–12.)

Each module has its own specific utility. Cloud brokers help CSPs to deploy their resources in the form of services, and users find resources throughout the service-oriented architecture. Cloud service customers can access CSPs by transcoding related tasks and matchmaking. Intercloud gateways provide interoperability and transcoding related tasks. The cloud broker assigns an appropriate CSP to a customer, and during the consumption of resources, it observes the utilization of resources by the user and also the QoS provided by the service provider.

12.2.5.3 Resource Estimation and Pricing Model

Brokers negotiate with CSPs. After the contract is settled, customers are provided with the service. In this regard, brokers have to predict the consumption of resources to allocate them in advance. It allows efficiency at the time of consumption. This preallocation is done on the basis of predictions depending on user behavior in the past and the probability of use in the future [10].

12.2.5.3.1 Resource Estimation Model

The required resource is calculated as [7] follows:

$$R = \sum_{i=0}^{n} \left\{ C_{pi} \left(\left(1 - \overline{P_i \left(\frac{L}{H} \right)_s} \right) - \sigma^2 \right) (1 - A_i) \right\} \qquad (12.1)$$

where
$R \in \{\text{CPU, Memory, Storage}\}$
C_{pi} is the basic price of the requested service
$P_i(L/H)_s$ is the average of the probabilities of giving the same resource currently requested
σ^2 is the variance of the service-oriented relinquish probability

Customers are of two types:

1. Low relinquish probability $P(L)$ where $0 < P(L) \leq 0.5$.
2. High relinquish probability $P(H)$ where $0.5 < P(H) \leq 1$.

When a customer requests for a service for the first time, the value of $\overline{P_i(L/H)}$ is set to 0.3, which is the average of 0.1 and 0.5, that is, low relinquish probability is set. Customers may have fluctuating behavior in utilizing resources, which may prove deceptive in making decision for resource allocation. A_i is the constant variable assigned to a user according to the history of average overall relinquish probability defined as

$$A_i = \overline{\left(\left(\sum_{k=0}^{n} P_i\left(\frac{L}{H}\right)_k \right), P_i\left(\frac{L}{H}\right)_{last} \right)} \quad \text{where } k > 0 \tag{12.2}$$

$$= 0.3 \quad \text{where } k = 0 \tag{12.3}$$

With the help of Equations 12.2 and 12.3, the cloud broker can estimate the requirement of resources. It helps in power consumption management. Resource allocation for different types of users is presented in Figure 12.9.

12.2.5.4 Pricing and Billing

The price for service "S" requested by a customer having low relinquish probability is given by Jrad et al. [8]

$$D_{SP(L)} = \int_{0}^{t} (C_{pi} + \mu_L + A * \beta) \tag{12.4}$$

where β is the ratio set by a broker (e.g., 15% of the total resources).

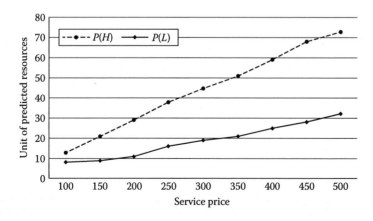

FIGURE 12.9
Resource prediction for different types of users. (From Aazam, M. and Huh, E.N., Advanced resource reservation and QoS based refunding in cloud federation, *Proceedings of 33rd IEEE GLOBECOM*, 2014, pp. 8–12.)

The price for service "S" requested by a customer having high relinquish probability is given by

$$D_{SP(H)} = \int_0^t (C_{pi} + \mu_H + A * \beta)$$ (12.5)

where the low and high relinquish probabilities μ_L and μ_H are given by

$$\mu_L = \frac{(C_{pi} * P_L)}{\delta}$$ (12.6)

$$\mu_H = \frac{(C_{pi} * P_H)}{\delta}$$ (12.7)

where δ is the total profit earned so far by CSPs from the current customer.

The more the profit earned from a customer, the lesser the value of μ added to the basic price for that customer; that is, the final price for the requested service will be less. The service price for different types of users is presented in Figure 12.10.

If a customer discontinues an ongoing service, the broker has to stop the service and refund the remaining amount to that customer. To calculate the refundable amount, the broker has to take into account a number of factors such as total service consumed by the customer, resources utilized by the customer, value of the service, and server downtime. Therefore, the refundable amount can be formulated as follows:

$$Y_t = Y_{un} + Y_{deg} + Y_{dt}$$ (12.8)

where

Y_t is the total amount to refund
Y_{un} is the refund amount for unutilized resources
Y_{deg} is the refund to be paid for service quality degradation
Y_{dt} is the refund to be paid for downtime

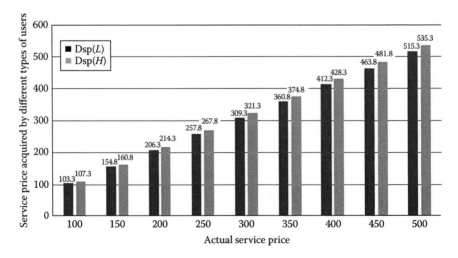

FIGURE 12.10
Service price for different types of users. (From Jrad, F. et al., SLA based service brokering in intercloud environments, *CLOSER*, 2012, pp. 76–81.)

The relation between the utilized resources (R_u) and unutilized resources (R_{un}) is given by

$$R_{un} = 1 - \left(\frac{R_u}{100} \right) \tag{12.9}$$

Therefore, the refund amount for unutilized resources is given by

$$Y_{un} = (R_{un} * C_{pi}) - \left(\frac{\beta}{R_u} \right) \tag{12.10}$$

As it is not always possible to deliver services as per the SLA, a CSP has to refund to the customer. The amount is calculated as follows:

$$Y_{deg} = \left[\left(\frac{Q_{SLA}}{Q_A} \right) * (C_{pi} - (Q_A * C_{pi})) \right] * \left(\frac{R_u}{100} \right)^2 \tag{12.11}$$

where
  Q_{SLA} is the quality promised in the SLA
  Q_A is the quality actually acquired

The value of Q_A varies from 0.1 to 0.9 where 0.1 denotes the worst service delivered and 0.9 means the best service delivered. The SLA always promises the best QoS, so the value of Q_{SLA} is always 0.9. Before introducing Y_{dt}, some capitalized terms used in the SLA are discussed as follows:

- *Affected customer ratio* (C_A): This is the ratio of unique visitors as measured by IP addresses affected by the unscheduled service outage to the total unique visitors as measured by IP addresses.
- *Customer planned downtime*: This is the downtime specified by customers, which is to be excluded from any calculation of outage period.
- *Unplanned downtime*: This is the total downtime minutes resulting from any condition or event beyond the company's reasonable control.
- *Availability* (M_A): This is calculated by adding the customer planned downtime and the unplanned downtime and subtracting the total value from the total number of minutes in the month.
- *Outage period* (M_{op}): This is equal to the number of downtime minutes resulting from an unscheduled service outage.

For this downtime inconvenience, CSPs have to refund the affected customers. The amount is calculated as follows:

$$Y_{dt} = C_{pi} * \left[\frac{(M_{op} * C_A)}{M_A} \right] \tag{12.12}$$

Hence, by adding the parameters Y_{un}, Y_{deg}, and Y_{dt}, the total refundable amount Y_t is determined and presented in Figure 12.11.

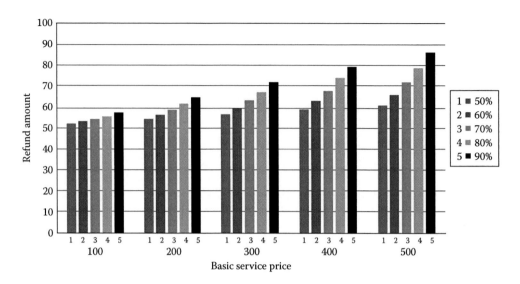

FIGURE 12.11
Refund amount for different resource consumption.

12.2.5.5 System Model and Assumption

A cloud provider offers computer resources such as CPU, memory, storage, and network bandwidth [6]. These resources are used to meet internal user demand.

When the internal demand is less than the capacity, additional resources are offered to the public cloud user after meeting the internal demand. To offer services to internal users, the cloud provider can create a logical pool of computing resources.

Here, the set of cloud provider can be represented by I. Let R_i be the set of resources by the cloud provider $i \in I$, where R_i = {CPU, Memory, Storage, Network-bandwidth}.

The available resource offered to the public cloud user, that is, the additional offer $r \in R_i$ of the provider $i \in I$ is denoted by R_i^r.

In virtual technology, all applications run on virtual machines. The set of applications is denoted by **A**, where $a \in A$ and a represents an instance of an application. Resource of type r required to support one instance of a virtual machine is denoted by Q_a^r. Resources of cloud providers are used not only by internal users but also by public cloud users. Cloud providers have to offer services in terms of the number of supplied virtual machines. For an application a, before actually knowing the internal demand, the number of virtual machines required is denoted by x_a. This type of service is called *committed offer*.

After knowing the actual internal demand, the unutilized resource can be offered to the public cloud users in terms of supplied virtual machines denoted by y_a for application a. This type of service is called *additional offer*.

Cloud providers have to decide the number of supplied virtual machines in the committed and additional offers. Therefore, the benefit in terms of revenue is *maximized*.

Without complete information when the decision of a committed offer has to be made, the internal demand (i.e., the number of required virtual machines) of provider i for application a is considered to be *random* and is denoted by a *random variable $D_{i,a}$*.

The internal demand for application a takes value from the set $D_{i,a} = \{D_{i,a}^1, ..., D_{i,a}^s, ..., D_{i,a}^{|D_{i,a}|}\}$, where $D'_{i,a} \in D_{i,a}$ and s denotes the scenario. The probability of demand $D_{i,a}^s$ is denoted by $P_r(D_{i,a}^s)$.

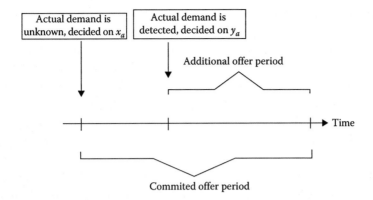

FIGURE 12.12
Activity diagram for cloud providers. (From Villegas, D., *J. Comput. Syst. Sci.*, 78, 1330, 2012.)

Let P_a^{comm} be the prices per virtual machine supplied to public cloud users for the committed offer and P_a^{add} be the prices per virtual machine supplied to public users for the additional offer.

It is assumed that $P_a^{comm} \geq P_a^{add}$ because the duration of service in the committed offer is longer than that of the additional offer, as shown in Figure 12.12.

Further, it is assumed that once the virtual machine is offered to the public user, it cannot be used by the internal user. Therefore, if an internal demand is not met (i.e., overdemand), the cost of overdemand incurs to the provider per an over-demanded virtual machine.

12.2.5.6 Resource and Revenue Sharing among Cloud Providers

Here, a stochastic linear programming game model is discussed to study resource and revenue sharing among cloud providers where the internal demand is random. The game is for a cooperative cloud provider I in the given coalition N, that is, i belongs to N, which is a subset of I.

The *price maximization* is given by [5]

$$\max_{x_a, y_a, z_a} \sum_{a \in A} \left(p_a^{comm} x_a \right) + \varepsilon_{\overline{d_i}} \left(\sum_{a \in A} \left(p_a^{add} y_a - c_a^{ovd} z_a \right) \right)$$

$$\sum_{a \in A} (x_a + y_a - z_a) Q_a^r \leq R_i^r - \sum_{a \in A} \overline{D_{i,a}} Q_a^r,$$

Subject to $r \in R_i, \overline{D_{i,a}} \in D_{i,a},$

$$x_a \geq 0, y_a \geq 0, z_a \geq 0$$

(12.13)

where
x_a is the number of virtual machines allotted for internal demand (i.e., committed offer)
y_a is the number of virtual machines allotted for public user demand (i.e., additional offer)
z_a is the number of virtual machines allotted for overdemand
P_a^{comm} is the price per virtual machine supplied to the public cloud user of the committed offer
P_a^{add} is the price per virtual machine supplied to the public user for the additional offer
c_a^{ovd} is the price per virtual machine supplied for overdemand
$\overline{D_{i,a}}$ is the expectation over the random variable $D_{i,a}$ representing the composite internal demand for all applications of provider i

12.3 Business Model of Mobile Computing Environment

Mobile business is a young promising industry created by the emergence of wireless data networks. The mobile business model consists of technology, application, network, services, regulation, and end users, which are illustrated in Figure 12.13 [3].

At the center of the mobile business world are users who have mobility-related needs. In order to fulfill their needs, the necessary and complementary supporting blocks that are required are as follows:

- *Technology*: This provides hardware and software infrastructure to end users for accessing mobile services. The primary actors in this area are device manufacturers and network equipment vendors. The secondary actors are device retailers, component makers, and operating system and micro-browser providers.

- *Services*: Users can access the content and application via their mobile devices. The primary actors of this domain are content and application providers.

- *Communication*: This provides communication service to end users, which enables access to mobile services. The primary actors are mobile network service providers and Internet service providers.

- *User needs:* The demands of end users or consumers are most important because ultimately they determine the success or failure of mobile businesses.

- *Regulation contexts:* They do not participate directly in the services but have a huge influence on other players. Players in this area include government, regulation authorities, and standardization groups [11].

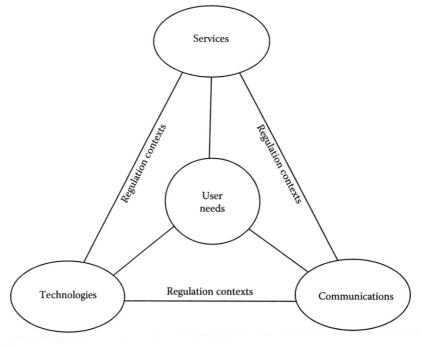

FIGURE 12.13
Business framework for mobile computing. (From Durkee, D., *Queue*, 8(4), 20, 2010.)

12.4 Cooperation among Service Providers

To enhance resource utilization and improve the performance of mobile applications, MCC is used. In the MCC environment (as represented in Figure 12.14), the service provider creates the resource pool cooperatively to support mobile applications [2]. To use the resource pool, an admission control mechanism is used for an optimization formula for an optimal decision, and consequently an optimal revenue-sharing model is introduced to share revenue among service providers. Here also a coalitional game model is developed.

12.4.1 Model of Mobile Cloud Computing

There is a set of areas (i.e., coverage areas of wireless access points) denoted by $[A] = \{1, ..., A\}$, where A is the total number of areas in an MCC environment. A set of access points is denoted by $[B] = \{1, ..., B\}$, where B is the total number of access points. The availability of an access point $b \in [B]$ to users in area $a \in [A]$ is denoted by $\alpha_{a,b}$, where $\alpha_{a,b} = 1$ if users in area a can connect and use bandwidth from access point b, and $\alpha_{a,b} = 0$, otherwise. There are P mobile applications offered in this MCC environment, and the set of mobile applications is denoted by $[P] = \{1, ..., P\}$. A set of data centers is denoted by $[D] = \{1, ..., D\}$, where D is the total number of data centers. The accessibility of a data center by a user using a mobile application is denoted by $\beta_{a,d,p}$, where $\beta_{a,d,p} = 1$ if users in area $a \in [A]$ using application $p \in [P]$ can run remote computing modules on a server in the data center $d \in [D]$, and $\beta_{a,d,p} = 0$ otherwise. In an MCC environment, as shown in Figure 12.14, N represents the set of S number of mobile CSPs; $K_{b,s}^{bw}$ represents the reserve bandwidth of provider $s \in N$ at access point b; $K_{d,s}^{cp}$ represents the number of servers reserved by the provider s at data center d; and R_p^{bw} is the required per instance of application p.

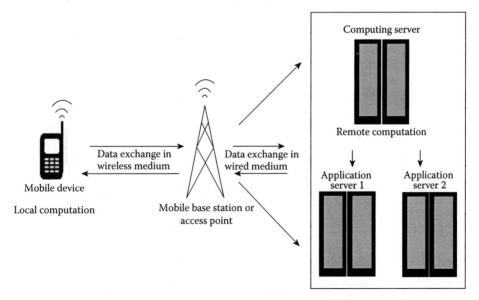

FIGURE 12.14
Mobile cloud computing model. (From Niyato, D. et al., Game theoretic modeling of cooperation among service providers in mobile cloud computing environments, *Wireless Communications and Networking Conference*, 2012, pp. 3128–3133.)

12.4.2 Model of Resource Pool

Multiple providers cooperate and create a resource pool to increase the amount of available resources. It is logically composed of reserved bandwidth from access points and servers in data centers to support mobile applications. Let $S \subseteq N$ denote a set of providers, that is, coalition cooperating to create a resource pool. $K_b^{bw}(S)$ and $K_d^{cp}(S)$ are the total reserved bandwidth and the total number of reserved servers at access point b and data center d in the given coalition S, respectively. The equations are given by [3]

$$K_b^{bw}(S) = \sum_{s \in S} K_{b,s}^{bw} \quad \text{and} \quad K_d^{cp}(S) = \sum_{s \in S} K_{d,s}^{cp}$$

The revenue obtained from a resource pool is aggregated for all cooperative providers in a coalition. A couple of issues arise with the cooperation among providers to create a resource pool. First, the revenue obtained from a resource pool must be shared among cooperative providers. Second, the providers have to decide the strategy of contribution to a resource pool, that is, to expand their capacity to gain higher profit or not.

12.5 Weblet-Based Mobile Cloud Computing Model

Cloud is the most promising service model for mobile devices. In MCC, the mobile device can perform its operations on the cloud. The weblet-based model has addressed the economic service provisioned for mobile cloud services [9].

In the weblet-based MCC, a service node manages loading and unloading of virtual images. A single service node handles one weblet at a time: either a transferred weblet request from an existing mobile cloud service domain or a new migrated request from the mobile device pool. Each virtual image can hold one weblet at a time. Service nodes handle three types of service requests, as shown in Figure 12.15:

1. *New weblet transfer*: A new weblet migration request is received or transferred from a mobile device or from other mobile cloud service provisioning domains.

2. *Intradomain transfer*: An existing weblet is transferred from one service node to another one within the same domain.

3. *Interdomain transfer*: A weblet is transferred from the current mobile cloud service provisioning domain to another. In the weblet-based mobile cloud service model [9], the cloud provides a large number of virtual images, though in reality the number is restricted to the cloud hardware configuration capacity. One virtual image is assigned to one CPU, and the CPU only handles one weblet at a time. This economic MCC model consists of three types of weblet migrations. These migrations are differentiated based on their economic gains in which an intra-domain weblet transfer usually generates higher economic gain than new weblet migration. The interdomain transfer migration means loss of revenue. Besides the economic gain, the economic MCC model needs to consider the cost due to occupation of CPUs. The model also has to consider the trade-off between the battery consumptions of mobile devices and the cloud service usage expenses. Thus, the total economic gain is determined considering all conditions.

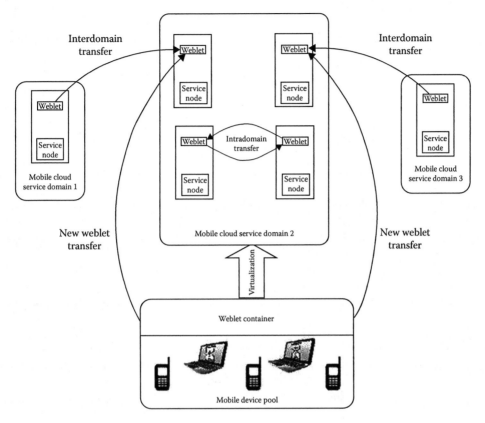

FIGURE 12.15
Weblet-based mobile cloud computing architecture. (From Huang, D. et al., Mobicloud: A secure mobile cloud framework for pervasive mobile computing and communication, *Proceedings of Fifth IEEE International Symposium on Service-Oriented System Engineering*, 2010.)

12.6 Mobile Cloud Service Insurance Brokerage

Mobile cloud insurance is a positive move toward risk management, which guarantees financial compensation for specific potential failures or downtime on the part of MCC service providers such as mobile network service providers or CSPs or application service providers. Mobile cloud service insurance is a novel approach presently used for mobile cloud brokerage [14]. In the mobile cloud service domain, an acceptable standard of QoS and quality of experience (QoE) must be maintained for subscribed services at the desired level. The performance measurement parameters of those mobile cloud services depend on the fulfillment of the customers' pre-approved QoS [15–24]. Variations in the QoS from that mentioned in the SLA result in the dissatisfaction of mobile cloud users. In the business environment of mobile cloud guaranteeing the runtime, QoS and QoE are essential for customer relationship management and retaining the potential customer pool. Sometimes, additional cost may be needed beyond the budget to maintain the desired level of quality. A consumer-level categorization is also required due to the limitation of resources and other runtime fault management. By this customer-level classification process, the service management for the premium consumers' QoS is ensured [25–38]. Adrija and Choudhury [14] have proposed inclusion of a novel service insurance tactic in the service broker as a component. The service insurances are expected to guarantee customer's satisfaction in the context of business aspects of the mobile cloud environment.

12.7 Business Aspects of Social Mobile Cloud Computing

In Chapter 8, we explored the architecture of social MCC. The convergence of the three major driving forces of mobile, social, and cloud has become the platform for new digital online business, which is the creation of new designs by blurring the offline world of business. Through big data analysis, business giants analyze the lifestyle patterns of users for selective advertisement. SMAC (social, mobile, analytics, and cloud) is the convergence of four technologies, which is currently driving innovations in the digital business world. Online business models for the social mobile cloud are used to transform conventional business using social media over mobile Internet and MCC. MCC exploits opportunities created by current technological advances to a major business advantage. The cooperative impacts of social technologies such as mobile Internet and cloud computing are creating new and incredible opportunities in business. They are also destroying unprepared or transforming industries and workers who are unwilling to or incapable of adapting. Business models for the social mobile cloud reveal a compelling view of how the social mobile cloud and new technology changes are playing key roles in the digital transformation of business and society, which are moving more quickly and cutting more deeply than any technology transformation ever seen [39].

12.7.1 Impact on Global Business

Social, mobile, analytics, and cloud create an ecosystem that allows a business to improve its operation and get closer to customers, which means maximum reach in minimum overhead. The structured or unstructured data generated by customers from mobile devices, sensors, social media, and web browsing have created new business models. The synergy created by social, mobile, analytics, and cloud together results in competitive advantage. The four components are shown in Figure 12.16.

12.7.2 Individual Roles of Components of SMAC

Each of the four technologies has its individual impact on businesses:

1. *Social*: Social media provides businesses with new ways to interact with customers.
2. *Mobile*: Mobile technologies have changed the way in which people communicate, shop, and work.
3. *Analytics*: Analytics provide the knowledge of how, when, and where people consume a certain service.
4. *Cloud*: Cloud computing provides new ways to access the technology and data a business needs to respond and solve business problems.

The convergence of these four technologies has built a disruptive force, which has created an entirely new business model for service providers.

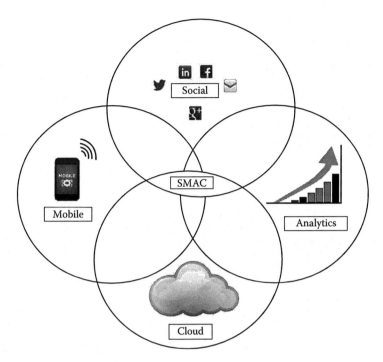

FIGURE 12.16
Components of SMAC for business advantage.

A media company named Netflix is a successful SMAC utilizer. Suppose a Netflix user streams a video clip for the Netflix cloud, say cricket match highlights, on his or her phone or tablet. That individual will have the option of signing into a Netflix account with Facebook's social login. After viewing a show, members are given multiple choices to provide remarks or feedbacks by rating the video with stars and writing reviews on social media. That data will be stored in the cloud. Netflix breaks down the analysis to a level such that its recommendation engine can provide suggestions for individual family members. This concept is known as "1:1 marketing." This marketing should be the ultimate goal of each and every SMAC initiative. There are some issues about which to be worried in the SMAC technology, and the most important among them is security. The 1:1 marketing initiative collects data from different sources. These data may be purchased from data brokers, which may violate customer privacy and data sovereignty [40].

12.8 Conclusion

In this chapter, various service models of MCC have been discussed in detail. In the MCC environment, it is assumed that some computing modules of mobile applications can run remotely on a powerful server in a cloud. Based on this assumption, the business model of cloud computing, cooperation among cloud providers, and the business model of cloud with federated version have been discussed. Finally, the economic models of MCC have been described along with the game theory based cooperation among mobile cloud providers. The insurance of mobile cloud-based environments is one of the key components of the success of BMCC. Broker-based fuzzy decision models are used nowadays based on the QoS and QoE of the mobile cloud user. Mobile cloud pricing depends on the QoE of the user and the security pricing of cloud-based mobile users. Dynamic pricing is desired over fixed pricing in the runtime environment. Price negotiation and auction in MCC are important for dynamic pricing to mitigate loss in real-time scenario. Mobile cloud insurance is a positive move toward risk management, which guarantees financial compensation for potential failures or downtime on the part of MCC service providers.

References

1. A. Osterwalder and Y. Pigneur, *Business Model Generation: A Handbook for Visionaries, Game Changers, and Challengers*, John Wiley & Sons, Hoboken, NJ, 2010.
2. D. Niyato, P. Wang, E. Hossain, W. Saad, and Z. Han, Game theoretic modeling of cooperation among service providers in mobile cloud computing environments, in *Wireless Communications and Networking Conference*, Paris, France, pp. 3128–3133, 2012.
3. D. Durkee, Why cloud computing will never be free, *Queue*, 8(4), 20, 2010.
4. M. Mihailescu and Y. M. Teo, Dynamic resource pricing on federated clouds, in *10th IEEE/ACM International Conference on Cluster, Cloud and Grid Computing*, Melbourne, Australia, pp. 513–517, 2010.
5. D. Niyato, A. V. Vasilakos, and Z. Kun, Resource and revenue sharing with coalition formation of cloud providers: Game theoretic approach, in *11th IEEE/ACM International Symposium on Cluster, Cloud and Grid Computing*, Washington, DC, pp. 215–224, 2011.
6. D. Villegas, N. Bobroff, I. Roderob, J. Delgado, Y. Liuc, A. Devarakonda, L. Fong, S. M. Sadjadi, and M. Parashar, Cloud federation in layered service model, *Journal of Computer and System Sciences*, 78, 1330–1344, 2012.
7. M. Aazam and E. N. Huh, Advanced resource reservation and QoS based refunding in cloud federation, in *Proceedings of 33rd IEEE GLOBECOM*, Austin, TX, pp. 8–12, 2014.
8. F. Jrad, J. Tao, and A. Streit, SLA based service brokering in intercloud environments, in *CLOSER*, Porto, Portugal, pp. 76–81, 2012.
9. D. Huang, X. Zhang, M. Kang, and J. Luo, MobiCloud: Building secure cloud framework for mobile computing and communication, in *The Proceedings of Fifth IEEE International Symposium on SOSE*, Nanjing, China, June 4–5, pp. 27–34, 2010.
10. H. Liang, D. Huang, and D. Peng, On economic mobile cloud computing model, in J. Y. Zhang, J. Wilkiewicz, and A. Nahapetian (eds.), *Mobile Computing, Applications, and Services*, Springer, Berlin, Germany, pp. 329–341, 2012.
11. G. Camponovo and Y. Pigneur, Business model analysis applied to mobile business, in *ICEIS*, Angers, France, vol. 4, pp. 173–183, 2003.
12. M. Aazam and E.-N. Huh, Inter-cloud media storage and media cloud architecture for inter-cloud communication, in *The Proceedings of the Seventh IEEE CLOUD*, Anchorage, AK, June 27–July 2, pp. 982–985, 2014.
13. Amazon Web Services, 2015. http://aws.amazon.com.
14. B. Adrija and S. Choudhury, Service insurance: A new approach in cloud brokerage, in R. Chaki et al. (eds.), *Applied Computation and Security Systems*, Springer, New Delhi, India, pp. 39–52, 2015.
15. C. M. Lawler, Cloud service broker model-sustainable governance for efficient cloud utilization. *Green IT Cloud Summit*, Washington, DC, Sheraton Premier, Tysons Corner, 18, 2012.
16. M. M. Hassan, B. Song, and E. N. Huh, A market-oriented dynamic collaborative cloud services platform, *Annals of Telecommunications*, 65, 669–688, 2010.
17. Cloud Vulnerabilities Working Group, Cloud computing vulnerability incidents: A statistical overview, Cloud Security Alliance, USA, UK, Singapore, 2013. https://cloudsecurityalliance.org/group/cloud-vulnerabilities/.

18. S. Pearson and A. Benameur, Privacy, security and trust issues arising from cloud computing, in *IEEE Second International Conference on Cloud Computing Technology and Science (CloudCom)*, Indianapolis, IN, pp. 693–702, 2010.

19. D. Catteddu, Cloud computing—Benefits, risks and recommendations for information security. European Network and Information Security Agency, Berlin, Germany, December 2012.

20. D. Gmach, J. Roliat, L. Cherkasovat, G. Belrose, T. Turicchi, and A. Kemper, An integrated approach to resource pool management: Policies, efficiency and quality metrics, in *IEEE International Conference on Dependable Systems & Networks*, Anchorage, AK, pp. 24–27, 2008.

21. A. N. Toosi, R. N. Calheiros, R. K. Thulasiram, and R. Buyya, Resource provisioning policies to increase IaaS provider's profit in a federated cloud environment, in *IEEE International Conference on High Performance Computing and Communications*, Banff, Alberta, Canada, pp. 279–287, 2011.

22. B. T. Ward and J. C. Sipior, The Internet jurisdiction risk of cloud computing, *Information Systems Management*, 27(4), 334–339, 2010.

23. R. Farrell, Securing the cloud—Governance, risk, and compliance issues reign supreme, *Information Security Journal: A Global Perspective*, 19(6), 310–319, 2010.

24. A. Cidon et al., Copysets: Reducing the Frequency of Data Loss in Cloud Storage, Presented at the *2013 USENIX Annual Technical Conference, USENIX*, San Jose, CA, pp. 37–48, 2013.

25. C. Wang et al., Privacy-preserving public auditing for data storage security in cloud computing, in *IEEE Proceedings of INFOCOM*, San Diego, CA, pp. 1–9, 2010.

26. M. Jensen et al., On technical security issues in cloud computing, in *IEEE International Conference on Cloud Computing, CLOUD'09*, Bangalore, India, pp. 109–116, 2009.

27. D. Jamil and H. Zaki, Security issues in cloud computing and countermeasures, *International Journal of Engineering Science and Technology (IJEST)*, 3(4), 2672–2676, 2011.

28. T. F. Fortis, V. I. Munteanu, and V. Negru, Steps towards cloud governance: A survey, in *IEEE Proceedings of the ITI 34th International Conference on Information Technology Interfaces (ITI)*, Cavtat, Dubrovnik, pp. 29–34, 2012.

29. M. Bezuidenhout, F. Mouton, and H. S. Venter, Social engineering attack detection model: SEADM, in *IEEE Information Security for South Africa (ISSA)*, Sandton, Johannesburg, pp. 1–8, 2010.

30. H. Raj et al., Resource management for isolation enhanced cloud services, in *Proceedings of the ACM Workshop on Cloud Computing Security*, New York, pp. 77–84, 2009.

31. A. Bakshi and B. Yogesh, Securing cloud from DDOS attacks using intrusion detection system in virtual machine, in *ICCSN'10, IEEE Second International Conference on Communication Software and Networks*, Singapore, pp. 260–264, 2010.

32. Cloud computing licensing: Buyer beware, 2015. http://searchcloudcomputing.techtarget.com/feature/Cloud-computing-licensing-Buyer-beware.

33. N. Cordell, *Intellectual Property in the Cloud*, Allen & Overy LLP, London, United Kingdom, May 2013.

34. M. Vrable, S. Savage, and G. M. Voelker, Cumulus: Filesystem backup to the cloud, *ACM Transactions on Storage (TOS)*, 5(4), 14, 2009.

35. A. Marinos and G. Briscoe, Community cloud computing, in M. G. Jaatun, G. Zhao, and C. Rong (eds.), *Cloud Computing*, Springer, Berlin, Germany, pp. 472–484, 2009.

36. C. Evangelinos and C. Hill, Cloud computing for parallel scientific HPC applications: Feasibility of running coupled atmosphere-ocean climate models on Amazon's EC2, Ratio 2, in *The First Workshop on Cloud Computing and its Applications*, Chicago, IL, pp. 2–34, 2008.

37. D. Niu et al., Quality-assured cloud bandwidth auto-scaling for video-on-demand applications, in *IEEE INFOCOM*, 2012.

38. H. Crowe, W. Chan, H. Leung, and H. Pili, Enterprise risk management for cloud computing, Committee of Sponsoring Organizations of the Treadway Commission, June 2012.

39. T. Shelton, *Business Models for the Social Mobile Cloud: Transform Your Business Using Social Media, Mobile Internet, and Cloud Computing*, John Wiley & Sons, Hoboken, NJ, 2013.

40. What mobile cloud means for your business, 2015. http://www.content-loop.com/mobile-cloud-means-business/.

13

Application of Mobile Cloud Computing

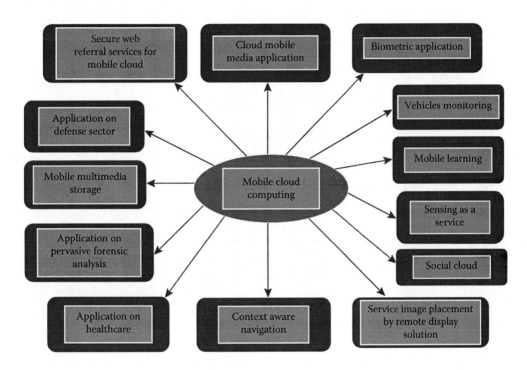

ABSTRACT Mobile cloud computing (MCC) is an emerging technology having many applications in mobile learning, monitoring of vehicles, digital forensic analysis, biometry, etc. With the help of MCC, storage and data processing are performed outside the mobile device and inside the cloud. MCC overcomes several disadvantages such as low bandwidth, limited speed, as well as limited storage capacity of traditional mobile learning, mobile health monitoring, and mobile gaming. In this chapter, various applications of MCC will be discussed.

KEY WORDS: *forensic cloud, mobile learning, biometric application, vehicles monitoring, context-awareness.*

13.1 Introduction

MCC is an integration of mobile computing into the cloud environment. With the rapid growth of mobile applications and the emerging concept of cloud computing, MCC has been pioneered as a potential technology for mobile services. MCC is a promising technology that allows data processing and storage outside mobile devices and in the centralized and powerful computation platform present in the cloud. MCC enables mobile devices to offload their applications to the cloud and greatly enriches the types of applications on mobile devices and enhances the quality of service (QoS) of the applications [1]. Cloud provides virtually endless resources that can be utilized by users in an on-demand fashion. The services in cloud can be accessed from the web browser of the client's mobile device over wireless network connection. This results in a considerable amount of energy saving and reduction in storage requirement for mobile devices, providing economic advantages for both cloud providers and users. The notions of user mobility and wireless access pattern give rise to certain challenges such as mobility management, QoS, energy management, and security and privacy issues to MCC. In this chapter, the different applications of MCC have been described.

The applications of MCC can be categorized on the following criteria:

- *Maintaining security and privacy of user data*: By using mobile cloud, a user can decide what information could be made public and what should be kept private.
- *Sensing capability*: A mobile device can be used as a sensor. Sensors can perceive the status of several environmental attributes such as temperature, humidity, and blood pressure. The sensor information can be sent to the cloud later. By accessing cloud, different users from different locations can access those data.
- *Health monitoring*: By using sensors, personal health information can be captured and sent to the cloud through mobile devices. Health centers accessing those data inside the cloud can give advice to users to improve their health status. Mobile devices can also act as sensors and can be used for personal health monitoring.
- *Reliability and data storage*: Mobile cloud protects users' data and recovers if failure occurs in the storage device.
- *Sensing as a service*: Mobile cloud provides platform, infrastructure, and software as a service. So, a user easily employs different applications with little compatibility issues.
- *Security of personal information*: Virtual machines present in the cloud ensure more security with the help of a secure search engine.

13.2 Cloud Mobile Media Application

MCC supports mobile applications with the features of cloud servers and storage along with the facility of mobile devices and mobile connectivity. Cloud mobile media (CMM) allow ubiquitous multimedia applications on mobile devices. CMM applications enable mobile users to access media from their mobile devices. Cloud mobile gaming (CMG) is one of the intensive CMM applications that requires high computation and mobile bandwidth. Multimedia libraries have gigantic volumes of information for communication in text, audio, graphics, video, and animation. With marvelous progression in technology, people are expecting more, and they want services anywhere and anytime.

13.2.1 Architecture

Heterogeneous applications of CMM can be provided by making use of cloud computing and its storage resources. They deal with different types of consumers. In this section, the overall architecture with end-to-end control and data flow between mobile devices and cloud server is described (Figure 13.1). CMM applications depend infrastructure as a service (IaaS) and platform as a service (PaaS) features located at public, private, or hybrid clouds. To access these applications, an instance of CMM is stored in the user's mobile device, which offers suitable user interface facilities with gesture, touch screen, voice, and text, allowing users to interact with the application running in the cloud [2]. From mobile devices, uplink is transmitted by control commands via the Wi-Fi access points or cellular radio access networks (RANs) to proper gateways such as packet data gateway (PGW) and service data gateway (SGW), situated at mobile core networks (CNs), after which data are finally sent to the Internet cloud. Transmitting downlink through CNs and RANs, the resultant multimedia data produced by the cloud are sent to the mobile device. The data can be the processing result of any application or retrieved from cloud storage resources. Thereafter, the CMM client decodes and displays the results on the display screen of mobile devices.

Services based on audio and video streaming are used in CMM applications. These are benefitted by cloud computing resources in executing demanding computation tasks of encoding, transcoding, and translation needed to adjust applications to different devices and networks [2]. Popular videos are cached at different resolutions and bit rates to reduce

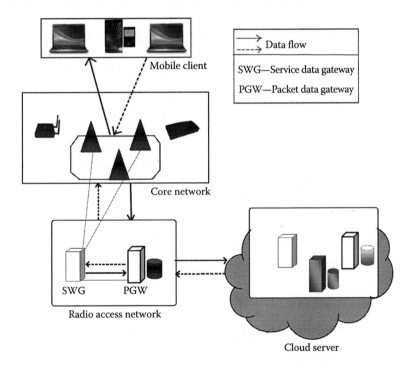

FIGURE 13.1
Cloud mobile media application architecture. (From Wang, S. and Dey, S., *IEEE Trans. Multimedia*, 15(4), 870, 2013.)

the computing cost. Cloud-based audio/video service not only reduces the initial capital expenses but also provides elasticity in cloud computing resources in order to deal with variable peak demands in a cost effective way.

An exceptional category of CMM applications is "cloud-based rendering." It can appreciably enhance the media practice of mobile users. On the basis of mobile capabilities, a gap is produced between the growing requirements of the latest 3D visualization and multiview rendering techniques. This can be improved with the support of cloud computing, and applications can run in the mobile devices such as tablets. Execution of all rendering techniques in the cloud instead of mobile devices permits users to run featured Internet games earlier available to high-end PC users only. This also allows users to participate in rich augmented reality and multiview immersive experiences, which are being developed for PC users. Multicore graphics processing units and suitable software are required in the cloud architecture to support the development of highly concurrent cloud-based rendering applications. In CMG, commands are sent from mobile devices. In response, the cloud renders and transmits the video via wireless networks to mobile devices.

Due to the transmission of video within the gaming session, significant additional bandwidth and fast response time result in high operating costs for the service provider. The network is overloaded with affluent traffic. Thus, video quality becomes low and the charge paid by mobile subscribers is increased due to tiered data plans.

An adaptive rendering approach is described by Wang and Dey [2]. The process of graphic rendering is to generate an image from the graphic scene file, which generally consists of geometry, texture, shading, lighting, and viewpoint information to describe the virtual scene. Rendering parameters are configured during the rendering setting phase. In CMG, communication complexity and computation complexity are introduced with each rendering setting. The first principle is to diminish object quantity in the graphic scene file as all the objects are not compulsory to play the game. In massively multiplayer online role-playing games, a participant needs to manipulate only one object: his avatar or representative in the virtual world of gaming. Removing objects such as flowers, rocks, and small animals will not hamper the game, and the load of graphic computation and complexity is reduced. The second principle is that rendering operations directly affect complexities. Finding the optimal rendering setting for the available cloud server computing resources and network bandwidth is time consuming. Also, the adaptive rendering setting should be performed in response to frequently changing network and server conditions. To overcome this conflict, the rendering adaptation technique is partitioned into two parts: online and offline steps. The offline steps will illustrate and predetermine the optimal rendering settings for different complexity levels. Hence in response to the variation of network and server resources, the online steps can select and vary the rendering settings in real time.

13.3 Biometric Application

Mobile devices can scan biometric evidence such as fingerprint, face, or iris and send the scanned file to the laboratory for processing. Information can be processed with the help of cloud computing, which supports resources for faster computation.

13.3.1 Face Recognition

Various techniques are classified according to how and what kind of features they extract and what learning algorithms are adopted. The MCC extensible system architecture for context-aware navigation by exploiting the computational power of resources is made available by cloud computing providers as well as the wealth of location-specific resources available on the Internet [3]. It includes a face recognizer in the cloud with input from a camera module integrated into sunglasses. The face recognizer is based on Viola-Jones AdaBoost features (VJ). This algorithm detects the presence of a face in real time to help in social interaction but cannot identify a particular face.

Face recognition is divided into three phases: face detection, extraction of suitable features from the file, and matching of features with the database for verification and identification. Mobile phones have applications that identify persons from a picture taken by the mobile camera if the person is in the contact list associated with his picture in the mobile phone. Cloud technology and social networking can be used to recognize people from a picture taken by the camera with contacts from Facebook accounts. Figure 13.2 shows the architecture of biometric applications.

13.3.2 Fingerprint Recognition

In forensic science, fingerprint recognition is an important phase of criminal justice. The crime branch of every country has made fingerprint datasets [3]. In the mobile biometric system, fingerprint identification is used to sign in for digital transaction. The technique of fingerprint verification is easier than fingerprint identification. Fingerprint is scanned with the help of mobile devices and sent to the cloud. In cloud computing, the fingerprint database is stored in the cloud server. The scanned fingerprint is checked against this database for identification. For fingerprint detection, a fingerprint orientation model based on 2D Fourier expansions (FOMFE) in the phase plane is proposed [3].

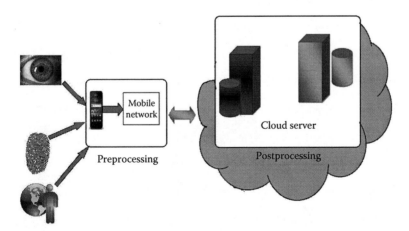

FIGURE 13.2
Architecture of biometric application. (From Stojmenovic, M., Mobile cloud computing for biometric applications, *Fifteenth International Conference on Network-Based Information Systems*, Melbourne, Australia, 2012, pp. 654–659.)

13.4 Vehicle Monitoring

13.4.1 Mobile Vehicular Cloud

The mobile vehicular cloud is used in the vehicular monitoring system, which consists of mobile nodes as vehicles and cloud computing as backend processing service. In the vehicle cloud, the leading applications are safe driving, urban sensing, content distribution, mobile advertising, and intelligent transportation. For example, vehicles pick up information via sensors (e.g., congestion, pavement conditions, surrounding cars, environment video clips, advertisements, etc.). They exchange and keep the data local, since local relevance and sheer volume of data make Internet upload unattractive. Other vehicles or Internet users can search for the data in the vehicle cloud with proper indexing and scope. There will also be significant computing on this cloud. For instance, computation of the full urban congestion picture, computation of the urban pollution map, collaborative reconstruction of pictures/video in accident or crime scenes, coordinated identification of possible terrorist threats, etc. are possible. Two applications are discussed in the following sections.

13.4.1.1 Route Tracking

A novel mobile sensor middleware, MobEyes supports proactive urban monitoring applications. Wireless vehicles are equipped with varieties of sensors and video cameras. Sensors detect an event and trigger the corresponding application. MobEyes processes and classifies the sensed data. Intervehicular routing is also one of its ingredients. The sensed data are stored locally, and onboard processors extract the required data, for example, license plate, coordinate position, and the like [4]. These summaries are dispersed at regular intervals in the cloud. A vehicle-tracking application can be implemented using MobEyes. In this application, police agents can reconstruct vehicle trajectories exploiting the collected summaries. Police nodes acquire information from vehicles through intervehicular routing. This application can be applied to monitor vehicles suspected of being used with terrorist intention. The application has to face challenges such as an increase in the number of targets, upload of fresh summaries periodically, and delivery of the information to the agent upon request.

13.4.1.2 Traffic Management

Information about the best route is conveyed to drivers so that they can avoid congestion. If all vehicles receive the same information about a route, they head toward the same route at the same time. This problem is called the route flapping problem. To overcome this problem, a navigator is placed in vehicles. The navigator is selected depending on the drivers' profiles and type of vehicles. Using mobile vehicular cloud, the instantaneous traffic flow can be learned by the navigator agency and differentiated route information is delivered to the drivers. It avoids the route flapping problem. Optimal routes are determined by the navigator and sent to the vehicles. In the cloud, the navigator server is implemented. It computes the traffic pattern from the destination ID carried by each onboard navigator message [4]. The server estimates road segment loads and delays and constructs the traffic load map and pattern matrix. It computes optimal incremental routes and dispatches such routes to the onboard navigators. The interaction between the onboard navigator and the navigator server helps in load balancing among numerous routes.

This application is a good example of synergy between vehicular cloud and Internet cloud. In particular, the sensing of segment traffic congestion is done in the vehicle cloud by means of reporting time and successive snapshots of GPS position, the route actuation, through instructions received by the onboard navigators from the navigator server. The navigator server, implemented in the Internet cloud, does the rest. It computes the traffic pattern from the destination ID carried by each onboard navigator message, determines optimal incremental routes, and dispatches such routes to the onboard navigators.

13.5 Pervasive Forensic Analysis

Digital forensics is a process to find legal evidence from computers and digital storage media. There are a number of commercial and open source tools for digital forensics investigation. Current forensic investigation requires correlative analysis of multiple devices and previous cases. Considering the benefits of MCC, the forensic service based on MCC could be a good solution to the problems faced by today's forensic tools. Typical forensic tools include a lot of functions for examining data on media, such as Windows registry reviewing, password cracking, and keyword searching. Today's tools running on a single system have limitations on forensic investigation because they need an extremely long time and computational power to analyze the growing size of digital devices. Additionally, the proliferation of operating systems and file formats causes another problem by dramatically increasing the complexity of data exploitation and the cost of tool development. Current forensic investigation requires correlative analysis of multiple devices and previous cases. Forensic cloud has been developed to show more feasibility in forensic analysis.

13.5.1 Pervasive and Collaborative Analysis

Many investigators apply numerous devices, for example, mobiles, personal computers, or notebooks, each of which is most prone to run on different operating systems and contain specific data and content. This makes it difficult for the investigators to easily access or share evidence data across devices or to maintain unified data. Productivity in forensic investigation can be improved by cooperating with other examiners and by sharing the evidence [5]. Mobile forensic analysis based on cloud computing enables collaborative investigation to be available anytime at any place. Providing data visualization based on behavior-based analysis and correlative analysis and supporting mobile-based forensic environment could be possible by developing a forensic service framework based on cloud computing.

13.5.2 Forensic Cloud

The technology of digital forensic investigation has been quickly changing, and today's digital forensic utensils have some problems in systemically answering the challenges. A new concept called forensic cloud has been developed as a new platform in digital forensics, as shown in Figure 13.3 [5]. The aim of forensic cloud is to enable forensic examiners to concentrate on the investigation process by separating technology from investigation. In order to implement this concept, the following requirements should be met:

- High-speed processing of basic forensic functions such as imaging, analyzing and searching, password cracking, etc.
- Supporting user mobility
- Secure data access

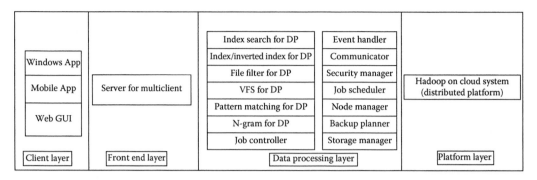

FIGURE 13.3
Forensic cloud architecture. (From Lee, J., Pervasive forensic analysis based on mobile cloud computing, *Third International Conference on Multimedia Information Networking and Security*, Shanghai, China, IEEE, 2011, pp. 572–576.)

13.5.2.1 Framework Design

The framework to implement mobile forensic analysis based on cloud computing is presented in this section. One of the important applications is the forensic index-based search. While it takes a long time to construct a database, it takes a short time for query processing and responding to the query. The framework has four layers: platform layer, data processing layer, front-end layer, and client layer.

13.5.2.1.1 Platform Layer

The platform layer lies at the bottom of the framework and is composed of distributed systems. In particular, Apache Hadoop is installed to manage the distributed systems. Hadoop provides various kinds of features for reliable and scalable distributed computing. An application developed on Hadoop can be easily migrated onto other Hadoop systems on cloud. For example, it is likely to run Hadoop on Amazon EC2 and S3. For the availability of Amazon Elastic Map Reduce services, applications based on Hadoop can be executed on that service without any alteration.

13.5.2.1.2 Data Processing Layer

This layer comprehends a lot of modules for establishing forensic analysis application. There are the N-gram tokenizer and pattern analyzer for creating index token. A file filtering component is used to extract plain text from various file formats such as .pdf, .docx, .zip, and the like. Additionally, a virtual file system library is essential to mount and deal with forensic imagery, which is obtained by the procedure in which the complete drive contents are imaged to a file and checksum values are considered to confirm the integrity of the image file. The inverted index is an index data structure where data are stored through mapping in a database file, or in a set of documents. Producing and navigating the inverted index structure is an important part of the index search procedure.

The inverted indexing component holding distributed computing is requisite to afford quick and full text searches on enormous data. The security resolutions, such as user authentication and access control, are offered because security might be a severe concern over digital forensic investigation on mobile devices. In addition, other components for exceptions handling and system protection are required to provide consistent distributed computation.

13.5.2.1.3 Front-End Layer

To support various applications, there are many servers present in the front-end layer. Clients send a request to the server for an application, and the server responds to the client with a suitable result.

13.5.2.1.4 Client Layer

There are three types of end users in the client layer: Windows application for the lab analysis, web application for the remote analysis, and mobile application for the pervasive analysis.

13.6 Mobile Learning

A mobile learning (M-learning) system consists of mobile devices and cloud computing and is implemented for supporting education systems. The main purpose of M-learning is that users can get information from centralized but shared resources anytime and anywhere they want to read without any cost. M-learning is an arrangement in which one can study any topic through any resource without the requirement of storing everything in their device. On the basis of payment, the services from the data centers of the cloud can be utilized for learning preferred topics over a mobile phone from any remote area [6].

13.6.1 Mobile Learning Procedure

People have to register first to access the M-learning system from cloud and study the regulations to use it via web. M-learning can also be downloaded and installed in the mobile as an application. People can access the information over the cloud through GPRS/Wi-Fi. Users can choose a particular topic among various topics. The data might be text documents, audio, and video files, which will be buffered from the cloud to the mobile user and downloaded in the mobile as per the availability of the memory capacity of the mobile. Two important sections of M-learning are cloud model and client model [6].

13.6.1.1 Cloud Model

The M-learning server associated with a huge database is situated at cloud. Registered users access the cloud through their PDAs. In this way, following an authentication check, the required information is shared from the server as per the client request. The process flow of mobile learning is shown in Figure 13.4. Storage is used to cache data and distribute the information among clients. Memory management refers to the organizing and managing of the data coming from the cloud to mobile subscribers. The process layer interacts with security firewalls and memory management.

13.6.1.2 Client Model

The client model depicts the user side. Registered users are allowed to download the application and install in their mobile phones. The users connect with the cloud with the help of GPRS/Bluetooth/Wi-Fi in their devices. They search for different subjects and select topics as per their area of interest. All the information related to that topic is downloaded to their

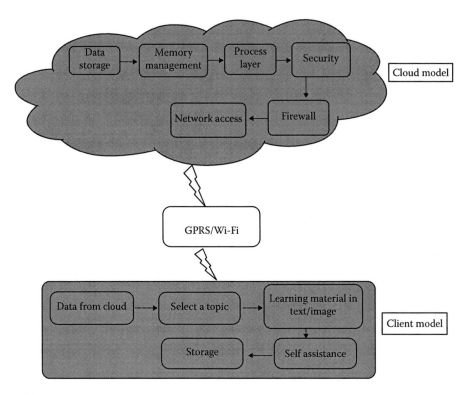

FIGURE 13.4
Process flow of mobile-learning cloud computing. (Reproduced from Rao, N.M. et al., *Int. J. Adv. Comp. Sci. Appl.*, Vol. 1, December 2010.)

mobile phone for reading. Figure 13.4 shows the data transfer from the data center of the cloud to mobile phones. The subscribers select a topic to retrieve from the data storage of the cloud and keep the downloaded or retrieved information in their mobile phones.

13.7 Remote Display

Remote display is a technology in which all higher processing parts of an application are performed on the cloud and users can access the ultimate result using their web browser in mobile devices. The main keywords for this technology are source and sink. Source is where the processing is done, and sink is the device where the result is displayed. As shown in Figure 13.5, video and audio are first played and encoded on the source side, and then they are sent to users' devices where the encoded part is decoded and displayed on the browser.

Remote display provides on-demand access in cloud instead of using mobile devices for processing data and complicated computing modules. It helps to expand the barrier of the mobile operating system. Since all applications are run in the cloud server, the users' mobile devices have no direct contact with the application. It helps developers in creating programs at lower cost. Only one version of any application is created, and it is employed by all users. The QoS of this technology depends on the quality of the video screen of the

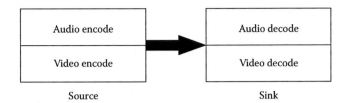

FIGURE 13.5
Remote display technology.

users' mobile devices. The price of programs and hardware cost are reduced. Developers get the chance to assemble programs at low cost. The mobile cloud provider builds a remote screen image (RSI) to run various services of different applications. These are known as virtual machines. Receiving results of related services later, RSIs collect and order those results on the screen. The screen is captured and sent to the users' mobile devices [7].

13.8 Context-Aware Navigation System

To navigate safely, a blind person must be able to detect nearby obstructions and curbs on the road and must also learn to detect stairs inside buildings, interpret traffic patterns, find bus stops, and know location. So, he or she must be fully aware of the context of their living environments. He or she must be able to track personal items. Existing navigation systems for the blind and visually impaired people fail to address the important aspects of context awareness and safety [8]. Most of these systems depend heavily on the underlying infrastructure. When the infrastructure is not available, they are useless. Another shortcoming of the existing navigation systems is the limitation or lack of contextual information provided to the user. The MCC environment can provide such a context-aware navigation system, the architecture of which is shown in Figure 13.6. In this system, the camera module integrated with the sunglasses captures visual context data. This integrated device is connected with the mobile phone with the help of Bluetooth. There is a speech device integrated with the mobile, and this device takes commands from the user.

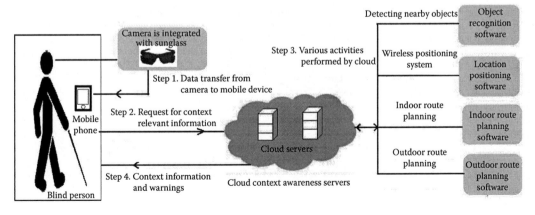

FIGURE 13.6
Context-aware navigation system architecture.

Context-aware guidance is achieved with computation in cloud and users' mobile devices. The mobile is integrated with compass and GPS locator. Cloud is reserved for performing different tasks such as refined location information, indoor route planning, outdoor route planning, and object recognition.

13.9 Cloud Computing Support for Enhanced Health Applications

For mobile health information delivery, access, and communication, mobile devices are being considered the main platforms [9]. However, mobiles devices face several challenges such as limitations in computation and power supply. Mobile devices are unable to run heavy multimedia and security algorithms as they have limited computational capacity and run on small batteries. Cloud computing provides functionality for managing information data in a distributed, ubiquitous, and pervasive manner. So, if mobile devices are integrated with cloud, several issues such as storage, remote access, and computation can be managed. This architecture can also be implemented in healthcare applications, as shown in Figure 13.7. Using the proposed system, an environment can be constructed for healthcare data storage, update, and retrieval. In the front-end interface of cloud computing, users communicate directly with the cloud, allowing the management of storage content. This architecture consists of three modules:

1. *Patient module*: Patients are the end users of this application. A valid patient must subscribe first to use all the features of this application. The patient needs to send necessary documents for registration in order to subscribe.

2. *Administrator module*: The administrator is the person who has a medical background. The functions of an administrator are as follows:

   a. *Generation of unique patient ID*: After verifying all the documents of a new user, the administrator registers the user as an authorized customer and sends a unique patient ID for future login.

   b. *Authentication with patient ID*: In the future whenever the patient wants to login again, the administrator first verifies the entered patient ID.

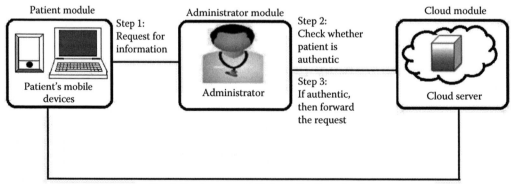

FIGURE 13.7
Architecture of cloud-based mobile healthcare system. (From Sutar, P.B. and Kulkarni, B.P., *Int. J. Eng. Innov. Technol.*, 2(1), 139, 2012.)

c. *Update the data in cloud*: The patient cannot update or perform any modification on the data stored in the cloud. Only the administrator has the rights to update the cloud's data to attain data security.

d. *Account management*: All registered customers accounts are managed by the administrator, who has the backup of all information related to a patient's account.

e. *Billing issues*: For accessing the cloud, the customer or patient has to pay charges. All billing issues are handled by the administrator.

3. *Cloud module*: Cloud stores patients records and performs computation.

The advantages of a cloud-based mobile healthcare system are as follows:

- *From patient's view*: Using this system, a patient does not need to carry all records and case history every time he or she visits another doctor in a different location. By using this application, the patient can access data from anywhere in the world at any time.

- *From hospital's view*: It is safe to store important data in cloud rather than in any local database as the cloud is a distributed system. Hence, healthcare centers can store all patient-related information in the cloud.

13.9.1 Femtocell-Based Health Monitoring Using MCC

In this architecture, a patient's health information is captured by body sensors and sent to mobile devices. From the mobile devices, this information is forwarded to the femtocell under which the mobile device is registered. The patient's health information is verified with the femtocell and checked whether the health status of the corresponding patient is normal or not. If any kind of abnormality is found, then that information is sent to the cloud. Health data stored in the cloud can be accessed by the medical team of the health center later. The team can immediately take proper action. The information is transmitted securely to the cloud, and the privacy of data is maintained through the use of femtocell [10]. The working model of this system is presented in Figure 13.8.

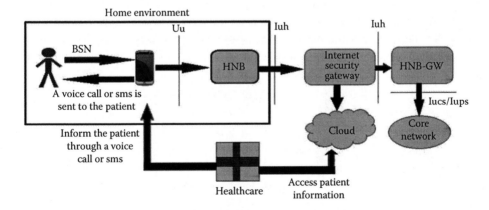

FIGURE 13.8
Femtocell-based health monitoring using MCC. (From De, D. and Mukherjee, A., Femtocell based economic health monitoring scheme using mobile cloud computing, *International Advance Computing Conference*, Gurgaon, India, IEEE, 2014, pp. 385–390.)

The main components of this system are body sensor network (BSN), femtocell, that is, home node base station (HNB), mobile station (MS), and Internet connection. As shown in Figure 13.8, the mobile users and HNB are connected via Uu interface. The HNB is connected to the Internet and HNB-Gateway (HNB-GW) via Iuh interface. HNB is connected with the core network with the help of HNB-GW. Iucs/Iups interface connects the HNB-GW and core network. A security gateway (SeGW) is maintained between the HNB and the HNB-GW for secure data transmission. First, a patient's health data such as body temperature, acceleration, heartbeat, blood sugar level, and blood pressure are captured by the BSN placed on the patient's body. Then those data are send to the MS. The MS sends those data to the HNB, which compares the received current data with the previously stored threshold data stored in the database inside the HNB. If any abnormality is found in the comparison, then the HNB sends those data to the cloud. All those data are stored safely inside the cloud. Later, the health center accesses those data and gives appropriate advice to the patient.

13.10 Sensing as a Service

Nowadays, a sensor can be attached inside any mobile device. Sensors attached to mobile devices can sense different types of information from different domains such as healthcare, environmental monitoring, and social networking. A new concept of sensing as a service can be introduced from this perspective, as shown in Figure 13.9. Here, sensing service is provided using mobile devices and the information is sent to the cloud. A sensing-as-a-service cloud must be able to support various mobile phone sensing applications on different mobile device platforms [11].

As shown in Figure 13.9, multiple sensing servers can be deployed to handle sensing requests from different users located in different places. First, a cloud user initiates a sensing request through an online form in a web server from any kind of mobile device. Then, the request is forwarded to a sensing server located in the cloud. The request is sent to a subset of mobile devices. The corresponding sensing task is fulfilled by those mobile devices. The sensed data will then be collected by a sensing server and stored in the cloud database. After that, the sensed data are returned to the requester. The most interesting part of this system is that a mobile device user is only be a cloud service user who can request sensing services from the cloud but also a service provider who can fulfill sensing tasks according to sensing requests from other cloud users.

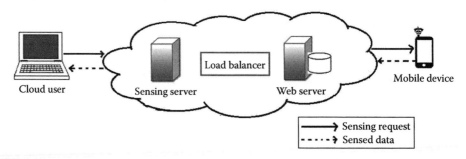

FIGURE 13.9
Architecture of sensing as a service system.

13.11 Secure Web Referral Services for Mobile Cloud Computing

World Wide Web and the hyper-linked web pages have made web browser a universal information portal for end users. An end user employs several services such as online chatting, electronic mail, and Internet services based on cloud computing. Users of mobile devices are easy targets for attackers or hackers. To cheat Internet users, attackers easily deploy malicious phishing websites to expose private information [12]. Cryptography-enhanced Internet protocols, for example, Secure Socket Layer (SSL) based HTTP, or HTTPS, have been widely used to protect Internet users from being attacked. However, those protocols are not enough to protect an Internet user. Two major security problems can occur due to human errors such as SSL Strip-based Man-in-the-Middle (MITM) attack and web-based phishing attack. In the SSL Strip attack, whenever a user types his username and password in the web page, the username and password are transmitted to the attacker in unencrypted format. In phishing, the human weakness can be found out using fake websites. To achieve good results against phishing websites and SSL Strip-based MITM attack, several techniques are present. All the existing security techniques rely on human factors. By using mobile cloud, a virtual machine can be provided to each user. This technique is called secure search engine (SSE). SSE consumes no power or computation on the client device. SSE acts as a personal security proxy. All the web traffics are redirected through SSE. Within the VM, the SSE can validate IP addresses by using web crawling technology. To check given URLs, a phishing filter is also used. SSE can remove potential human errors by returning deterministic answers of security inspections to either accept or reject the web request. The system can also protect user privacy and improve performance by using private and anonymously shared caches.

13.12 Mobile Multimedia Storage

Following the footsteps of technological advances, the use of mobile phones is increasing day by day, and so the importance of the mobile database has also rapidly increased. This type of database is generally called mobile database system, which stores an enormous amount of data and information in the cloud and can be accessed through the Internet. A multimedia database consists of image, video, audio, and text, so it requires a large storage space and high-speed data transferring [13]. This database is known as mobile multimedia storage.

There are various applications of mobile multimedia storage. The required data from mobile multimedia storage can be accessed anywhere and anytime through the mobile network. MCC is more or less dependent on the device and network mobility and promptness of the Internet. Nowadays, people can use their mobile phones as a small personal computer because of the robustness of applications and use of MCC. We can listen to music, backup any photos, see videos, access information anytime and anywhere since all these are stored in the cloud of mobile multimedia storage. The data can be easily accessed from the browser by typing the desired URL. These data are also well protected by the user's password and various encryption algorithms. This is very helpful in the development of mobile applications and their utilization.

A bigger challenge that MCC faces is the space constraint in which the mobile client and the server communicate with each other. Figure 13.10 illustrates a mobile multimedia database (MMDB) and the interaction of mobile cloud with other components such as a mobile device, broadcasting station, and network provider.

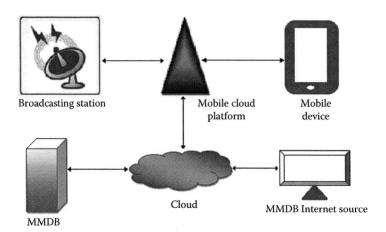

FIGURE 13.10
Interaction of MMDB with mobile cloud.

13.13 Application of Mobile Cloud Computing in Defense Sector

The defense department wants to use handheld mobile devices in the battlefield. By using these devices, a lot of tasks, such as image recognition, natural language processing, decision making, and mission planning, can be performed. For example, to easily identify a key target with high accuracy, a soldier can wear night vision goggles having the capabilities of object recognition. In object recognition applications, the accuracy of identifying the target and the time latency are important factors. The accuracy is controlled by an object recognition algorithm. This algorithm requires high computation power [14], but the mobile devices are not capable of doing large computation, so identifying the target object with high accuracy is a big challenge as time is critical in warfare. Various challenges in this field are as follows:

1. When we want to access the sheer volume of information (i.e., big data), the computation time is increased significantly.
2. Mobile devices in battlefield are connected to remote servers via satellite links. The latency of those links is far from negligible.
3. As mobile devices have limited storage capacity, they can utilize only locally available resources.

All these challenges can be dealt with the help of MCC, as shown in Figure 13.11.

In this architecture, a mobile device can take a picture of an observed object and send a request to the cloud server by attaching the image with the request. Then, the server can compare this image with images stored in its database. After that, it returns a matched result to the mobile device. The results are transferred via satellite. To reduce the delay, cloudlets can be used between mobile devices and cloud for data preprocessing and caching. The mobile device would send the picture to the cloudlet with the help of a high-speed link and the cloudlet carries out some or all the required processing. If the cloudlet is not

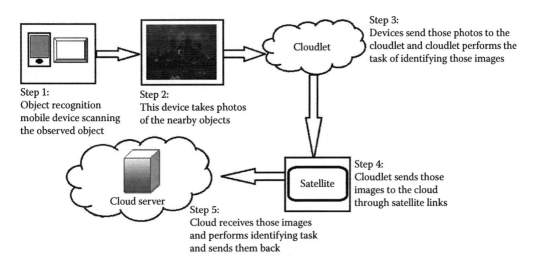

FIGURE 13.11
Application of MCC in defense sector.

able to perform all the processing due to lack of data, it sends the remaining computation request to the cloud through satellite. Cloud processes the request and sends response back to the cloudlet. Cloudlet then sends that request to the mobile device.

13.14 Application in Social Cloud

Nowadays, the Internet is the main medium that connects millions of people sitting in their rooms. Many of these Internet users maintain their relationships online through social networking sites such as Facebook, LinkedIn, Twitter, and others. For sharing information and communication between people, social networks have become an excellent platform. They can also reflect real-world relationships. Social networking plays an important part in daily lives. Facebook is one of the common examples of social networking sites.

Social networking sites serve as a vital platform for ideation and e-commerce. For example, some organizations and integrated applications use Facebook credentials for the purpose of authentication instead of their own credentials [15]. As a huge number of users are connected and they share information through those sites, a large trustable storage space is needed. Cloud can provide this type of storage space, which is known as social cloud. Cloud environments provide abstractions of computation. Storage clouds extend the capabilities of storage-limited devices as well as provide transparent access to data anytime, anywhere [15]. Figure 13.12 illustrates the social cloud architecture.

As shown in Figure 13.12, Internet users access various types of social networking sites from their mobile devices. They communicate among themselves through those sites and also share various types of important information. Since one site is accessed by a large number of users at a time, the volume of the data is very large. It is called big data, which must be kept secure. After that, those information go to the web and application server, which forwards them to the cloud server. Commercial cloud providers such as Amazon EC2/S3, Microsoft Azure, Google App Engine, and smaller open clouds such as Eucalyptus and Nimbus provide access to scalable virtualized resources.

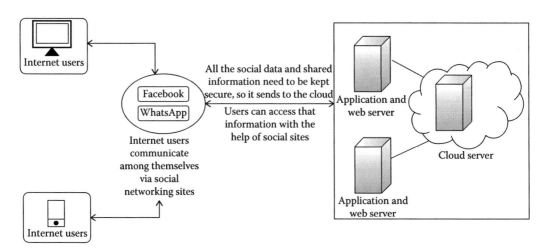

FIGURE 13.12
Social cloud architecture. (From Dhale, V.D. and Mahajan, A.R., Review of cloud computing architecture for social computing, *Proceedings of National Conference on Innovative Paradigms in Engineering and Technology,* Nagpur, India, 2012, pp. 15–19.)

13.15 Conclusion

MCC is a significant approach to enhance the features of real-time applications as it combines mobile devices and cloud computing technology. Mobile devices are used to collect realistic data from the environment, and data processing and distribution are performed inside the cloud. In this chapter, various applications of MCC, such as M-learning, mobile gaming, forensic analysis, mobile health monitoring, and so forth, were discussed. The challenges to be faced in the MCC applications were also studied.

Questions

1. Describe the architecture of cloud mobile media application?
2. Describe the biometric application of MCC?
3. Describe the architecture of mobile vehicular cloud?
4. What is forensic cloud?
5. How can MCC be applied to mobile learning?
6. What is a context-aware navigation system?
7. Describe briefly the applications of MCC in healthcare?
8. What are social cloud applications?
9. How is MCC applied in the defense sector?
10. What is sensing as a service?

References

1. H. T. Dinh, C. Lee, D. Niyato, and P. Wang, A survey of mobile cloud computing: Architecture, applications, and approaches, *Wireless Communications and Mobile Computing*, 13(18), 1587–1611, 2011.
2. S. Wang and S. Dey, Adaptive mobile cloud computing to enable rich mobile multimedia applications, *IEEE Transactions on Multimedia*, 15(4), 870–883, 2013.
3. M. Stojmenovic, Mobile cloud computing for biometric applications, in *15th International Conference on Network-Based Information Systems*, Melbourne, Australia, pp. 654–659, 2012.
4. M. Gerla, Vehicular cloud computing, in *11th Annual Mediterranean Ad Hoc Networking Workshop*, Ayia Napa, Cyprus, IEEE, pp. 152–155, 2012.
5. J. Lee, Pervasive forensic analysis based on mobile cloud computing, in *Third International Conference on Multimedia Information Networking and Security*, Shanghai, China, IEEE, pp. 572–576, 2011.
6. N. M. Rao, C. Sasidhar, and V. S. Kumar, Cloud computing through mobile-learning, *International Journal of Advanced Computer Science and Applications*, Vol. 1, December 2010.
7. T. D. Nguyen, M. V. Nguyen, and E. N. Huh, Service image placement for thin client in mobile cloud computing, in *IEEE Fifth International Conference on Cloud Computing*, Honolulu, HI, pp. 416–422, 2012.
8. P. Angin, B. Bhargava, and S. Helal, A mobile-cloud collaborative traffic lights detector for blind navigation, in *11th International Conference on Mobile Data Management*, IEEE, pp. 396–401, 2010.
9. P. B. Sutar and B. P. Kulkarni, Cloud computing support for enhanced health applications, *International Journal of Engineering and Innovative Technology*, 2(1), 139–141, 2012.
10. D. De and A. Mukherjee, Femtocell based economic health monitoring scheme using mobile cloud computing, in *International Advance Computing Conference*, IEEE, pp. 385–390, 2014.
11. X. Sheng, X. Xiao, J. Tang, and G. Xue, Sensing as a service: A cloud computing system for mobile phone sensing, *Proceedings of the IEEE Sensors Conference*, Taipei, Taiwan, pp. 1–4, 2012.
12. L. Xu, L. Li, V. Nagarajan, D. Huang, and W. T. Tsai, Secure web referral services for mobile cloud computing, in *IEEE Seventh International Symposium on Service Oriented System Engineering*, Redwood City, CA, pp. 584–593, 2013.
13. M. S. Khan, N. Qamar, M. A. Khan, F. Masood, M. Sohaib, and N. Khan, Mobile multimedia storage: A mobile cloud computing application and analysis, *International Journal of Computer Science and Telecommunications*, 3(3), 68–71, 2012.
14. P. Hu, J. Shen, and S. Fang, Application of mobile cloud computing in operational command training simulation system, in *IEEE 12th International Conference on Computer and Information Technology*, Chengdu, China, pp. 532–535, 2012.
15. S. P. Ahuja and B. Moore, A survey of cloud computing and social networks, *Network and Communication Technologies*, 2(2), 11, 2013.
16. V. D. Dhale and A. R. Mahajan, Review of cloud computing architecture for social computing, in *Proceedings of National Conference on Innovative Paradigms in Engineering and Technology*, Nagpur, India, pp. 15–19, 2012.

14

Future Research Scope of Mobile Cloud Computing

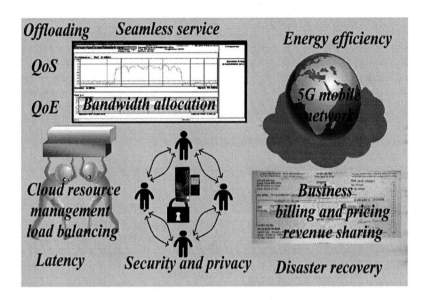

Will Densification be the Death of 5G?
Will Cellular Networks Eventually Become Interference-Overloaded?

Jeffrey Andrews
(Professor, The University of Texas at Austin ComSoc Technology News,
May 2015, IEEE Solution is MCC)

ABSTRACT Mobile cloud computing has overcome several disadvantages of mobile computing such as low bandwidth, limited speed, limited storage of mobile devices, etc. Still, there are various issues where research is needed. In this chapter, various applications of mobile cloud computing are discussed. Energy efficiency, latency minimization, efficient resource management, billing, and security are challenging areas of mobile cloud computing. In this chapter, we have focused on these fields with recommended solutions.

KEY WORDS: *energy, latency, pricing, resource, security.*

14.1 Introduction

Mobile cloud computing (MCC) has become an essential part of human life today. With the explosion of mobile applications and the support of cloud computing for a variety of services for mobile users, MCC is introduced as an integration of cloud computing into the mobile environment. MCC brings new types of services and facilities to mobile users to take full advantages of cloud computing [1]. MCC efficiently reduces energy and time consumption of application execution, enhances data storage capabilities of mobile devices as well extends the battery lifetime of these devices. Yet, there are several fields in MCC where more improvements, such as bandwidth, network access management, quality of service (QoS), pricing, service convergence, standard interface, energy efficiency, latency, resource management, and application migration, are required. Considering these factors, future research scopes in MCC are discussed in this chapter.

14.2 Efficient Bandwidth Allocation

Efficient resource allocation technique has become an emerging area of interest due to the explosive increase in the number of mobile users. Here, two technologies are considered to increase the low bandwidth: (1) fifth-generation (5G) network [2,3] and (2) femtocell.

14.2.1 Use of 5G Network

Fifth-generation network increases the bandwidth capacity for subscribers [2]. Recent MCC environment consists of third-generation (3G) and fourth-generation (4G) mobile networks. To increase the speed of execution, higher bandwidth is required. In every aspect, the 5G network is useful for MCC as it deals with small cell network with a cloud environment [3]. Due to the use of small cell, the signal strength as well as bandwidth is better.

14.2.2 Mobile Cloud Computing Architecture with 5G Network

Figure 14.1 shows the MCC architecture with 5G network. It shows four different networks—3G, 4G/LTE, WLAN, and GPRS/Edge—connected to the cloud.

This architecture may improve the quality of voice call and Internet services and expand the coverage area of network and link. But the 5G network involves heterogeneity. Densification arises due to the use of small cells [3], which in turn causes interference management problems. Research is required in this field to solve the problems of densification and interference management.

14.2.3 Mobile Cloud Computing Architecture with Femtocell Base Station

Femtocell is a small, low-power cellular base station [4–6]. It has a small coverage of approximately 10 m [7–9]. Hay Systems Ltd has combined femtocells [11] and cloud computing to offer a scalable, secure, and economical network service for mobile operators [1]. Femtocell can remove various obstacles such as low signal strength and low bandwidth. Inclusion of femtocell in the MCC environment is shown in Figure 14.2.

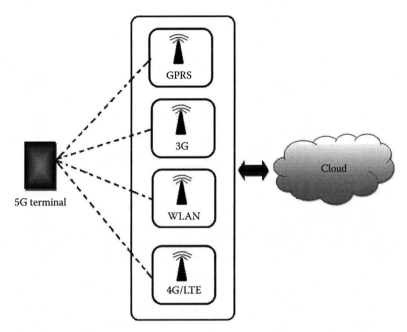

FIGURE 14.1
Mobile cloud computing architecture with 5G network.

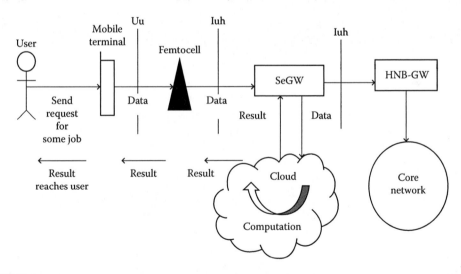

FIGURE 14.2
Mobile cloud computing architecture with femtocell base station.

Figure 14.2 pictorially depicts the proposed architecture where all the computations required in the mobile terminal are offloaded to the cloud [10] via femtocell. When a mobile user wants to perform some computations in the mobile device, the necessary data are sent to the cloud via the femtocell base station. The mobile user and femtocell are connected via Uu interface. The femtocell is connected with the core network by the femtocell gateway or the home node base station gateway (HNB-GW), which connects several femtocells or HNBs [11]. A security gateway provides a proper security mechanism between femtocell and HNB-GW [11] over the Internet. An IPsec tunnel is established between the femtocell and the security gateway; hence, all Iuh interface traffic is tunneled through this connection. The femtocell and HNB-GW are connected with the Iuh interface through a broadband connection. All the data required for the computation reach the cloud via the security gateway [11]. The results of the computations thereafter reach the mobile terminal through the security gateway and the femtocell. If the femtocell is used in MCC, the call dropping and blocking probability may be reduced. Offloading computation to the cloud reduces the power consumption of the mobile device. Moreover, using femtocell base stations instead of macrocell or microcell base stations, the power consumption of a mobile network can be reduced. These techniques are used in MCC for efficient bandwidth allocation and to reduce power consumption.

14.2.4 Efficient Spectrum Utilization Using Cognitive Radio

Improved link performance and optimized bandwidth use require efficient network access management [1]. Wireless access management is performed by cognitive radio in mobile communication [12]. Cognitive radio is a wireless communication technology where communication between nodes takes place through an optimal wireless system based on the availability of radio resources in a heterogeneous wireless communication environment [12]. This system utilizes the spectrum efficiently and allows unlicensed users to employ the spectrum allotted to licensed users. Cognitive radio is integrated into MCC for efficient spectrum utilization. In this case, mobile users should be able to detect the radio resource availability by spectrum sensing.

14.2.4.1 Mobile Cloud Computing with Cognitive Radio Technology

Cognitive networks promise reliable and timely data communication for accessing the cloud [12]. Both cognitive radio and cloud have computing and storage capacity, and cognitive radio efficiently utilizes the bandwidth. Hence, if we merge these two technologies, both of them overcome the performance bottlenecks of each other. An MCC environment based on cognitive radio [12] consists of heterogeneous environment. These technologies are used to utilize the bandwidth efficiently. But sometimes due to user mobility, many obstacles such as signal distortion and communication failure have become more critical to maintain seamless connectivity. So, it is mandatory to increase the QoS of the MCC environment. A new technology called cloudlet can be merged to the MCC environment to get high QoS.

14.3 Use of Cloudlet in Mobile Cloud Computing

A cloudlet is a computer or a group of computers connected to the Internet and accessible to nearby mobile devices. If the mobile devices do not wish to offload to the cloud due to cost and delay, a nearby cloudlet can be used [13]. Hence, mobile users can meet the demand for interactive response by reduced-delay, single-hop, and high-bandwidth wireless access to the cloudlet [13]. If no cloudlet is found nearby, the mobile device may access the distant cloud or, in the worst possible case, make use of its own resources. Despite the fact that cloudlets successfully deal with the limitations of high WAN latency, they still have two disadvantages [13]. First, mobile users remain dependent on the service provider for providing such cloudlet infrastructure in LAN networks. To alleviate this constraint, a more dynamic cloudlet concept is proposed by Satyanarayanan et al. [13], in which all devices in the LAN network can cooperate in the cloudlet.

The second shortcoming of virtual machine (VM)-based cloudlets [13] is the coarse granularity of VMs as an element of allotment. Instead of executing the whole application remotely in the VM, improved performance can be realized by dynamically partitioning the application into components. Moreover, as cloudlet resources are limited, there is a strong probability that the cloudlet runs out of resources when many users run their applications entirely in the cloudlet infrastructure. This limitation can be dealt with if the applications are offloaded in components rather than as a whole. Satyanarayanan et al. [13] have proposed a new cloudlet architecture in which applications are managed at component level. These application components are distributed between the cloudlets. This cloudlet is not fixed; mobile devices can join or leave the cloudlet at runtime. These features eliminate the disadvantages of conventional cloudlets as well as provide a solution to the high WAN latency problem associated with the cloud.

14.4 Cross-Cloud Communication

The competition between cloud service providers (CSPs) can lead to a differentiation of cloud services according to type, cost, and availability. When a single cloud cannot meet user needs, a composite of clouds can be developed to meet user demands in a unified fashion. Mobile sky computing can meet user satisfaction.

14.4.1 Mobile Sky Computing

Mobile sky computing [14] is an integration of mobile computing and sky computing. It enables providers to support a cross-cloud communication. Users can also implement mobile services and applications. Figure 14.3 shows the architecture of a mobile sky computing environment.

Mobile sky computing may overcome some shortcomings of the current MCC systems, for example, the difficulty of comparing offerings coming from different providers. Mobile sky environments would be greatly enhanced by the ability to negotiate specific latency and bandwidth between various resources provisioned in the clouds. This is the direction of cross-cloud communication.

14.5 Standard Interface in Mobile Cloud Computing

When mobile users communicate with the cloud, interoperability becomes a critical factor [1]. The web interface is the current interface between the mobile and the cloud. But several problems arise while using web interfaces:

- The web interface is not designed specifically for mobile devices; hence, the overhead is higher.

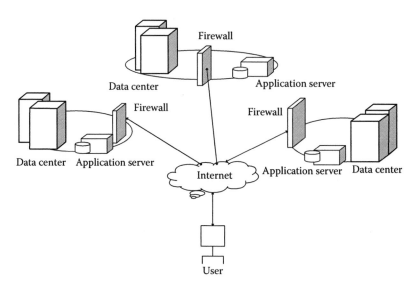

FIGURE 14.3
Architecture of mobile sky computing.

- Compatibility among mobile devices is another issue in which standard protocol, signaling, and interface are required for interaction between mobile users and cloud for seamless service. So, it is necessary to develop a suitable interface for MCC environment. A developer should emphasize on Android and windows for creating operating systems for an MCC environment.

14.6 Cloud Resource Access Mechanism

Heterogeneous networks consist of many technologies such as 3G, LTE, Wi-Fi, and WiMax [14]. MCC needs an always-active, on-demand available wireless links with a scalable bandwidth [12]. Since users in MCC are always moving, they need seamless connectivity. But when users move from one network to another, connectivity with the old network is snapped and a fresh connection is made with the new network. The operation takes some time in which running application processes disconnect. So to maintain seamless connectivity, the developer should create a protocol that provides global connectivity anytime, anywhere. IPV6 may be the protocol that will give this type of connectivity.

14.7 Elastic Application Model

An elastic application model that enables flawless use of cloud resources to enhance the capability of resource-constraint mobile devices is proposed by Amin et al. [15]. The model partitions a single application into multiple components called weblets and gives a variation of weblet implementation. A weblet can be either platform independent (like Java) or platform dependent, and its location of execution is transparent; that is, it can be executed on the mobile device itself or transferred to the cloud. Thus, an elastic application can augment the potential of a mobile device together with computation power, storage, and network bandwidth [15]. An elastic application has two general properties:

1. Elastic applications are partitioned for partial execution inside the mobile device and cloud. The partition is done such that each component or weblet has minimum dependency on the other. This increases robustness and reduces the communication overhead of the weblet.
2. The execution configuration, which is an assignment of weblet to the execution units of these applications, is not static. Rather, it is determined at a runtime when these applications are launched and modified.

Figure 14.4 shows the possible execution configurations for an application using two weblets.

An elastic application is shown in Figure 14.5. It contains weblets, a manifest, and a UI component. Weblets are independent software units that run on either the cloud or the device and illustrate web service interfaces through HTTP. The manifest is a static XML file, which consists of metadata for the application.

Manifest is used to specify constraints for the application and weblets. On the device side, the primary component is the device elasticity manager (DEM). DEM configures applications at the launch time and modifies the configurations during the runtime. The construction of an application includes the location of the weblets, information about whether or not the components are replicated or shadowed, and selection of paths used for communication with the weblets. The router forwards the requests from UI components to the weblets. Cloud elasticity service consists of the application manager, cloud manager, and sensing information collection. The cloud manager allocates and releases resources to and from the underlying cloud nodes. The application manager installs and maintains applications on behalf of elastic devices. The collection of operational data is called sensing information. A node manager on each cloud node supervises resources associated with the server. It directly communicates with the application manager and the cloud manager. Weblet containers run on each node.

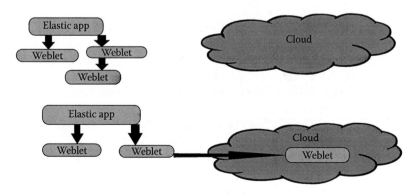

FIGURE 14.4
Execution configurations of elastic applications.

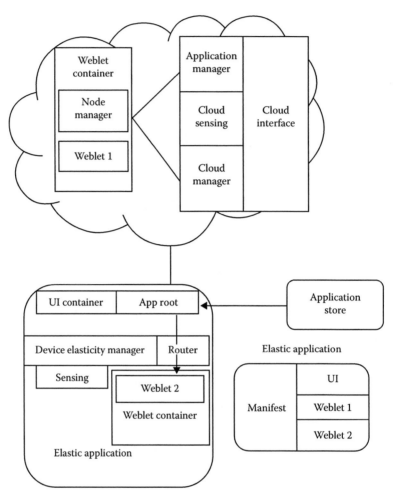

FIGURE 14.5
Reference architecture for elastic application.

The elastic application has the following advantages:

1. Elastic applications are independent of the computational abilities of the recent mobile platforms and can be organized to use the available multiple processing cores. When more computations are required to perform with a large amount of storage, the cloud can be used. The resources of mobile devices are reserved and consume less power. The complex applications can obtain cloud resources. The ability to assign cloud resources and migration makes the device more flexible. Moreover, the partitioned application components can be replicated. If an instance of the replicated component fails, the application is not affected.

2. This model provides a test bed for future mobile devices. The applications executed inside the cloud are moved to the future devices. This extends the application lifetime.

14.8 Security and Privacy in Mobile Cloud Computing

MCC can suffer from different attacks on the mobile device due to low power and slow CPU execution. So, it is difficult to restrict hackers from hacking the mobile device. So, multilevel security should be provided in the MCC environment. Another issue is authentication. Every time users should be re-authenticated by different providers.

14.8.1 Secured Mobile Cloud Computing

In the MCC environment, no data are stored in the mobile device during execution. The data go directly to the cloud, where the execution occurs. CSPs mainly work with third-party vendors, as there is no safeguard with data. Another case of privacy violation is "tracking" of individuals based on the location-based navigation data offloaded to the cloud. Furthermore, the data stored at a particular location may not be secured at another location due to different access rules. Since the user does not know the storage locations, it is hard to determine what laws to be applied for safeguarding data. Hence, serious security concerns may arise for a company that stores its trade secrets inside the cloud or a user uploading a unique concept to the cloud. Encrypting the data before storage can be a solution to the problem [16,17]. Figure 14.6a and b shows two encryption scenarios.

In Figure 14.6, the encrypted data are stored inside the cloud to prevent an unauthorized user to access the data. In Figure 14.6a, the cloud vendor operates on the encrypted data, whereas in Figure 14.6b, the cloud vendor decrypts the data to operate on it. Another possible solution is stenography [18], which is used to transform the data so that operations can be performed without exposing them.

14.9 Enterprise Mobile Cloud

14.9.1 Identity Security

Many applications are attached to the identity of the user [16]. When another user wants to access the application from the same resource, then he or she has to provide the identity sample to the device. After the identity is matched, the user can access the application.

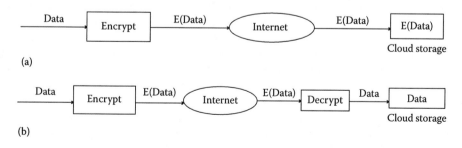

FIGURE 14.6

Two encryption scenarios in mobile cloud computing. (a) Operation on encrypted data and (b) operation on decrypted data.

If the security is password based, the password can be easily hacked and then anyone can access the application, so it needs some strong authentication techniques. Biometric authentication may be the most suitable solution for this kind of scenario. Another challenge to MCC is carrying personal or official data on the device or cloud. More users are sharing information using mobile devices. They either share information using the mobile device itself or download from the cloud storage and then send the information to another user. When the device is lost or stolen, both personal and official data can be compromised. Here, biometric authentication can ensure a more secure mobile cloud environment.

14.9.2 Disaster Recovery

Mobile cloud service must be developed to include any type of disaster recovery plan. Disaster recovery may be developed by employing redundant data, but this method is not so suitable. It is more suitable for implementing the cloud service at different locations, which are also near the user. However, this strategy does not indicate proper disaster recovery implementation and needs more research.

14.9.3 Pricing in Mobile Cloud Computing

Mobile service providers (MSPs) and CSPs have different methods of service management, customer management, payment, and pricing. Consequently, various issues arise: How would price division take place among different entities? How would the price be set and how would consumers make payment? For example, when a user plays a mobile game on the cloud, the service charge of the game is divided among the MSP, CSP, and the game service provider in such a way that each of them is satisfied. The CSP [19] provides cloud services to the mobile user via a data network like the Internet. The CSP receives its payment from the MSP as per the revenue collected by the mobile operator from the user. There exists an intermediate proxy between the MSP and the Internet service provider (ISP) [19], which is used to maintain information on the amount of data transferred by the CSP to the mobile user as well as to manage the revenue-sharing agreement between the MSP and the CSP. The proxy is defined as an intermediate agent acting as a communication link between the elements of two remote networks to perform pre-configured functions. Monitoring the revenue-sharing agreement between the CSP and the MSP is a major function of the proxy and can be performed by monitoring the IP address or the domain name of the CSP. For data transmission in a packet-switched network, the amount of data can be checked by calculating the number of transmitted packets on the basis of the peers' IP addresses. In the case of a circuit-switched network, the same activity can be performed by calculating the time duration of the active call.

As shown in Figure 14.7, the communication network has a number of mobile operators providing services to a pool of users. The mobile operators are linked to a data network to offer data services to end users. Figure 14.7 shows how the CSPs are connected to the data network through a number of service providers. It also shows the connection between an intermediate proxy and a mobile network to check the amount of transmitted data to end users from the CSP. Apart from this, the intermediate proxy contains the domain name or IP address of each CSP for keeping track of the agreement with the CSP on revenue sharing. To use the CSP's offered data services, each mobile device contains a web browser to view the provided contents. The mobile network is associated with a finance and billing system containing the address of each content provider and the account number of each

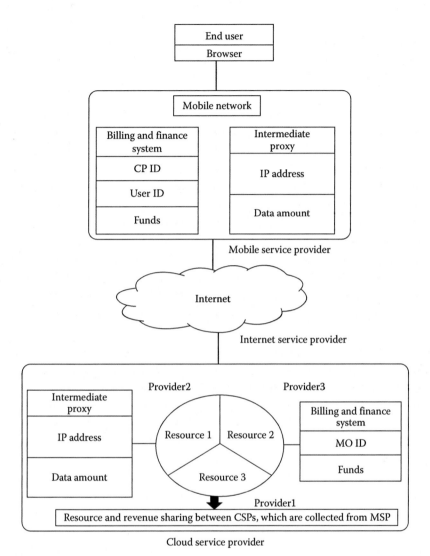

FIGURE 14.7
System model of revenue sharing of MSP and CSP.

consumer who has agreed to the payment of the service charge. The finance and billing system can track the revenue that the mobile operator charges from consumers for the cloud data services and the amount of payment made to each CSP. The amount of payment made to the content provider, when the service provided by it is used by the end user, is proportional to the revenue generated by the MSP. On the basis of a fixed percentage, the revenue can be shared. There can also be a maximum or minimum monetary amount that the mobile operator has to pay to the content provider every time a customer uses the cloud data services. Moreover, if a huge number of packet transmissions occur within a short time, a "limiting" factor should be considered by the intermediate proxy such that if the packets rapidly follow each other, the intermediate proxy may not be able to count all the packets. For updating bills, the intermediate proxy provides the details of the packet count to the finance and billing system of the mobile operator at preconfigured intervals.

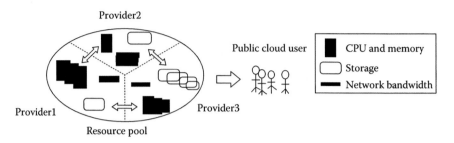

FIGURE 14.8
Model showing pricing and revenue sharing in MCC.

The collected information can be used to pay the CSP based on the revenue that the mobile operator gets for the data services. For ensuring that the invoice coming from the CSP is correct, the finance and billing system can be used. Alternatively, the CSP is paid by the mobile operator on a regular or automatic basis. The payment can also be made according to the agreed term without waiting for a billing statement from the cloud.

As the CSPs may consist of one or more databases to serve end users, each content provider has an intermediate proxy to keep track of the established agreement between the MSP and the CSP, but it is not cost effective. If we create a coalition of CSPs, one intermediate proxy is sufficient to keep track of the data amount transferred from the CSP to the end user through the data network. After collecting revenue from the MSP, it is the duty of CSPs to share the revenue among themselves. But in this case, another issue arises: how to form a coalition among CSPs and how to share the revenue collected from the mobile operator. For CSPs, the computing resource is not only accommodated to the internal user but also to public user. Therefore, the CSP can generate more revenue from the available resources. So, the CSP can form a coalition to serve the public cloud user. Coalition formation is one of the important approaches to improve efficiency and resource utilization. In Figure 14.8, three CSPs cooperate to offer services to public cloud users. As shown in Figure 14.8, provider1 has a large internal demand for data storage and the storage capacity left for public cloud user is small; provider2 experiences large internal demand for CPU core and data storage; and provider3 has a large internal demand for network bandwidth. If these CSPs offer service to public cloud users separately, the benefit will be small as the number of VMs supplied by each provider will be limited by the smallest available resource; for example, the number of VMs from provider1 is limited by the small storage capacity due to the internal demand. However, if these providers cooperate to create a resource pool, the available resources of each provider can be integrated; for example, the available CPU cores from provider1 can be used with the available storage from provider3 to offer more VMs to public cloud users and gain better benefit.

14.9.4 Service Execution and Delivery

Mobile cloud users need an efficient monitoring system, which measures the QoS they receive. Service location agent is a system or contract used to measure quality against a fee. The term of service is defined during the negotiation phase, and real-time performance is done during the monitoring phase. This service location agent can be developed in a dynamic mobile cloud environment with more powerful flexibility.

14.10 Reducing Energy Consumption of Offloading

Recently, the use of mobile devices, especially smartphones, has increased rapidly. With the increased usage of these devices, the demand for better QoS and quality of experience (QoE) has also increased. Mobile devices, though efficient in terms of fulfilling user demands, require methodologies to enhance their battery backups [1,10]. The storage of high volume of data and execution of complex applications in these devices require the device to be charged periodically [1]. This hampers QoE. MCC allows offloading of data storage and application execution from the mobile devices to the cloud servers [1]. The offloading efficiently solves the problem of battery backup, but it does not come free of charge and incurs certain overheads such as the cost and energy required for communication between the mobile devices and the cloud. For establishing communication with the cloud, some message transfers are required between the cloud and the mobile device [20]. These message transfers incur additional cost as well as energy consumption. Therefore, studying the overhead incurred for both data storage and application execution is needed.

14.10.1 Femtocell-Based Offloading Strategy for Mobile Cloud Computing

It has been observed that with offloading computation from the mobile device to cloud servers, energy can be saved if the time required for data transfer for communicating with the cloud is less than the execution time of the same application within the mobile device and the speed of the cloud server is amply greater than that of the mobile device [2]. MCC can reduce energy consumption of mobile devices, and the savings from offloading the computation are more than the energy cost of the additional communication [21]. The introduction of femtocell in the MCC architecture would reduce the energy consumption to a greater extent. As offloading data to the cloud requires Internet and femtocells are connected to the service provider's network via broadband, by introducing femtocell base stations, improved QoS can be obtained in terms of signal strength and access to the Internet along with secured data transmission due to the presence of security gateway.

14.10.2 Cloud Path Selection for Offloading Data Storage and Application Execution

MCC has gained immense popularity in the recent years. It facilitates data storage and application execution to be offloaded from the mobile device to the cloud [20]. Offloading helps reduce energy consumption of the mobile device as well as extends the battery life [20]. Recently, a variety of clouds, which provide similar type of services, have come into existence. For example, Dropbox, MS Skydrive, Google Drive, and Box provide similar services [22]. It is, therefore, of utmost importance to select an optimum cloud for offloading. Selection of the cloud path consumes a lot of time and energy since many criteria, such as speed, bandwidth, price, security, and availability, need to be considered for making final decisions [22]. Once an optimal cloud is found for executing a certain type of application corresponding to a mobile device, it is unnecessary to rerun the path selection algorithm again when a similar application is generated from the same mobile device [20].

The path selection method based on ant colony optimization can be used to prevent the waste of energy in selecting a path for the same mobile device while offloading similar applications [22–24]. At the beginning, ants move about in a random manner, and on finding food, they come back to their colony by laying down pheromone trails. Other ants follow these pheromone trails to find food, rather than moving at random.

With passing time, the pheromone trail evaporates [22–24], which is advantageous as it eliminates the chance that every ant would eventually follow the same path instead of exploring new paths [22–24]. A case can be considered in which similar types of applications are to be offloaded from a mobile device to the cloud. As the applications are similar, the same cloud selected for executing the first application can be used to execute the others. In this case, rather than executing the path selection algorithm to choose an optimal cloud, it is proposed to cache the path and use it whenever similar applications are generated.

14.10.3 Offloading Strategy for Mobile Devices

The availability of desktop-based rich applications in smartphones via web-based cloud services leads to increased computation and resource need in smartphones. This increases energy consumption in smartphones [25]. To overcome the limitations of smartphones, an intelligent offloading management algorithm is highly desirable. This algorithm needs to be smart enough to learn all the computation needs in smartphones and make energy optimal decisions based on the gained knowledge [25] and will initiate computation offloading to the cloud server from mobile devices only when necessary. It should also be smart enough to consider all the possible user contexts and circumstances.

14.10.4 Offloading Overhead in Mobile Cloud Computing

An application can be designed for mobile devices that would differentiate offloading events into data storage offloading and application execution offloading [21]. The application would keep an account of the number of data storage and application execution events being offloaded from a mobile device to the cloud. From the information recorded by the application, an average estimation of the overhead incurred for offloading data storage and application execution can be made [21]. The information recorded gives an account of the cost incurred in terms of message transmission as well as the energy required for communication and accessing the cloud in both the events of offloading. The recorded information would help prevent unnecessary message transmission involved in offloading. Thus, unnecessary resource and cost utilization can be avoided with the design of such applications.

14.11 Improving Quality of Service with Cloudlets

The basic idea of MCC is to run computation-intensive applications in the resource-rich cloud servers rather than in the resource-limited mobile devices. This trait of offloading applications from mobile devices to the cloud via the Internet results in significant energy saving of the mobile devices. However, long WAN latency is a major disadvantage with the employment of MCC [26]. WAN delays in the crucial course of user interaction can impair usability and affect the QoS. As the latency increases, interactive response suffers. In order to obtain the benefits of MCC without being WAN limited, cloudlets can be used [13,26]. When mobile devices do not want to offload to the cloud because of delay and cost, a nearby cloudlet can be used [26].

14.11.1 Load Balancing among Cloudlets

A cloudlet may be overloaded at some point in time and may share its load with other cloudlets. An overloaded cloudlet may experience delay in providing service or may even provide incorrect results. This situation may hamper the QoS and in turn the QoE. Thus, load balancing among cloudlets is very essential to avoid such circumstances. For load balancing it is important to maintain a load threshold value, which would indicate a maximum load value a cloudlet may have based on its processing capacity of the CPU. A cloudlet cannot accept any load if its current load is at or above the load threshold. Cloudlets with a load below the threshold can accept more load. In this way, the load of each cloudlet can be balanced.

14.11.2 Load Sharing among Cloudlets

The cloudlets can also share the load with other cloudlets. Hence, a cloudlet to which an application is offloaded can split the application into multiple components in such a way that the split components have negligible interdependency. Interdependency refers to the condition in which the output of one component is the input of another component. It is minimized so that the communication between the cloudlets is minimized, and in turn the cost of communication in terms of number of messages transferred between the cloudlets will be less. In addition, the energy required for transferring and receiving messages between cloudlets will be reduced as well. Each component thus obtained will be offloaded to a nearby cloudlet whose load is below the threshold. Each cloudlet will send the result of computation to the cloudlet to which the application was offloaded. This cloudlet will finally combine all the intermediate results to generate the final result and then send it to the requesting device.

14.11.3 Provisioning of Seamless Mobile Cloud Services

Analogous to the mobile users of this generation, future mobile users would demand extremely competent application services in mobile clouds [25]. Although the need for high-end mobile devices is rising, MCC will still be highly dependent on low-end mobile devices because of their low cost and energy saving [25]. This is a major issue in providing mobile cloud services to users with rich mobile experience using low-end devices. At times when the cloud server is engaged or not available due to network error or link failure, a mobile device can form groups with other mobile devices to deliver a service. This would not cause any of the devices to be overburdened, and service would be delivered seamlessly to the user. Since energy saving and improvement in user experience conflict with each other, reducing energy consumption without affecting user experience is a challenge.

14.12 Reduction in Cloud Data Center Energy Consumption

Cloud servers are always switched on irrespective of whether they are providing service or not. This causes unnecessary energy consumption in the idle state [25]. To save energy, some of the servers are generally turned off during the period of low demand, posing a threat of upsetting customer requirements during peak time if enough servers are not available to serve increased demand. Thus, a special server can be designated to monitor network traffic. During the period of high demand, it would generate some sort of message to the server administrator indicating that the demand is high and more servers are needed to be activated [25]. The server administrators would then switch on the idle servers. In this way, servers would consume minimum energy. Thus, instead of all servers being switched on all the time, only one server that monitors network traffic would be switched on throughout, and the threat of lack of servers during high demand will be reduced.

14.13 Resource Management

Lack of consciousness in resource management is another reason for excessive energy consumption. To overcome this, the resource management system should be improved for both cloud servers and mobile devices [25]. The resource management system must be such that resources are allocated in an on-demand fashion as the process progresses its execution. There should be no provisioning of allocation of resources prior to execution since it is difficult to correctly predict the resource needs before execution and unnecessary resource utilization may happen with allocation before the execution begins.

14.14 Application Migration Schemes for Data Center Servers

The influx of workload to the cloud data center server is changeable. Thus, to save energy consumption by running servers, a dynamic application migration scheme, which would learn the nature of workload arrival and dynamically migrate applications to an energy-efficient node, is necessary [25].

14.15 Conclusion

MCC is an emerging and fast growing field of cloud computing. It is a technology that combines the benefits of both mobile computing and cloud computing and provides efficient services to the mobile industry. The main motivation of this technology is to use service, whether storage, software, or application, on the cloud by mobile users. The number of mobile devices is increasing every day, and most people use smartphones now. So, the mobile cloud environment is used by a maximum number of users. Hence, it should not be bound to a specific field but should be broader. Some specific research directions regarding bandwidth allocation, billing system, security, energy efficiency, load balancing, and resource management have been discussed. If they are developed, the mobile cloud environment may become more powerful and end users would get much help from the MCC environment.

Questions

1. Explain the methods of efficient bandwidth allocation in MCC.
2. What is a cloudlet? Why is it used?
3. Draw and describe the cloudlet-based MCC architecture.
4. Describe the reference architecture of elastic application.
5. Write a short note on mobile sky computing.
6. Explain the femtocell-based offloading strategy in MCC.
7. How would femtocell help in the densification of 5G mobile network?
8. Describe the load-balancing and load-sharing strategies of MCC.
9. How can a secured MCC environment be achieved?
10. Explain the problem of densification of 5G mobile network.

References

1. H. T. Dinh, C. Lee, D. Niyato, and P. Wang, A survey of mobile cloud computing: Architecture, applications, and approaches, *Wireless Communications and Mobile Computing*, 13(18), 1587–1611, 2013.
2. A. Tudzarov and T. Janevski, Functional architecture for 5G mobile networks, *International Journal of Advanced Science and Technology*, 32, 65–78, 2011.
3. J. Andrews and A. Gatherer, Will densification be the death of 5G?, *IEEE ComSoc Technology News*, CTN Issue, May 2015. http://www.comsoc.org/ctn/will-densifcation-be-death-5g.
4. A. Mukherjee, P. Gupta, and D. De, Mobile cloud computing based energy efficient offloading strategies for femtocell network, in *IEEE Applications and Innovations in Mobile Computing*, Kolkata, India, pp. 28–35, 2014.
5. A. Mukherjee and D. De, A cost-effective location tracking strategy for femtocell based mobile network, in *IEEE International Conference on Control, Instrumentation, Energy and Communication*, Kolkata, India, pp. 533–537, 2014.
6. D. De, B. Mahata, and S. Jana, Effective load distribution algorithm considering power and interference mitigation in a dense femtocell based network, in *Second International Conference on Computer, Control and Information Technology*, Adi Saptagram, India, pp. 692–697, 2012.
7. D. De and A. Mukherjee, Femtocell based economic health monitoring scheme using mobile cloud computing, in *IEEE Fourth International Advance Computing Conference*, Gurgaon, India, pp. 385–390, 2014.
8. A. Mukherjee, S. Bhattacherjee, S. Pal, and D. De, Femtocell based green power consumption methods for mobile network, *Computer Networks*, Elsevier, 57(1), 162–178, 2013.
9. A. Mukherjee and D. De, Congestion detection, prevention and avoidance strategies for an intelligent, energy and spectrum efficient green mobile network, *Journal of Computational Intelligence and Electronic Systems*, American Scientific Publishers, 2(1), 1–19, 2013.
10. K. Kumar and Y. H. Lu, Cloud computing for mobile users: Can offloading computation save energy? *Computer*, 43(4), 51–56, 2010.
11. J. Boccuzzi and M. Ruggiero, *Femtocells: Design & Application*, McGraw-Hill, New York, 2011.
12. F. Ge, H. Lin, and A. Khajeh, Cognitive radio rides on the cloud, in *IEEE Military Communications Conference*, San Jose, CA, pp. 1448–1453, 2011.

13. M. Satyanarayanan, P. Bahl, R. Caceres, and N. Davies, The case for VM-based cloudlets in mobile computing, *IEEE Pervasive Computing*, 8(4), 14–23, 2009.
14. K. Keahey, M. Tsugawa, A. Matsunaga, and J. Fortes, Sky computing, *IEEE Internet Computing Magazine*, 13(5), 43–51, 2009.
15. M. A. Amin, K. Bin Abu Bakar, and H. Al-Hashimi, A review of mobile cloud computing architecture and challenges to enterprise users, in *IEEE Seventh GCC Conference and Exhibition*, Doha, Qatar, pp. 240–244, 2013.
16. Y. Li, H. Dai, and B. Yang, Identity-based cryptography for cloud security, *IACR Cryptology ePrint Archive*, p. 169, 2011.
17. L. Yan, C. Rong, and G. Zhao, Strengthen cloud computing security with federal identity management using hierarchical identity-based cryptography, in *Cloud Computing, Lecture Notes in Computer Science*, Springer, Berlin, Germany, pp. 167–177, 2009.
18. A. Mukherjee and D. De, Symmetric key based audio stenography for mobile network, *IJEIR*, 1(3), 271–277, 2012.
19. D. Niyato, A. V. Vasilakos, and Z. Kun, Resource and revenue sharing with coalition formation of cloud providers: Game theoretic approach, in *Proceedings of the 11th IEEE/ACM International Symposium on Cluster, Cloud and Grid Computing*, Newport Beach, CA, pp. 215–224, 2011.
20. M. V. Barbera, S. Kosta, A. Mei, and J. Stefa, To offload or not to offload? The bandwidth and energy costs of mobile cloud computing, in *Proceedings of INFOCOM IEEE*, Turin, Italy, pp. 1285–1293, 2013.
21. P. A. Miettinen and J. K. Nurminen, Energy efficiency of mobile clients in cloud computing, in *Proceedings of the Second USENIX Conference on Hot Topics in Cloud Computing*, New York, USENIX Association, p. 4, 2010.
22. H. Wu, Q. Wang, and K. Wolter, Methods of cloud-path selection for offloading in mobile cloud computing systems, in *IEEE Fourth International Conference on Cloud Computing Technology and Science*, Taipei, Taiwan, pp. 443–448, 2012.
23. S. Roy and S. R. Sahoo, Path planning of mobile agents using AI technique, PhD dissertation, National Institute of Technology, Rourkela, India, 2007.
24. T. Stützle and M. Dorigo, ACO algorithms for the traveling salesman problem, in K. Miettinen et al. (eds.), *Evolutionary Algorithms in Engineering and Computer Science*, Wiley, Chichester, United Kingdom, pp. 163–183, 1993.
25. M. Rahman, J. Gao, and W. Tsai, Energy saving in mobile cloud computing, in *Proceedings of IEEE International Conference on Cloud Engineering*, Redwood City, CA, pp. 285–291, 2013.
26. T. Verbelen, S. Pieter, D. T. Filip, and D. Bart, Cloudlets: Bringing the cloud to the mobile user, in *Proceedings of the Third ACM Workshop on Mobile Cloud Computing and Services*, New York, pp. 29–36, 2012.

Index